TECHNOLOGICAL INTERVENTIONS IN DAIRY SCIENCE

Innovative Approaches in Processing, Preservation, and Analysis of Milk Products

Innovations in Agricultural and Biological Engineering

TECHNOLOGICAL INTERVENTIONS IN DAIRY SCIENCE

Innovative Approaches in Processing, Preservation, and Analysis of Milk Products

Edited by

Rupesh S. Chavan, PhD
Megh R. Goyal, PhD

Apple Academic Press Inc.
3333 Mistwell Crescent
Oakville, ON L6L 0A2 Canada

Apple Academic Press Inc.
9 Spinnaker Way
Waretown, NJ 08758 USA

© 2018 by Apple Academic Press, Inc.

First issued in paperback 2021

Exclusive worldwide distribution by CRC Press, a member of Taylor & Francis Group
No claim to original U.S. Government works

ISBN 13: 978-1-77-463637-4 (pbk)
ISBN 13: 978-1-77-188609-3 (hbk)

Library and Archives Canada Cataloguing in Publication

Technological interventions in dairy science : innovative approaches in processing, preservation, and analysis of milk products / edited by Rupesh S. Chavan, PhD, Megh R. Goyal, PhD.

(Innovations in agricultural and biological engineering)
Includes bibliographical references and index.
Issued in print and electronic formats.
ISBN 978-1-77188-609-3 (hardcover).--ISBN 978-1-315-16940-8 (PDF)

1. Dairy products industry--Technological innovations. 2. Dairy processing. 3. Dairy products. 4. Milk. I. Chavan, Rupesh S., editor II. Goyal, Megh Raj, editor III. Series: Innovations in agricultural and biological engineering

| SF250.5.T43 2018 | 637 | C2017-907543-8 | C2017-907544-6 |

CIP data on file with US Library of Congress

Apple Academic Press also publishes its books in a variety of electronic formats. Some content that appears in print may not be available in electronic format. For information about Apple Academic Press products, visit our website at **www.appleacademicpress.com** and the CRC Press website at **www.crcpress.com**

CONTENTS

LIST OF CONTRIBUTORS

A. K. Agrawal
Department of Dairy Engineering, College of Dairy Science and Food Technology Chhattisgarh Kamdhenu Vishwavidyalaya (CGKV), Raipur 492012, India. E-mail: akagrawal.raipur@gmail.com

Usman Ali
National Agricultural Food Biotechnology Institute, C-127, Industrial Area, Phase 8, Mohali, Punjab, India. E-mail: get4usman@gmail.com

Md. Irfan A. Ansari
Department of Agricultural Engineering, Birsa Agricultural University, Ranchi 834006, Jharkhand, India. E-mail: irfan26200@yahoo.com

Santanu Basu
Dr. S. S. B. University Institute of Chemical Engineering & Technology, Panjab University, Chandigarh 160014, India. E-mail: santbasu@gmail.com

Shraddha Bhatt
Department of Biotechnology, Junagadh Agricultural University, Junagadh 362001, Gujarat, India. E-mail: shraddha.bhatt761@gmail.com

Rupesh S. Chavan
Department of Quality Assurance, Mother Dairy at Junagadh, Junagadh 362001, Gujarat, India. E-mail: rschavanb_tech@rediffmail.com

Jatin Gol
GCMMF, Anand, Gujarat, India; E-mail: jatin_gol@yahoo.com

Megh R. Goyal
Agricultural and Biomedical Engineering from General Engineering Department, University of Puerto Rico, Mayaguez Campus; Agriculture Sciences and Biomedical Engineering, Apple Academic Press Inc., USA. E-mail: goyalmegh@gmail.com

A. S. Hariyani
Dairy Chemistry Department, College of Dairy Science, Kamdhenu University, Amreli 365601, Gujarat, India. E-mail: ashariyani@ku-guj.com

Tanmay Hazra
Dairy Chemistry Department, College of Dairy Science, Kamdhenu University, Amreli 365601, Gujarat, India. E-mail: tanmayhazra08@gmail.com

Shaik Abdul Hussain
Dairy Technology Division, National Dairy Research Institute (NDRI), Karnal 132001, Haryana, India. E-mail: abdulndri@gmail.com

Akruti Joshi
Shree P. M. Patel Institute of P.G. Studies and Research, Anand 388001, Gujarat, India. E-mail: akruti.joshi@gmail.com

Shubhneet Kaur
Department of Food and Nutrition, Lady Irwin College, Sikandra Road, New Delhi 110001, India.
E-mail: shubhneetkaur3@gmail.com

Anit Kumar
Department of Food Science and Technology, National Institute of Food Technology
Entrepreneurship and Management (NIFTEM), Kundli, Haryana 131028, India.
E-mail: aks.kumar6@gmail.com

Nichal Mayur
GCMMF Ltd., Anand 388001, Gujrat, India. E-mail: mayurnichal619@gmail.com

Vijendra Mishra
National Institute of Food Technology Entrepreneurship and Management, HSIIDC, Kundli 131028,
Sonepat, India. E-mail: vijendramishra.niftem@gmail.com

Ashwini S. Muttagi
DRDC Foods, Dabur India Limited, Plot No 22, Site 4, Sahibabad, Ghaziabad 201301,
Uttar Pradesh, India. E-mail: ashumuttagi@gmail.com

Madhav R. Patil
University Department of Dairy Chemistry, Biochemistry & Food Technology Division, College of
Dairy Technology, Udgir, Latur 413517, Maharashtra, India. E-mail: madhavpatil34@gmail.com

Prasad S. Patil
Dairy Microbiology Division, National Dairy Research Institute (NDRI), Karnal 132001, Haryana,
India. E-mail: patilprasad11@gmail.com

Chakraborty Purba
Dr. S. S. B. UICET, Panjab University, Chandigarh 160014, Chandigarh, India.
E-mail: purba.chakraborty@yahoo.co.in

Shalini Gaur Rudra
Division of Food Science and Postharvest Technology, Indian Agricultural Research Institute,
New Delhi 110012, India. E-mail: gaurshalini@gmail.com

Prabin Sarkar
Dairy Chemistry Division, National Dairy Research Institute (NDRI), Karnal 132001, Haryana,
India. E-mail: prabinndri@gmail.com

Rachna Sehrawat
Department of Food Engineering, National Institute of Food Technology Entrepreneurship and
Management (NIFTEM), Kundli, Haryana 131028, India. E-mail: sehrawatrachna@gmail.com

Uma Shanker Shivhare
Dr. S. S. B University Institute of Chemical Engineering & Technology, Panjab University,
Chandigarh 160014, India. E-mail: usshiv@yahoo.com

Ashish Kumar Singh
Dairy Technology Division, National Dairy Research Institute (NDRI), Karnal 132001, Haryana,
India. E-mail: aksndri@gmail.com

Geetesh Sinha
Department of Dairy Engineering, College of Dairy Science and Food Technology, Chhattisgarh
Kamdhenu Vishwavidyalaya (CGKV), Raipur 492012, India. E-mail: geeteshsinha20@gmail.com

A. J. Thesiya
Dairy Technology Department, College of Dairy Science, Kamdhenu University, Amreli 365601, Gujarat, India. E-mail: ajthesiya@kuguj.com

Sudhir Kumar Tomar
Dairy Microbiology Division, NDRI, Karnal 132001, Haryana, India.
E-mail: sudhirndri@gmail.com

Ashutosh Updhyay
Department of Food Science and Technology, National Institute of Food Technology Entrepreneurship and Management (NIFTEM), Kundli, Haryana 131028, India.
E-mail: ashutosh.ft@gmail.com

Rajesh Kumar Vishwakarma
ICAR-CIPHET, Punjab Agricultural University, Ludhiana 141004, India.
E-mail: rkvciphet@gmail.com

Akanksha Wadehra
Dairy Technology Division, National Dairy Research Institute (NDRI), Karnal 132001, Haryana, India. E-mail: smartakanksha@gmail.com

Devbrat Yadav
Dairy Chemistry Division, National Dairy Research Institute (NDRI), Karnal 132001, Haryana, India. E-mail: dev.007.yadav@gmail.com

LIST OF ABBREVIATIONS

ASDs	autistic spectrum disorders
ATP	adenine triphosphate
BF	Burkina Faso
BOD	biochemical oxygen demand
BSA	bovine serum albumin
CAP	colloidal calcium phosphate
CC	cream cheese
CCP	colloidal calcium phosphate
CD	Crohn's disease
CIP	cleaning in place
CLSM	confocal laser scanning microscopy
CMP	caseinomacropeptide
CO	cheese with oregano
COP	cleaning-out-of-place
CPO	crude palm oil
CR	cheese with rosemary
CRC	colorectal cancer
CSEM	cryoscanning electron microscopy
CWP	cell wall particles
DP	degree of polymerization
ED	electrodialysis
EM	electromagnetic
EOs	essential oils
EPS	exopolysaccharide
ETP	effluent treatment plant
FA	ferulic acid
FAAs	free amino acids
FAE	ferulic acid esterase
FC	fat content
FDA	Food and Drug Administration
FFAs	free fatty acids
FISH	fluorescent in situ hybridization
FRAD	fine returns air disperser

GC-MS	gas chromatography–mass spectrometry
GIT	gastrointestinal tract
GRAS	generally recognized as safe
HD	hydrodistillation
HP	high pressure
HPLC	high-performance liquid chromatography
HPP	high-pressure processing
HSG	highly sensitized group
IBD	inflammatory bowel diseases
IBS	irritable bowel syndrome
IDF	International Dairy Federation
IGGs	immunoglobulins
ISO	International Organization for Standardization
LAOS	large amplitude oscillatory shear
LOX	lipoxygenase
LP	light pulses
LPC	leaf protein concentrate
LPM	Lambert–Pearson model
LPS	lipopolysaccharides
LVR	linear viscoelastic region
MAP	modified atmosphere packaging
MD	membrane distillation
MF	microfiltration
MFGM	milk fat globule membrane
MHB	Muller–Hinton broth
MHG	microwave hydrodiffusion and gravity
MIC	minimum inhibitory concentration
MS	manosonication
MSD	microwave steam distillation
MTS	manothermosonication
MVR	mechanical vapor recompression
MW	microwave
MWSD	microwave steam diffusion
NB	nutritive broth
NCIMS	National Conference on Interstate Milk Shipments
NIC	noninhibitory concentration
OD	osmotic distillation
OMF	oscillating magnetic fields

PCR	polymerase chain reaction
PD	pressure difference
PEF	pulsed electric field
PEOs	plant essential oils
PF	preterm infant formula
PLC	programmable logic controller
PMO	pasteurized milk ordinance
RA	rheumatoid arthritis
RCT	rennet coagulation time
RF	radio frequency
RFEF	radio frequency electric field
RO	reverse osmosis
SAOS	small angle oscillatory shear
SC	supercritical
SCC	spinning cone column
SCFA	short-chain fatty acids
SCFE	superficial fluid extraction
SEM	scanning electron microscopy
SF	standard infant formula
SFP	soy-fortified paneer
SG	sensitized group
SI	solubility index
SIP	sterilizing-in-place
SMP	skimmed milk powder
SODD	soybean oil deodorizer distillate
SS	stainless steels
TCA	tricarboxylic acid
THD	turbohydrodistillation
TPA	texture profile analysis
TSB	tryptone soya broth
TVR	thermovapor recompression
UAE	ultrasound-assisted extraction
UC	ulcerative colitis
UF	ultrafiltration
UHT	ultrahigh temperature processing
UTM	universal testing machine
UV	ultraviolet
UVP	ultrasound velocity profiling

VAO	vanillyl alcohol oxidase
WHC	water-holding capacity
WIP	wash-in-place
WMP	whole milk powder
WPA	whey protein aggregates
WPC	whey protein concentrate

PREFACE 1 BY RUPESH S. CHAVAN

Milk is nutritious, appetizing, and nature's perfect food and is recommended by nutritionists for development of sound body and consumed by all sectors of people. Demand for milk is greater in developed countries as compared to developing countries, but the gap has narrowed due to an increase in urbanization, population, and consumption. India ranks first in milk production, accounting for 18.5% of world production, achieving an annual output of 146.3 million tons during 2014–15 as compared to 137.69 million tons during 2013–14 recording a growth of 6.26%, whereas the Food and Agriculture Organization (FAO) has reported a 3.1% increase in world milk production from 765 million tons in 2013 to 789 million tons in 2014. Pasteurized milk and ultrahigh temperature (UHT) milk market share in 2014 increased to 83.7% and 6.9% (in volume), respectively, as compared to 81.9% and 5.2% in 2013.

Milk is a perishable commodity, and being rich in nutrients it acts as the perfect substrate for growth of microflora, and to reduce this, different thermal and nonthermal techniques are implemented. Thermal treatments are the common techniques used for extending the shelf life of milk, for example, pasteurization, sterilization, and UHT, but the loss of nutrients is a concern associated with these treatments. Nonthermal treatments like high-pressure processing, pulse electric field, ultrasonication, and irradiation are also explored to process milk to minimize the loss of nutrients as compared to thermal treatment. Postprocess contamination is a major factor that can affect the shelf life of milk, and to maintain the product, safe packaging plays an important role when the milk and milk products are stored at refrigeration or ambient temperature.

An attempt to cover all of these important interventions in dairy science is made and the outcome of the efforts is the present book the book is divided in four sections and have tried to cover topics including novel technological interventions in dairy science, nonthermal technologies for processing of milk and milk products, determining the physical and rheological properties of milk, and milk products and factors affecting the same, fouling of milk and cleaning-in-place in the dairy industry, novel methods to detect adulteration of ghee or milk fat.

This book provides updated information to advancements in the field of dairy science and technology. I hope this book will fulfil the requirements of industrial professionals, scientists, regulatory personnel, consultants, academics, students, and field related personnel. The book will also help to bridge the gaps between the industrial applications of recent techniques, which are mentioned in the existing books in the market for the dairy professionals.

—**Rupesh S. Chavan, PhD**
Editor

PREFACE 2 BY MEGH R. GOYAL

With farms of hundreds of cows producing large volumes of milk, the larger and more efficient dairy farms are more able to weather severe changes in milk price and operate profitably, while "traditional" very small farms generally do not have the equity or cash flow to do so. The common public perception of large corporate farms supplanting smaller ones is generally a misconception, as many small family farms expand to take advantage of economies of scale, and incorporate the business to limit the legal liabilities of the owners and simplify such things as tax management.

Worldwide, the largest milk producer is the European Union with its present 27 member countries, with more than 153,000,000 metric tons in 2009 (more than 95% cow milk). Countrywise, the largest producer is India (more than 55% buffalo milk), the largest cow milk exporter is New Zealand, and the largest importer is China.

In the United States, the top six dairy states are, in order by total milk production: California, Wisconsin, New York, Idaho, and Pennsylvania. Dairy farming is also an important industry in Florida, Minnesota, Ohio, and Vermont. There are 65,000 dairy farms in the United States. Pennsylvania has 8500 farms with 555,000 dairy cows. Milk produced in Pennsylvania yields annual revenue of about US$1.5 billion. Herd size in the United States varies between 1200 on the West Coast and Southwest, where large farms are commonplace, to roughly 50 in the Midwest and Northeast, where land-base is a significant limiting factor to herd size. The average herd size in the United States is about 100 cows per farm but the median size is 900 cows with 49% of all cows residing on farms of 1000 or more cows.

A *milk product* is a food produced from the milk of mammals. Dairy products are usually high-energy-yielding food products. A production plant for the processing of milk is called a dairy or a dairy factory. Apart from breastfed infants, the human consumption of dairy products is sourced primarily from the milk of cows, water buffaloes, goats, sheep, yaks, horses, camels, domestic buffaloes, and other mammals. Dairy products are commonly found throughout the world.

I recall my childhood. I was breastfed by mother until I was 8 years old. It calmed me down from my nervousness and hypertension, and imparted to me security in the lap of my mother. Up to fifth grade, I never tasted dairy milk except few drops in an Indian tea. In sixth grade, once a day we were asked to sit on a 1 m × 50 m mat and were served one 16 oz glass of milk by American Peace Corps volunteers. I enjoyed not only drinking the milk but also the dedication of these volunteers. This milk was prepared from powdered milk that was imported from Europe. What a gesture to the undernourished students in the developing countries!

At the 49th annual meeting of the Indian Society of Agricultural Engineers at Punjab Agricultural University (PAU) during February 22–25 of 2015, a group of ABEs and FEs convinced me that there is a dire need to publish book volumes on focus areas of agricultural and biological engineering (ABE). This is how the idea was born on new book series titled Innovations in Agricultural & Biological Engineering. This book, Technological Interventions in Dairy Science, is 17th volume under this book series, and it contributes to the ocean of knowledge on dairy engineering.

The contribution by all cooperating authors to this book volume has been most valuable in the compilation. Their names are mentioned in each chapter and in the list of contributors. This book would not have been written without the valuable cooperation of these investigators, many of whom are renowned scientists who have worked in the field of dairy engineering throughout their professional careers.

I am glad to introduce Dr. Rupesh S. Chavan, who worked as Assistant Professor at the National Institute of Food Technology Entrepreneurship and Management, Kundli, under the Ministry of Food Processing Industries, India; and In-charge of the International Bakery Research and Training Center. Presently, he is a Senior Executive in the Department of Quality Assurance, Mother Dairy at Junagadh. He is a professor/researcher and specializes in dairy and bakery products. Without his support and leadership qualities as lead editor of this book volume and his extraordinary work on dairy engineering applications, readers will not have this quality publication.

I would like to thank editorial staff, Sandy Jones Sickels, Vice President, and Ashish Kumar, Publisher and President at Apple Academic Press, Inc. (AAP) for making every effort to publish the book. Special thanks are also due to the AAP Production Staff.

I request readers offer constructive suggestions that may help to improve the next edition. Please send your comments to goyalmegh@gmail.com.

I express my deep admiration to my family for understanding and collaboration during the preparation of this book volume. As an educator, there is a piece of advice to one and all in the world: "Permit that our almighty God, our Creator, provider of all and excellent Teacher, feed our life with Healthy Milk and Milk Products and His Grace—and Get married to your profession."

—Megh R. Goyal, PhD, PE
Senior Editor-in-Chief

FOREWORD

Milk is known for nourishing mankind from ancient time, and with increasing incomes and rising production, milk and other dairy products are gaining importance in the human diet. In the modern era, milk production has become a large industry, and various value-added products, such as concentrated milk, butter, cheese, yogurt, ice cream, and other dairy products, are being made and marketed, which continuously increases demand in the market.

In the last three decades, world milk production has increased by more than 50%, from 500 million tons in 1983 to 769 million tons in 2013. With 18% of global production, India is the world's largest producer of milk, followed by the United States, China, Pakistan, and Brazil. Most of the increase in milk production has been in South Asia since 1970s, which is the main driver for growth in milk production in developing world, according to the Food and Agricultural Organization.

Higher consumption of milk and other dairy products is being observed in Asia, Latin America, and the Caribbean region. Milk and dairy products are providing opportunities for better livelihood for farmers, processors, and related persons in addition to fulfilling nutritional requirement of consumers. To accelerate the growth in this industry and to accomplish better processing operations, knowledge of milk properties, innovative nonthermal processing methods, the techniques for milk preservation and safety methods in dairy science such as cleaning-in-place in milk industry and detection methods of adulteration in milk/milk fat are essential.

This book, *Technological Interventions in Dairy Science,* is unique in presenting these important aspects in a common platform. This book provides an in-depth look at selected topics, from physical properties of milk and other milk products to nonthermal processing of milk and it also discusses safety methods in dairy science, which include cleaning-in-place in the dairy industry and techniques to determine adulteration in milk. The book has 12 chapters presented in four sections that can be referred to as needed or read from start to end for a full explanation of linkage in various processes.

This book provides valuable information to a variety of audiences—from academic and research to industry and consumers. I hope that the information presented in this publication will prove its usefulness to students, academicians, researchers, and industry professionals and will help to fulfill their specific requirements as well as to consumers for general information about the overall processes and operations in the milk industry.

Khursheed A. Khan, MTech
Assistant Professor
Department of Agricultural Engineering
R.V.S. Agriculture University, College of Horticulture
Mandsaur 458001, Gwalior, Madhya Pradesh, India
E-mail: khan_undp@yahoo.ca

WARNING/DISCLAIMER

PLEASE READ CAREFULLY

The goal of this volume, *Technological Interventions in Dairy Science: Innovative Approaches in Processing, Preservation, and Analysis of Milk Products,* is to guide the world community on how to efficiently manage the technology available for different processes in dairy engineering.

The editors and the publisher have made every effort to make this book as accurate as possible. However, there still may be grammatical errors or mistakes in the content or typography. Therefore, the contents in this book should be considered as a general guide and not a complete solution to address any specific situation.

The editors and publisher shall have neither liability nor responsibility to any person, organization, or entity with respect to any loss or damage caused, or alleged to have caused, directly or indirectly, by information or advice contained in this book. Therefore, the purchaser/reader must assume full responsibility for the use of the book or the information therein.

The mention of commercial brands and trade names are only for technical purposes. It does not mean that a particular product is endorsed over another product or equipment not mentioned. The editors and publisher do not have any preference for a particular product.

All weblinks that are mentioned in this book were active at the time of publication. The editors and the publisher shall have neither liability nor responsibility if any of the weblinks are inactive at the time of reading of this book. The information in this book is not meant for legal use; the information herewith is meant to facilitate clear and smooth communication among the professionals in bioengineering and biology.

ABOUT THE LEAD EDITOR
DR. RUPESH S. CHAVAN

 Rupesh S. Chavan, PhD, is a dairy technologist by profession with an expertise in the implementation of food safety management systems in dairy and food industries. He is a PhD holder with specialization in Dairy Science and Technology along with masters and bachelors degree in Dairy Technology with a MBA in Operations Management.

Currently, Dr. Chavan is associated with Mother Dairy Fruits and Vegetable Pvt. Ltd., Junagadh, Gujarat, India. In his previous assignment, he has worked as an Assistant Professor at the National Institute of Food Technology Entrepreneurship and Management, Kundli, under the Ministry of Food Processing Industries, India; and in-charge of the International Bakery Research and Training Center.

He is a professor/researcher and specializes in dairy and bakery products. He is the author/coauthor for more than 50 scientific publications. He has worked for industries including the Gujarat Cooperative Milk Marketing Federation and ITC, Bangalore for different roles that include production, quality assurance, research, and development, food safety for about 5 years in the field of milk and milk products and biscuits. In his PhD, research, he worked on the manufacturing of whey-based spray-dried tomato soup powder.

Currently, Dr. Chavan is working and development of FSSC systems and upgradation of the same for Mother Dairy, Junagadh. He is also actively involved in projects such as the effect of microfluidizer on the quality of fruit-flavored and low-fat yogurt; antibiotic resistance in *lactobacilli* of food and fecal origin; detection of genes, influence of stress and horizontal transfer; and interpretation of mixed milk (buffalo and cow) coagulation process and assessment of product quality.

He is a life member of the Indian Dairy Association (IDA), Society of Indian Bakers (SIB), and SASNET-FF. He was also engaged in education in Dairy Technology for BTech (Food Technology and Management) and MTech students. He has attended subject-specific skill-oriented courses, trainings, seminars, conferences, and workshops. He is a Principal

Investigator and Co-principal Investigator for projects on milk and milk products funded by different esteemed agencies. He was a university topper during BTech (Dairy Technology). Dr. Chavan also holds vast experience of research, manufacturing, quality assurance, and food safety systems in biscuit and milk industry.

Readers may contact him at: rschavanb_tech@rediffmail.com

ABOUT THE SENIOR EDITOR-IN-CHIEF

Megh R Goyal, PhD, PE
Retired Professor in Agricultural and Biomedical Engineering, University of Puerto Rico, Mayaguez Campus; and Senior Acquisitions Editor, Biomedical Engineering and Agricultural Science, Apple Academic Press, Inc.

Megh R. Goyal, PhD, PE, is a Retired Professor in Agricultural and Biomedical Engineering from the General Engineering Department in the College of Engineering at University of Puerto Rico–Mayaguez Campus; and Senior Acquisitions Editor and Senior Technical Editor-in-Chief in Agriculture and Biomedical Engineering for Apple Academic Press Inc.

He has worked as a Soil Conservation Inspector and as a Research Assistant at Haryana Agricultural University and Ohio State University. He was the first agricultural engineer to receive the professional license in Agricultural Engineering in 1986 from the College of Engineers and Surveyors of Puerto Rico. On September 16, 2005, he was proclaimed as "Father of Irrigation Engineering in Puerto Rico for the twentieth century" by the ASABE, Puerto Rico Section, for his pioneering work on micro irrigation, evapotranspiration, agroclimatology, and soil and water engineering. During his professional career of 45 years, he has received many prestigious awards. A prolific author and editor, he has written more than 200 journal articles and textbooks and has edited over 50 books. He received his BSc degree in engineering from Punjab Agricultural University, Ludhiana, India; his MSc and PhD degrees from Ohio State University, Columbus; and his Master of Divinity degree from Puerto Rico Evangelical Seminary, Hato Rey, Puerto Rico, USA. Readers may contact him at: goyalmegh@gmail.com.

OTHER BOOKS ON AGRICULTURAL & BIOLOGICAL ENGINEERING BY APPLE ACADEMIC PRESS, INC.

Management of Drip/Trickle or Micro Irrigation
Megh R. Goyal, PhD, PE, Senior Editor-in-Chief

Evapotranspiration: Principles and Applications for Water Management
Megh R. Goyal, PhD, PE, and Eric W. Harmsen, Editors

Book Series: Research Advances in Sustainable Micro Irrigation
Senior Editor-in-Chief: Megh R. Goyal, PhD, PE

Volume 1: Sustainable Micro Irrigation: Principles and Practices
Volume 2: Sustainable Practices in Surface and Subsurface Micro Irrigation
Volume 3: Sustainable Micro Irrigation Management for Trees and Vines
Volume 4: Management, Performance, and Applications of Micro Irrigation Systems
Volume 5: Applications of Furrow and Micro Irrigation in Arid and Semi-Arid Regions
Volume 6: Best Management Practices for Drip Irrigated Crops
Volume 7: Closed Circuit Micro Irrigation Design: Theory and Applications
Volume 8: Wastewater Management for Irrigation: Principles and Practices
Volume 9: Water and Fertigation Management in Micro Irrigation
Volume 10: Innovation in Micro Irrigation Technology

Book Series: Innovations and Challenges in Micro Irrigation
Senior Editor-in-Chief: Megh R. Goyal, PhD, PE

- Micro Irrigation Engineering for Horticultural Crops: Policy Options, Scheduling and Design
- Micro Irrigation Management: Technological Advances and Their Applications

- Micro Irrigation Scheduling and Practices
- Performance Evaluation of Micro Irrigation Management: Principles and Practices
- Potential of Solar Energy and Emerging Technologies in Sustainable Micro Irrigation
- Principles and Management of Clogging in Micro Irrigation
- Sustainable Micro Irrigation Design Systems for Agricultural Crops: Methods and Practices
- Engineering Interventions in Sustainable Trickle Irrigation: Water Requirements, Uniformity, Fertigation, and Crop Performance

Book Series: Innovations in Agricultural & Biological Engineering
Senior Editor-in-Chief: Megh R. Goyal, PhD, PE

- Dairy Engineering: Advanced Technologies and Their Applications
- Developing Technologies in Food Science: Status, Applications, and Challenges
- Engineering Interventions in Agricultural Processing
- Engineering Practices for Agricultural Production and Water Conservation: An Inter-disciplinary Approach
- Emerging Technologies in Agricultural Engineering
- Flood Assessment: Modeling and Parameterization
- Food Engineering: Emerging Issues, Modeling, and Applications
- Food Process Engineering: Emerging Trends in Research and Their Applications
- Food Technology: Applied Research and Production Techniques
- Modeling Methods and Practices in Soil and Water Engineering
- Processing Technologies for Milk and Dairy Products: Methods Application and Energy Usage
- Soil and Water Engineering: Principles and Applications of Modeling
- Soil Salinity Management in Agriculture: Technological Advances and Applications
- Technological Interventions in the Processing of Fruits and Vegetables
- Technological Interventions in Management of Irrigated Agriculture
- Engineering Interventions in Foods and Plants
- Technological Interventions in Dairy Science: Innovative Approaches in Processing, Preservation, and Analysis of Milk Products

- Novel Dairy Processing Technologies: Techniques, Management, and Energy Conservation
- Sustainable Biological Systems for Agriculture: Emerging Issues in Nanotechnology, Biofertilizers, Wastewater, and Farm Machines
- State-of-the-Art Technologies in Food Science: Human Health, Emerging Issues and Specialty Topics
- Scientific and Technical Terms in Bioengineering and Biological Engineering

EDITORIAL

Apple Academic Press Inc. (AAP) is publishing various book volumes on the focus areas under book series titled *Innovations in Agricultural and Biological Engineering*. Over a span of 8 to 10 years, Apple Academic Press will publish subsequent volumes in the specialty areas defined by American Society of Agricultural and Biological Engineers (asabe.org).

The mission of this series is to provide knowledge and techniques for agricultural and biological engineers (ABEs). The series aims to offer high-quality reference and academic content in Agricultural and Biological Engineering (ABE) that is accessible to academicians, researchers, scientists, university faculty, and university-level students and professionals around the world. The following material has been edited/ modified and reproduced below [From: *Goyal, Megh R., 2006. Agricultural and biomedical engineering: Scope and opportunities. Paper Edu_47 Presentation at the Fourth LACCEI International Latin American and Caribbean Conference for Engineering and Technology (LACCEI' 2006): Breaking Frontiers and Barriers in Engineering: Education and Research by LACCEI University of Puerto Rico – Mayaguez Campus, Mayaguez, Puerto Rico, June 21 – 23*]:

WHAT IS AGRICULTURAL AND BIOLOGICAL ENGINEERING (ABE)?

"Agricultural Engineering (AE) involves application of engineering to production, processing, preservation and handling of food, fiber, and shelter. It also includes transfer of technology for the development and welfare of rural communities", according to <isae.in>. *"ABE is the discipline of engineering that applies engineering principles and the fundamental concepts of biology to agricultural and biological systems and tools, for the safe, efficient and environmentally sensitive production, processing, and management of agricultural, biological, food, and natural resources systems"*, according to <asabe.org>. *"AE is the branch*

of engineering involved with the design of farm machinery, with soil management, land development, and mechanization and automation of livestock farming, and with the efficient planting, harvesting, storage, and processing of farm commodities", definition by: <http://dictionary.reference.com/browse/agricultural+engineering>.

"*AE incorporates many science disciplines and technology practices to the efficient production and processing of food, feed, fiber and fuels. It involves disciplines like mechanical engineering (agricultural machinery and automated machine systems), soil science (crop nutrient and fertilization, etc.), environmental sciences (drainage and irrigation), plant biology (seeding and plant growth management), animal science (farm animals and housing) etc.*", by http://www.ABE.ncsu.edu/academic/agricultural-engineering.php.

"According to https://en.wikipedia.org/wiki/Biological_engineering: "*BE (Biological engineering) is a science-based discipline that applies concepts and methods of biology to solve real-world problems related to the life sciences or the application thereof. In this context, while traditional engineering applies physical and mathematical sciences to analyze, design and manufacture inanimate tools, structures and processes, biological engineering uses biology to study and advance applications of living systems.*"

SPECIALTY AREAS OF ABE

Agricultural and Biological Engineers (ABEs) ensure that the world has the necessities of life including safe and plentiful food, clean air and water, renewable fuel and energy, safe working conditions, and a healthy environment by employing knowledge and expertise of sciences, both pure and applied, and engineering principles. Biological engineering applies engineering practices to problems and opportunities presented by living things and the natural environment in agriculture. BA engineers understand the interrelationships between technology and living systems, have available a wide variety of employment options. The <asabe.org> indicates that "*ABE embraces a variety of following specialty areas*". As new technology and information emerge, specialty areas are created, and many overlap with one or more other areas.

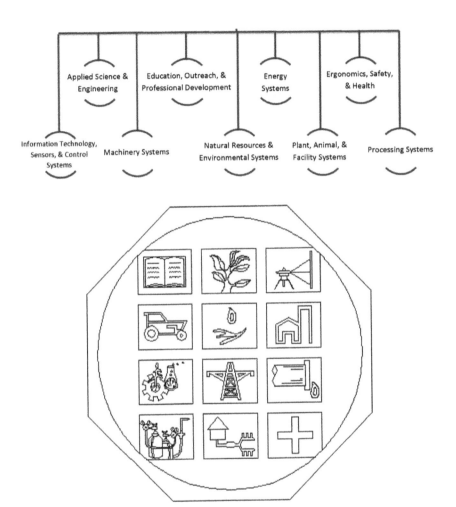

1. *Aquacultural Engineering*: ABEs help design farm systems for raising fish and shellfish, as well as ornamental and bait fish. They specialize in water quality, biotechnology, machinery, natural resources, feeding and ventilation systems, and sanitation. They seek ways to reduce pollution from aquacultural discharges, to reduce excess water use, and to improve farm systems. They also work with aquatic animal harvesting, sorting, and processing.

2. *Biological Engineering*: It applies engineering practices to problems and opportunities presented by living things and the natural environment. It also includes applications of nanotechnology in agricultural and biological systems.

3. *Energy*: ABEs identify and develop viable energy sources—biomass, methane, and vegetable oil, to name a few—and to make these and other systems cleaner and more efficient. These specialists also develop energy conservation strategies to reduce costs and protect the environment, and they design traditional and alternative energy systems to meet the needs of agricultural operations.

4. *Farm Machinery and Power Engineering*: ABEs in this specialty focus on designing advanced equipment, making it more efficient and less demanding of our natural resources. They develop equipment for food processing, highly precise crop spraying, agricultural commodity and waste transport, and turf and landscape maintenance, as well as equipment for such specialized tasks as removing seaweed from beaches. This is in addition to the tractors, tillage equipment, irrigation equipment, and harvest equipment that have done so much to reduce the drudgery of farming.

5. *Food and Process Engineering*: Food and process engineers combine design expertise with manufacturing methods to develop economical and responsible processing solutions for industry. Also, food and process engineers look for ways to reduce waste by devising alternatives for treatment, disposal, and utilization.

6. *Forest Engineering*: ABEs apply engineering to solve natural resource and environment problems in forest production systems and related manufacturing industries. Engineering skills and expertise are needed to address problems related to equipment design and manufacturing, forest access systems design and construction; machine–soil interaction and erosion control; forest operations analysis and improvement; decision modeling; and wood product design and manufacturing.

7. *Information and Electrical Technologies Engineering*: It is one of tthe most versatile areas of the ABE specialty areas because it is applied to virtually all the others, from machinery design to soil testing to food quality and safety control. Geographic information systems, global positioning systems, machine instrumentation and controls, electromagnetics, bioinformatics, biorobotics, machine vision, sensors, and spectroscopy. These are some of the exciting information and electrical technologies being used today and being developed for the future.

8. *Natural Resources*: ABEs with environmental expertise work to better understand the complex mechanics of these resources, so that they can be used efficiently and without degradation. ABEs determine crop water requirements and design irrigation systems. They are experts in agricultural hydrology principles, such as controlling drainage, and they implement ways to control soil erosion and study the environmental effects of sediment on stream quality. Natural resources engineers design, build, operate, and maintain water control structures for reservoirs, floodways, and channels. They also work on water treatment systems, wetlands protection, vertical farming, and other water issues.

9. *Nursery and Greenhouse Engineering*: In many ways, nursery and greenhouse operations are microcosms of large-scale production agriculture, with many similar needs—irrigation, mechanization, disease and pest control, and nutrient application. However, other engineering needs also present themselves in nursery and greenhouse operations: equipment for transplantation; control systems for temperature, humidity, and ventilation; and plant biology issues, such as hydroponics, tissue culture, and seedling propagation methods. And sometimes the challenges are extraterrestrial: ABEs at NASA are designing greenhouse systems to support a manned expedition to Mars!

10. *Safety and Health*: ABEs analyze health and injury data, the use and possible misuse of machines, and equipment compliance with standards and regulation. They constantly look for ways in which the safety of equipment, materials, and agricultural practices can be improved and for ways in which safety and health issues can be communicated to the public.

11. *Structures and Environment*: ABEs with expertise in structures and environment design animal housing, storage structures, and greenhouses, with ventilation systems, temperature and humidity controls, and structural strength appropriate for their climate and purpose. They also devise better practices and systems for storing, recovering, reusing, and transporting waste products.

Career in ABE

One will find that university ABE programs have many names, such as biological systems engineering, bioresource engineering, environmental engineering, forest engineering, or food and process engineering. Whatever the title, the typical curriculum begins with courses in writing, social sciences, and economics, along with mathematics (calculus and statistics), chemistry, physics, and biology. Students gain a fundamental knowledge of the life sciences and how biological systems interact with their environment. One also takes engineering courses such as thermodynamics, mechanics, instrumentation and controls, electronics and electrical circuits, and engineering design. Then student adds courses related to particular interests, perhaps including mechanization, soil and water resource management, food and process engineering, industrial microbiology, biological engineering, or pest management. As seniors, engineering students work in a team to design, build, and test new processes or products.

For more information on this series, readers may contact:

Ashish Kumar, Publisher and President
Sandy Sickels, Vice President
Apple Academic Press, Inc.
Fax: 866-222-9549
E-mail: ashish@appleacademicpress.com
http://www.appleacademicpress.com/publishwithus.php

Megh R. Goyal, PhD, PE
Book Series Senior
Editor-in-Chief
Innovations in Agricultural and Biological Engineering
E-mail: goyalmegh@gmail.com

PART I
Nonthermal Technologies for Processing of Milk and Milk Products

CHAPTER 1

APPLICATIONS OF NONTHERMAL PROCESSING IN THE DAIRY INDUSTRY

PRASAD S. PATIL[1], AKANKSHA WADEHRA[2], MADHAV R. PATIL[3], RUPESH CHAVAN[4,*], and SHRADDHA BHATT[5]

[1]*Dairy Microbiology Division, National Dairy Research Institute (NDRI), Karnal 132001, Haryana, India*

[2]*Dairy Technology Division, National Dairy Research Institute (NDRI), Karnal 132001, Haryana, India*

[3]*University Department of Dairy Chemistry, Biochemistry & Food Technology Division, College of Dairy Technology, Udgir, Latur 413517, Maharashtra, India*

[4]*Department of Quality Assurance, Mother Dairy at Junagadh, Junagadh 362001, Gujarat, India*

[5]*Department of Biotechnology, Junagadh Agricultural University, Junagadh 362001, Gujarat, India*

Corresponding author. E-mail: rschavanb_tech@rediffmail.com

CONTENTS

ABSTRACT

Nowadays, prime goal of milk processors is extending the shelf life of milk and milk products without compromising their quality and safety. In general, use of heat is still a common practice in dairy industries to guarantee the microbiological safety of milk and its subproducts. However, the processing of milk through heating had a noticeable evolution during the twentieth century, which has continued until the present time. Technological improvements together with the efforts and diligence of processors, technologists, and dairy researchers in bringing superior quality products to consumers, have been triggering the investigation and development of thermal and nonthermal approaches for milk processing capable of substituting the traditional well-established preservation processes. Novel thermal processing technologies are dealing with the change in temperature as opposed to the novel nonthermal processing technologies such as pulsed electric fields, high pressure, pulsed light, ultrasound, and gamma radiation among others, where temperature may also change but is not the major parameter responsible for food processing.

1.1 INTRODUCTION

Processing and preservation of food has been a challenging task for mankind throughout the ages. Many processing techniques have been developed and adopted since decades. Processes such as curing, pickling, fermentation, etc. have been carried out for generations throughout the world. Further, it can be noted that most of the production processes adopted in food industries require the application of heat. Though these products have become established throughout the whole world and are important in their own right; however, the products resulting from these processes are radically changed in comparison to the fresh produce in terms. Application of such high-energy treatments usually diminishes cooking flavor, and causes loss of vitamins, essential nutrients, and food flavors in the product. Consumers' demand for safe (due to mistrust for artificial preservation) and "fresh-like" product has given impetus to research and led to studies for the development of alternative processing methods. Hence, a wide range of novel processes have been studied over the last 100 years. Many of these technologies remain much in the research arena; however, others have come onto the brink of commercialization.

In dairy industry, presently we are using thermal processing operations such as pasteurization, sterilization, drying, and evaporation. Even many novel technologies have been developed in the thermal processing such as ohmic heating, microwave (MW) heating, ultrahigh temperature treatment, etc. All these technologies use heat as a source of treatment because of which we have disadvantages. To reduce or avoid the negative impact or disadvantages of the thermal processing technologies, novel nonthermal technologies such as high-pressure processing (HPP), pulsed electric field (PEF), membrane processing, active or smart packaging, and ultrasound, have been developed for the food and dairy industry.

Nonthermal methods allow processing of foods below the temperatures used during thermal pasteurization, thus flavors, essential nutrients, and vitamins undergo minimal or no changes during processing.

This chapter discusses about the different nonthermal techniques used for processing of milk and milk products including: (1) irradiation, (2) HPP, (3) ultrasound, (4) membrane processing, and (5) electrical methods such as PEF, light pulses (LP), radio frequency electric field (RFEF), and oscillating magnetic fields (OMF).[41,43]

1.2   IRRADIATION

Irradiation is the process of applying low levels of radiant energy to a material to sterilize or preserve it. Radiation is a physical means of food processing that involves exposing the prepackaged or bulk foodstuffs to radiation. The products are exposed to the radiation rays and kept inside a shielded structure for a specific time to deliver the desired dose. Foodstuffs are generally irradiated with gamma radiation from a radioisotope source such as ^{60}Cobalt (^{60}Co) or ^{137}Caesium (^{137}Cs), or with electrons or X-rays generated using an electron accelerator. These rays are termed as the ionizing radiations. Most food irradiation facilities utilize the radioactive element cobalt-60 (^{60}Co) as the source of high-energy gamma rays. These gamma rays have sufficient energy to dislodge electrons from some food molecules, thereby converting them into ions (electrically charged particles). Gamma rays do not have enough energy to affect the neutrons in the nuclei of these molecules; therefore, they are not capable of inducing radioactivity in the food.

The effectiveness of a given radiation dose varies depending on the density, antioxidant levels, moisture, and other components or characteristics of the foods. External factors, such as temperature, the presence or absence of oxygen, and subsequent storage conditions, also influence the effectiveness of radiation.

1.2.1 RADIATION DOSE FOR FOOD PROCESSING

When ionizing, radiations penetrate a food, and energy is absorbed. This is the "absorbed dose" and is expressed in Grays (Gy), where 1 Gy is equal to an absorbed energy of 1 J/kg. The absorbed dose varies in relation to the intensity of radiation, exposure time, and composition of the food. The dosage is classified as below:

a) Low-dose applications (<1 kGy).
b) Medium-dose application (1–10 kGy).
c) High-dose application (above 10 kGy).

1.2.1.1 RADURIZATION

It is used to treat foods that are high in pH and which have high water activity. Targeted organisms are usually Gram-negative bacteria and psychrotrophs, and can also be used to target yeasts and molds in foods that are low in pH and water activity. The treatment dose for radurization is approximately 1 kGy.

1.2.1.2 RADICIDATION

This treatment is used to destroy pathogenic organisms on the food product and is able to kill vegetative cells. The treatment dose for this treatment ranges from 2.5 to 5.0 kGy.

1.2.1.3 RADAPERTIZATION

This process is not recommended for most foods as it involves treating the product to levels of radiation of approximately 30 kGy. This high level of radiation kills all vegetative cells and also destroys spores from organisms such as *Clostridium botulinum*.

1.2.2   MODE OF ACTION

The mode of action of ionizing radiation can be considered in three phases:

a) The primary physical action of radiation on atoms
b) The chemical consequences of these physical actions
c) The biological consequences of living cells in food or contaminating organisms

When the radiation rays pass through the foods, there are collisions between ionizing radiations and food particles at the atomic and molecular levels, resulting in the production of ion pairs and free radicals. Because most foods contain substantial quantities of water and contaminating organisms contain water in their cell structure, the radiolytic products of water are particularly important. Ionizing radiation splits water to produce hydrogen, hydrogen radicals (H^+, OH^-), H_2, H_2O_2, and H_3O^+. These species are highly chemically reactive and react with many substances. These results in activation of microorganisms or insects as the chemical changes cause alteration of cell membrane structure, reduced enzyme activity, reduced nucleic acid synthesis, affects energy metabolism through phosphorylation, and produce compositional changes in cellular DNA. In radiation, inactivation of microorganism is achieved without the need for high temperatures. Hence, it is also called "cold sterilization."[6]

1.2.3   APPLICATION OF IRRADIATION IN DAIRY INDUSTRY

It is well known that despite substantial efforts in avoidance of contamination, an upward trend in the number of outbreaks of foodborne illnesses caused by nonspore-forming pathogenic bacteria have been reported in many countries. Good hygienic practices can reduce the level of contamination but the most important pathogens can presently neither be eliminated from most farms nor it is possible to eliminate them by primary processing, particularly from those foods which are sold raw. Several decontamination methods exist but the most versatile treatment among them is the processing with ionizing radiation. Decontamination of food by ionizing radiation is a safe, efficient, environmentally clean, and energy efficient process. Irradiation is particularly valuable as an end product decontamination procedure. Radiation treatment at doses of 2–7 kGy, depending on condition of irradiation and the milk and milk product can effectively eliminate potentially pathogenic nonspore-forming bacteria including both longtime recognized pathogens such as *Salmonella* and *Staphylococcus aureus* as well as emerging or "new" pathogens such as *Campylobacter*, *Listeria monocytogenes*, or *Escherichia coli* O157:H7 from suspected milk and milk products without affecting sensory, nutritional, and technical qualities.

Irradiation of fluid milk also results in unacceptable flavor scores. Off-flavors and browning originating from chemical reactions involving lactose were identified. Sterilization of yogurt bars, ice cream, and nonfat dry milk by gamma irradiation using a dose of 40 kGy at 78°C resulted in an overall decrease in acceptance. Preservation of yogurt using irradiation was studied and it was reported that the nonirradiated plain yogurt reached a population of 10^9 cfu/g after 6 days and irradiated plain yogurt using a dose of 1 kGy, the microbial load was found to be significantly lower as compared to nonirradiated even after the end of 18 days. Irradiation combined with refrigeration further extended the shelf life of yogurt to 29–30 days.

The major application of irradiation in milk and milk products is in cheese sector. However, it is reported by several scientists that dosage rate as low as 0.5 kGy can induce off-flavors into cheese products rendering them nonacceptable by the consumers. Cheeses such as cheddar, Gouda, Camembert, and others are often subjected to irradiation to increase the shelf life of the product using lower dosage rates, while higher doses are required for sterilization purposes. The product once treated has an indefinite shelf life from a microbiological standpoint, provided of course that sterility is maintained.

1.3   HIGH-PRESSURE PROCESSING

High-pressure technology is increasingly being used to produce value-added food products. Although the fact that "high pressure kills microorganisms and preserves food" was discovered way back in 1899 and has been used with success in chemical, ceramic, carbon allotropy, steel/alloy, composite materials, and plastic industries for decades, yet it was only in the late 1980s that its commercial benefits became available to the food processing industries.[36]

Hite[24] studied the effect of high pressure on foodborne pathogens in milk and reported that there was a significant reduction in viable microbes when subjecting the milk to a pressure of 650 MPa. The use of HPP has come about so quickly due to the ability of HPP to deliver foods with fresh-like taste without added preservatives. Till date, a variety of food products such as jams, fruit juice, tomato juice, meat, oysters, ham, fruit jellies, pourable salad dressings, salsa, and poultry have been processed using this technology.[38] Figure 1.1 shows a simplified flow diagram.

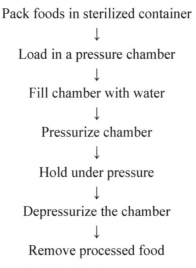

FIGURE 1.1   Flow diagram for HPP of foods.

There are two general principles of the application of HPP in food processing: *Le Chatelier's Principle and Isostatic Rule*. The former states that any phenomenon that is accompanied by a decrease in volume will be enhanced by pressure; and the latter indicates that pressure is instantaneously and uniformly transmitted throughout a sample.

1.3.1   EFFECTS OF CHARACTERISTICS OF FOOD ON EFFICACY OF HPP

The primary aim of treating foods with HPP in most cases is to reduce or eliminate the relevant foodborne microorganisms. The molecular composition of the food substrate, however, can significantly affect the extent to which HPP kills the inhabitant microorganisms. The application of HPP may inactivate enzymes or alter the physical properties of the food material (e.g., denature structural proteins or densify texture). It is important to consider both the effects of the food substrate on slowing the microbial inactivation kinetics and the effect of the HPP on the properties of the foodstuff, when optimizing processing conditions for specific foods. It is reported that microorganisms appear to be less sensitive to the lethal effects of high pressure when treated in certain foods, such as milk. The reasons for these effects are not clear, but it may be that certain food constituents such as proteins and carbohydrates or cations such as Ca^{2+} contribute to this effect. The pH and water activity (a_w) of foods can also significantly affect the inactivation of microorganisms by HPP. Most microorganisms tend to be more susceptible to pressure in lower pH environments, and pressure-damaged cells are less likely to survive in acidic environments.

1.3.2 MICROBIAL ASPECTS OF HPP

High pressures processing can lead to microbial inactivation, probably due to two reasons: protein denaturation and cell injury. Proteins integrated in membranes surrounded by lipids are of fundamental importance in membrane-bound activity. In bacteria, the integrity of the membrane is very sensitive to pressure. Hence, a destabilization effect may occur when applying high pressure to bacteria due to denaturation of enzyme(s) important in the metabolism of microorganisms. On the other hand, cells under pressure can shrink due to change in volume. Therefore, the change in volume might affect the membrane structure and cell content leading to the protein effect explained above. Also after depressurization, the cell membrane can be injured and/or disrupted, causing leakage of the cell content and then to cell death.[29]

The response of different types of microorganisms to pressure varies significantly in the following order of increasing resistance: vegetative bacteria < yeasts and molds < viruses < bacterial spores. The resistant of microorganisms to pressure varies considerably depending on the pressure range applied, temperature and treatment duration, and type of microorganisms. In HPP, the food product is subjected to pressure usually within the range of 300–700 MPa for inactivation of vegetative microorganisms.

Bacterial endospores commonly require pressures applied at elevated treatment temperatures. Gram-positive bacteria tend to be more resistant than Gram-negative bacteria.[5,20] A cell in the exponential phase of growth have been found to be less resistant than those in the stationary phase.[31] Yeast and mold are more sensitive to pressure, whereas viruses are very resistant.

In HPP, bacterial spores were found to be more resistant than vegetative bacteria, surviving up to 1200 MPa at room temperature.[39] It is well understood that by improving HPP conditions or combination with other treatments and agents, spores can be inactivated. It was found that relatively low pressure, for example, 100 MPa, or even less, could cause germination of some spores. Suspensions of *Bacillus cereus, Bacillus subtilis,* and *Bacillus pumilus* spores in water or potassium phosphate buffer were germinated by pressures of 32.5 and 97.5 MPa, respectively. After germination, spores were more sensitive to agents such as heat, pH, and pressure than they were in dormant state. Another approach is to combine heat treatment with HPP. A mild heat treatment can encourage spores to germinate, making them more susceptible to HPP. Therefore, a preheat treatment followed by HPP can be more effective for inactivating spores than heating during pressurization.[16]

1.3.3   COMBINATION OF HPP WITH OTHER TREATMENTS

HPP has also been studied in combination with other treatments to achieve better microbial inactivation and shelf life extension. The combined effect of HHP and heat on microbial inactivation has been widely investigated among various microorganisms suspended in different media, and in solids, and liquids foods. Inactivation of *L. monocytogenes* in ultrahigh temperature processing (UHT) milk can be achieved after 15 min of treatment at 200 MPa and 55°C. As mentioned earlier, combined effect of pressure and heat also enhances the inactivation of spores which otherwise is difficult to inactivate when pressure or heat is applied solely. The effect of HHP on enzymes varies widely for different enzymes, probably due to the structural differences of individual enzymes. In general, combinations of pressure with moderate temperatures increase the level of enzyme inactivation, but in some cases, an increase in enzyme activity has been reported.[23] Temperatures between 45°C and 55°C and pressures between 600 and 900 MPa can inactivate pectinesterase, lipase, polyphenol oxidase

(PPO), lipoxygenase, peroxidase (POD), lactoperoxidase, phosphatase, and catalase in different extensions.

1.3.4 APPLICATIONS OF HPP IN DAIRY INDUSTRY

HHP processing can be applied successfully at room temperature, reducing the thermal energy needed for heat processing[20,27] to deliver food products without undesirable changes such as in nutritional and sensory characteristics.[39] HHP is already applied in the food industry to pasteurize guacamole, meats, and seafood but not for sterilization. In 1990, the first HHP processed pasteurized jam (low acid product) appeared on the Japanese retail market, and in 1991 fruit yogurts, jellies, salad dressing, fruit sausages, and citrus juices also appeared.[22] Thermal processing has been shown to reduce the antioxidant activity of fruits. A study showed that HPP (450 MPa) revealed better retention of antioxidant activity and color compounds in fruit smoothies as compared to thermal treatment.[26] The detailed applications of HPP in milk and milk products are dealt in separate chapter in the book.

1.4 ULTRASOUND

Ultrasound is a form of energy generated by sound waves of frequencies above 16 kHz. Until recently the majority of applications and developments of ultrasound in food technology involves the noninvasive analysis with particular reference to quality assessment. More recently, food technologists have discovered that it is possible to use ultrasound to aid processing. The type of ultrasound used in these applications is more powerful (>5 W/cm^2) and at lower frequency (<40 kHz) than that used in diagnostic applications and generally referred to as powerful ultrasound.[30]

1.4.1 PRINCIPLE OF OPERATION

The low-frequency high-power ultrasound has sufficient energy to break intermolecular bonds and along with intensities, cavitation effect can be generated which can alter physical properties as well as enhance or modify many chemical reactions. When ultrasound passes through a liquid medium, cavitation occurs causing alternate compressions and rarefactions. These regions of pressure change cause bubble to form in medium and expand. A point is reached where the ultrasonic energy provided is not sufficient to retain the vapor phase in the bubble, therefore the bubble collapses creating shock waves. This creates region with high temperature and pressure reaching up to 5500°C and 50 MPa.[15]

The cavitation bubbles have variety of effects within the liquid medium depending upon the type of system in which it is generated. These systems can be broadly classified into: homogeneous liquids (as in water), heterogeneous solid–liquid (as in cleaning of the surface with water with the aid of ultrasound, extraction of oil from oilseeds, etc.), and heterogeneous liquid–liquid (aids disruption and mixing resulting in the fine emulsion).[30]

A power ultrasound consists of three basic parts: (1) generator; (2) transducer; and (3) coupler. Ultrasonic generators transform electrical energy into ultrasound energy (a type of mechanical energy) via a transducer and then transmitted to the medium through a coupler.[15]

1.4.2   ULTRASOUND AND COMBINED TECHNOLOGIES

Ultrasound is often more effective when combined with other antimicrobial methods. Hence, different authors have attempted to use it in combination with other antimicrobial methods to increase its effect on microbial and enzyme inactivation. Beneficial combinations include thermosonication (heat and ultrasound), manosonication (MS; pressure and ultrasound), and manothermosonication (MTS; pressure, heat, and ultrasound).[34]

The application of ultrasound and heat has been termed thermosonication. Heat combined with ultrasound is considered to reduce process temperatures and processing times, for pasteurization or sterilization processes that achieve the same lethality values as with conventional processes.[30,43] Reduction of the temperature and/or processing time should result in improved food quality.[44] The combined use of heat and ultrasound markedly increases the lethality of heat treatments and consequent reductions in time and/or temperature of heat process. Combined heat and ultrasound treatments have been reported to lower maximum processing temperatures by 25–50%. After treatment, changes in color and vitamin C were minimal. Different authors have investigated combinations of heat and ultrasound to decrease the intensity of heat treatments. The heat resistance of *B. cereus*, *Bacillus licheniformis*, *Bacillus stearothermophilus*, and *Thermoduric streptococci* decreased following ultrasonication treatment at 20 kHz.[15] The simultaneous application of ultrasound with an external hydrostatic pressure of up to 600 kPa (MS) increases substantially the lethality of the treatment. It is reported that MS treatments sensitize spores of *B. subtilis* to lysozyme.[35]

MTS has proved to be an efficient tool to inactivate microorganisms, especially in those conditions under which their thermos tolerance is higher and deactivation of enzymes such as POD, lipoxygenase, lipase, and protease, and tomato or orange pectin methylesterase.[15] Unfortunately, ultrasound does not inactivate alkaline peroxidize or lactoperoxidase enzymes. These enzymes thus cannot be used to indicate a successful ultrasonic treatment. If ultrasonication is to be used as an alternative to thermal pasteurization, a need exists to find a quick and efficient method to indicate whether ultrasonication was sufficient in terms of ensuring a microbiologically safe product.

1.4.3 APPLICATION OF ULTRASOUND IN DAIRY INDUSTRY

One of the earliest uses of power ultrasound in processing was in emulsification. If bubbles collapse near the phase boundary of two immiscible liquids, the resultant shock wave can provide a very efficient mixing of the layers.[30] Trials have shown that fat globules undergo a substantial reduction in size (up to 80%) following ultrasonication and emulsions produced are often more stable than those produced conventionally. Ultrasound has been used industrially in the manufacture of salad cream, tomato ketchup, peanut butter, some cream soups, and fruit juices. Ultrasonic cutting is another application of ultrasound in the food industry. This technique helps improve the quality of food segments by a reduction of the cutting force. This improvement is achieved by a modification of tool work piece interactions as the movement of the cutting device is super-positioned with ultrasonic excitation. D'Amico et al.[14] reported that ultrasound treatment (sonifier probe at 20 kHz, 100% power level, 150 W acoustic power intensity) with or without the effect of mild heat (57°C) resulted in 5 log reduction of *L. monocytogenes* in UHT milk, a 5 log reduction in total aerobic bacteria in raw milk. Zenker et al.[44] revealed that ultrasound treatment has a positive influence on shelf life, improvement in surface color stability (lightness), and L-ascorbic acid retention in fruit and vegetable juice besides reduction in the microbial count in the medium.

Ultrasound is also widely used to enhance the extraction of different components, including herbal extracts, protein, oil, bioactives, etc.[14,40] The mechanical effect of ultrasound provides greater penetration of solvent into cellular materials and improves mass transfer. There is an additional benefit for the use of power ultrasound in extractive processes, which result from the disruption of biological cell walls to facilitate the release of contents.[30] Vilkhu et al.,[42] reviewed that ultrasound assisted extraction (UAE) process provides an opportunity for extraction of heat sensitive bioactive and food components at lower processing temperatures. In freezing, ultrasound assists in shortening of freezing process and control crystal size distribution in the product, less cell damage, thereby leads to a product of better quality, for example, in the bakery-based ice creams.[28]

The application of ultrasound during filtration can eliminate the negative effects of membrane fouling and concentration polarization, thus enhancing the filtration efficiency.[40] When ultrasound is introduced during filtration, cake layer and concentration polarization on the membrane surface can be disrupted, whereas the intrinsic permeability of the membrane is not affected by ultrasound. As a consequence, the flow resistance decreases and the flux increase.[13] In the dairy, ultrasound has been successfully employed to enhance the dairy whey filtration.[30,32]

1.5 PULSED ELECTRIC FIELD

PEF processing is a nonthermal method of food processing that uses short bursts of electricity for microbial inactivation and causes minimal or no detrimental effect on food quality attributes.[1,2,8,21] High-intensity PEF processing involves the application of pulses of high voltage (typically 20–80 kV/cm) for short time periods (less than 1 s) to fluid foods placed between two electrodes.[7] During this process, mainly microbial and also some enzyme inactivation is achieved, while minimizing the heating of foods, therefore, reducing detrimental changes in the sensory and physical properties of food.

1.5.1 PULSED ELECTRIC SYSTEM

The high-intensity PEF processing system consists of five major components: a high-voltage power supply, an energy storage capacitor, a treatment chamber(s), a pump (to conduct food through the treatment chamber(s), a cooling device, voltage current, and temperature controlling device), and a computer to control operations.[21] Generation of PEF requires a fast discharge of electrical energy within a short period of time which is accomplish by the pulse forming network, an electrical circuit consisting of one or more power supplies with the ability to charge voltages (up to 60 kV), switches, capacitors (0.1–10 μF), inductors (30 μH), resistors (2–10 Ω), and treatment chambers.

1.5.2 PRINCIPLE OF OPERATION

Inactivation of microorganisms exposed to high-voltage PEF is related to the electromechanical instability of the cell membrane. Several theories have been proposed to explain microbial inactivation by PEF; the most accepted being electroporation. Thus, electroporation brings about through the action of an intense electric field loss of membrane integrity, inactivation of proteins, and leakage of cellular contents of microorganisms. Electric field strength and treatment time are two most important factors involved in PEF processing.[11,25] When exposed to high electrical field pulses, the cell membranes develop pores either by enlargement of existing pores or by the creation of new ones. The pores increase membrane permeability, allowing the loss of cell content or intrusion of surround media, either of which can cause cell death.[12]

The microbial cell membranes are affected by factors such as: electric field intensity, pulse width, treatment time, treatment temperature, pulse waveshapes, type of microorganism, concentration and growth stage of microorganism, pH, presence or absence of antimicrobials and ionic compounds, medium conductivity, and medium ionic strength.

1.5.3 APPLICATION OF PEF IN DAIRY INDUSTRY

Application of PEF is restricted to dairy products that can withstand high electric fields, have low electrical conductivity, and do not contain or form bubbles.[11] The products are minimally affected by the process since damage occurs on a cellular level and flavor and enzyme activity are not significantly diminished. Among all liquid products, PEF technology has been most widely applied to apple juice, orange juice, milk, liquid egg, and brine solution.[33] Homogenized milk inoculated with *Salmonella dublin* when treated with 36.7 kV/cm and 40 pulses for 25 min had no growth after storage at 7–9°C for 8 days. Similarly, PEF treated (40 kV/cm, 30 pulses, time of 2 µs) skim milk (0.2% milk fat) had a shelf life of 15 days. However, treatment of skim milk with 80°C for 6 s followed by PEF treatment at 30 kV/cm, 30 pulses and pulse width of 2 µs increased the shelf life to 22 days. PEF treated skim milk inoculated with *E. coli* was able to achieve a 4.0 log reduction when treated with 63 pulses of 2.5 µs pulse width.

PEF technology has recently been used in alternative applications including drying enhancement, enzyme activity modification, preservation of solid and semisolid food products, and wastewater treatment, besides pretreatment applications for improvement of metabolite extraction.[2,21,36]

1.6 PULSED LIGHT AND UV TECHNOLOGY

Pulsed light (PL) involves the use of intense and short-duration pulses of broad spectrum "white light" (ultraviolet (UV) to the near-infrared region). UV processing involves the use of radiation from the UV region of the electromagnetic (EM) spectrum. The germicidal properties of UV irradiation (200–280 nm) are due to DNA mutations induced by DNA absorption of the UV light. Traditionally, this technology, mainly due to its poor penetration level, has been used to sterilize or reduce the microbial population on food surfaces, packaging material, or processing plants. Other new technologies based on the treatment with light at different wavelengths have been lately developed to inactivate microorganisms in foods and beverages. A recently patented process described the sanitation of fresh foods and beverage products using multiple stages of exposure at different wavelengths of UV, near-IR, and IR light.

1.6.1 PRINCIPLE OF OPERATION

Susceptibility of microorganisms to PL is in order of: Gram-negative bacteria < Gram-positive bacteria < fungal spores. The lethal action of PL is due to a photothermal and/or a photochemical mechanism. It is possible that both mechanisms coexist, and the relative importance of each one would depend on the target microorganism.[37]

One disadvantage in PL treatment is the possibility of shadowing, which makes the microorganisms in lower layers very hard to destroy. It should be noted that more is opaque and thicker is the food item, the lower is the inactivation of microorganisms below the surface.[9] Also, the entire surface of the food should be flashed in order to achieve the decontamination of its whole surface. There should not be any irregularities of the food surface as it complicates the decontamination process due to shadowing. Also, if many food items are treated together, each piece can shadow the other which again complicates the process. Factors influencing the efficacy of a PL treatment are: distance of food from a source of light (nearer the sample from the lamp, higher is the efficacy of PL treatment); sample thickness; and food composition (e.g., the presence of proteins and oil decreases the efficacy of PL).[18]

1.6.2 APPLICATIONS OF PULSE LIGHT IN DAIRY INDUSTRY

PL treatment of foods is approved by the FDA[17] under the code 21CFR179.41. Continuous flow PL technique was used for inactivation of *S. aureus* in milk. Milk temperature increased up to 38°C, depending on the residence time as well as the distance of the product from the light source. This temperature increase caused a fouling effect as well as possible changes in milk quality.

1.7 RADIO FREQUENCY HEATING OF FOODS

Radio frequency (RF) heating is a promising technology for dairy applications because of the associated rapid and uniform heat distribution, large penetration depth, and lower energy consumption. RF heating has been successfully applied for drying, baking, and thawing of frozen meat, and in meat processing. However, its use in continuous pasteurization and sterilization of foods is rather limited.

During RF heating, heat is generated within the product due to molecular friction resulting from oscillating molecules and ions caused by the applied alternating electric field. RF heating is influenced principally by the dielectric properties of the product when other conditions are kept constant. It is evident that frequency level, temperature, and properties of food, such as viscosity, water content, and chemical composition, affect the dielectric properties and thus the RF heating of foods. Therefore, these parameters should be taken into account when designing a RF heating system for foods.

1.7.1   PRINCIPLE OF RADIO FREQUENCY (DIELECTRIC) HEATING

There are three types of electronic heating: ohmic, dielectric (RF and MW), and induction.

Ohmic or *electric* resistant heating relies on direct ohmic conduction losses in a medium and requires the electrodes to contact the medium directly. Ohmic heating gives a direct heating because the product acts as an electrical resistor. The heat generated in the product is the loss in resistance. Since the heating effect depends on the eddy current induced in the material, this type of heating works well with conductors. In food processing, ohmic heating is used mainly for liquid products, as it is possible to establish the necessary electrical contact between electrodes and the media.

Induction produces heat by Joule effect in a conductor by inducing eddy currents in a manner similar to that of a transformer when a current in the secondary windings is generated by induction from a current in the primary windings. Heating can be confined to that part of the work piece or material, which is directly opposite the coil inducing the current. In this system, the secondary coil is made of stainless steel pipe through which the products flow. Fluids are heated instantaneously as they pass through the coiled pipe.

RF heating involves the heating of poor electrical conductors. It is also characterized by freedom from electrical and mechanical contact. RF heating involves the application of a high voltage alternating electric field to a medium sandwiched between two parallel electrodes. One electrode is grounded to set up capacitor configuration. This causes the operation to resemble a condenser in which some energy loss occurs across the plates causing a heating effect in the material spacing the plates. In condensers, the interplate material may be paper, air or plastic, or other electrical insulators with the ability to separate the potential difference of the plates without breakdown as the voltage increases. There is practically no possibility that compounds of metal of the RF heater will migrate into the food.

Typically RF heaters generate heat by means of an RF generator that produces oscillating fields of EM energy and comprises a power supply and control circuitry, a hydraulic press and a parallel plate, and a system for supporting processed material. The walls of the plates do not provide the electrical field since it is produced by the generator, but they do offer the boundaries of the field. Three frequencies are commonly employed for RF heating: 13.56, 27.12, and 40.68 MHz.

RF heaters that are poorly designed or shielded can emit stray RF EM energy that could be hazardous to health. Of particular concern is that RF heaters operate at frequencies in or near the region of human absorption (60–100 MHz). Hence, the workers near RF heaters can absorb considerable amounts of stray energy. The problem of stray fields can be relatively easily solved by proper design and appropriate location of shields for the RF fields. RF heating differs from the higher frequency MW heating (915 and 2450 MHz) in which the wavelength is comparable or smaller than the dimensions of the sample and heating occurs in a metal chamber with resonant EM standing wave modes, as in a MW oven, or in a waveguide. MW energy is generated by special oscillator tubes magnetrons or klystrons, and it can be transmitted to an applicator or antenna through a waveguide or coaxial transmission line. The output of such tubes tends to be in a range from 0.5 to 100 kW and requires a power supply.

MWs are primarily radiation phenomenon; they are able to radiate into a space that could be the inside of the oven or cavity. MW ovens incorporate a waveguide to deliver MW energy to cook food in a cavity. In the MW frequency range, the dielectric heating mechanism dominates up to moderated temperatures. The water content of the foods is an important factor for the MW heating performance. For normal wet foods, the penetration depth from one side is approximately 1–2 cm at 2450 MHz. Because RF uses longer wavelengths than MW, EM waves in the RF spectrum can penetrate deeper into the products so there is no surface overheating or hot or colds spots, common problems with MW heating. The RF heating also offers simple uniform field patterns as opposed to the complex nonuniform standing wave patterns in a MW oven.

1.7.2 APPLICATIONS OF RF AND MW HEATING IN DAIRY INDUSTRY

There are many promising applications of RF heating in the dairy industry because of the potential to improve the quality of food products. Continuous-flow MW treatment has been proposed for milk pasteurization due to its potential advantages over the conventional tubular and plate heat exchangers. Continuous milk pasteurization has been successful at 2450 MHz using a simple waveguide heat exchanger. The complete inactivation of *Yersinia enterocolitica, Campylobacter jejuni,* and *L. monocytogenes* occurred at 8, 3, and 10 min when the cells were heated at a constant temperature of 71.1°C using MWs with initial microbial loads of 106–107 K/mL.

1.8   MEMBRANE PROCESSING TECHNOLOGY

Membrane separations are techniques used industrially for removal of solutes and emulsified substances from solutions by application of pressure onto a very thin layer of a substance with microscopic pores, known as a membrane. Membrane separation processes include reverse osmosis (RO), ultrafiltration (UF), microfiltration (MF), dialysis, and electrodialysis (ED). Of the membrane separation techniques, RO, UF, MF, and ED are commercially used. UF has found many applications in food processing and has been successfully employed in a number of liquid "cold sterilization" and clarification applications. For the case of UF, membrane pore size ratings are generally in the range of 0.001–0.020 μm.

1.8.1   ELECTRODIALYSIS

The main principle of ED functioning is the separation of ions from an aqueous solution through the application of membranes and electricity. Anion and cation exchange membrane systems are stabilized between negative and positive charged electrodes. During the passage of a substance comprising ions from the channels of the system, the Na^+ ions move toward the negatively charged electrode, whereas the Cl^- ions move toward the positive. Since membranes are impermeable to ions, the latter are accumulated at both sides of the system, depending on their load, while the ionic charge is reduced at the remaining partitions. A typical example is the demineralization of cheese whey products.

1.8.2   REVERSE OSMOSIS

The diffusion of substances through RO membranes is achieved by pressure application. The filtration efficiency of various carbonate or noncarbonate solutes can reach up to 95–99% resulting in the production of water with exceptional qualitative characteristics. RO is widely applied for processing raw and wastewater as it is economical and can be carried out at room temperature. The application of RO in the dairy industry focuses mainly on the separation of waste solids from the water used for the different operations. RO is used to achieve to reduce the biochemical oxygen demand (BOD) of by-products in order to reduce the cost of water treatment and secondly to the export of high added value solids which can be dehydrated and disposed for feeding requirements.

1.8.3 NANOFILTRATION

In recent years, nanofiltration (NF) has been extensively used in industrial scale for separating particles that bear or not electric charge in water solutions. NF is often applied in the following separation processes; desalting of brackish and seawater, retention of minerals from wastewater, effluents processing of industries correlated to the manufacturing of clothing and removal of contaminants from companies producing medicines and edible and organic products. Prescribing these separation processes is extremely important for these industries. As a result, several models have been tested for parallel determination of steric results and control of membranes' permeability.[10] NF procedure can be effectively used to separate different substances from liquid streams of organic nature. Through this process, both cost and environmental pollution can be minimized. Furthermore, NF can be applied for the removal of undesirable or hazardous substances of low molecular weight.[19]

1.8.4 ULTRAFILTRATION

UF is a membrane process the driving force of which is pressure. In this method, membranes with suitable properties are used for the separation of large molecules and colloidal substances from liquids. The size of the molecules that can pass through the membranes depends on the diameter of membrane pores. Substances with a molecular weight lower than 300 are easily dissolved in liquid solutions and characterized by increased osmotic pressure. During the application of UF, these substances penetrate the membrane as filtration depends on the size of the pores and not on osmotic processes. As a result, UF can be applied at low-pressure levels (usually 1–5 atm). UF is effectively used in a wide range of applications for the manufacture of various foods such as dairy products, products of plant origin, alcoholic beverages, sweets, etc. This technology finds important application in the treatment of water and effluents as well.[32]

1.8.5 MICROFILTRATION

MF can be easily compared to UF and traditional filtration. Pores of these membranes can be manufactured from various substances. Thus, ceramic, glass, metallic or even membranes made from metal oxides, graphite, or different polymers are widely used. MF can be applied in various industries either to protect products from microorganisms or for particles separation. MF membranes are highly applied for carrying out solid–liquid size separation because they are characterized by several technical advantages such as moderate conditions, no phase alterations, decreased energy requirements, absence of additives, and effective design. The most important disadvantage of MF membranes is the existence of fouling phenomena. Specifically, at increased flow rates a layer of particles forms on the surface of the membranes. However, at decreased concentration levels no layer formation normally occurs. Another major disadvantage of this kind of membranes is the pore blocking due to the accumulation of foulants.

1.8.6 OSMOTIC DISTILLATION

Osmotic distillation (OD) is a concentration method widely used in the processing industry of several of liquid foods (e.g., fruit or vegetable juices). It can effectively take place at low temperatures, maintaining the qualitative characteristics of these perishable products. OD can be characterized as a direct osmosis, in which removal of water from a feed can occur by means of a hypertonic solution (mainly concentrated brine) flowing downstream a membrane. It can be distinguished from the common direct osmosis due to the specific membrane used, which has pores and hydrophobic properties, and is mainly fabricated from polypropylene.[3]

1.8.7 MEMBRANE DISTILLATION

Membrane distillation (MD) is a thermally based procedure used for the separation of vapor molecules which can penetrate the microporous membrane layer. The membrane is characterized by high hydrophobicity and the filtration process takes place due to vapor pressure difference (caused by temperature changes) across the membrane system. It can be applied to cover various needs such as desalination, wastewater processing, etc. and is a very useful technique for the food industry.[4]

1.8.8 DISADVANTAGES OF MEMBRANE TECHNOLOGY

Fouling is a process resulting in loss of performance of a membrane due to the deposition of suspended or dissolved substances on its external surfaces, at its pore openings or within its pores. During membrane operation, fouling can occur when certain compounds in the filtered substance are plugged or coated on membrane pores limiting the ability to filter out and disabling membrane. Concentration polarization phenomenon is one of the most severe problems occurring in the operation of membrane processes because it can cause significant decrease of the transmembrane flux. During the filtration process, a few substances in the filtered materials can be rejected by the membranes. Occasionally, some of these substances can be accumulated on the membranes' surface and trigger the concentration polarization phenomenon. Concentration polarization incidents are characterized by complexity and occur when separation of substances contained in the filtered product clog membrane pores or reduce flux rate and gradually rise to qualitative degradation of the final product.

1.8.9 APPLICATION OF MEMBRANE TECHNOLOGY IN DAIRY INDUSTRY

Membrane processes have found a continuously increasing number of applications in the protein industry (whey/milk protein standardization/ concentration, soy protein isolates), alcoholic and nonalcoholic drinks (wine, beer, fruit juice both cloudy and clear), and egg products. Membrane processes have a crucial advantage over other technologies, that is, they can potentially support sustainable industrial growth by saving energy, decreasing capital costs, lessening environmental impact, and optimizing the exploitation of raw materials. Membrane processes (MPs) have already replaced their conventional energy-intensive techniques thus leading to substantial drop in energy, cost, and environmental impact.

1.9 SUMMARY

Many of the novel technologies described in this chapter (e.g., high hydrostatic pressure, PEF, PL, ultrasound, etc.) are able to produce more nutritive, fresher, and minimally processed dairy foods. However, they are still in the research arena and their effectiveness on microorganisms and enzyme inactivation still has to be established with any possible health risks. Therefore, additional research is still needed on these techniques before implementing these technologies at the industrial scale.

KEYWORDS

- dielectric heating
- high-pressure processing
- manosonication
- manothermosonication
- membrane distillation
- membrane processing
- oscillating magnetic fields
- osmotic distillation
- pulsed electric field
- pulsed light

REFERENCES

1. Ade-Omowaye, B. I. O.; Angersbach, A.; Taiwo, K.; Knorr, D. Uses of Pulsed Electric Field Pretreatment to Improve Dehydration Characteristics of Plant Based Foods. *Trends Food Sci. Technol.* **2001**, *12*(8), 285–295.

2. Ade-Omowaye, B. I. O.; Eshtiaghi, N.; Knoor, D. Impact of High Intensity Electric Field Pulses on Cell Permeabilization and as Pre-processing Step in Coconut Processing. *Innov. Food Sci. Emerg. Technol.* **2000,** *1*(3), 203–209.

3. Ali, F.; Dornier, M.; Duquenoy, A.; Reynes, M. Evaluating Transfers of Aroma Compounds During the Concentration of Sucrose Solutions by Osmotic Distillation in a Batch-type Pilot Plant. *J. Food Eng.* **2003,** *60,* 1–8.

4. Alkhudhiri, A.; Darwish, N.; Hilal, N. Membrane Distillation: A Comprehensive Review. *Desalination* **2011,** *287,* 2–18.

5. Alpas, H.; Kalchayanand, N.; Bozoglu, F.; Sikes, A.; Dunne, C. P.; Ray, B. Variation in Resistance to High Hydrostatic Pressure Among Strains of Food Borne Pathogens. *Appl. Environ. Microbiol.* **1999,** *65,* 4248–4151.

6. Arvanitoyannis, I. S.; Stratakos A. C.; Tsarouhas, P. Irradiation Applications in Vegetables and Fruits: A Review. *Crit. Rev. Food Sci. Nutr.* **2009,** *49,* 427–462.

7. Barsotti, L.; Merle, P.; Cheftel, J. C. Food Processing by Pulsed Electric Felds, I: Physical Aspects. *Food Rev. Int.* **1999,** *15,* 163–180.

8. Bazhal, M. I.; Ngadi, M. O.; Raghavan, G. S. V.; Smith, J. P. Inactivation of *Escherichia coli* O57:H7 in Liquid Whole Egg Using Combined Pulsed Electric Field and Thermal Treatments. *Lebensm. Wiss. Technol.* **2006,** *39,* 419–425.

9. Beuchat, L. R. Vectors and Conditions for Preharvest Contamination of Fruits and Vegetables with Pathogens Capable of Causing Enteric Diseases. *Br. Food J.* **2006,** *108,* 38–53.

10. Bowen, W. R.; Welfoot, J. S. Modelling of Membrane Nanofiltration Pore Size Distribution Effects. *Chem. Eng. Sci.* **2002,** *57,* 1393–1407.

11. Calderon-Miranda M. L.; Barbosa-Canovas, G. V.; Swanson, B. G. Inactivation of *Listeria innocua* in Skim Milk by Pulsed Electric Fields and Nisin. *Int. J. Food Microbiol.* **1999,** *51,* 19–30.

12. Castro, A. J.; Barbosa-Canovas, G. V.; Swanson, B. G. Microbial Inactivation of Foods by Pulsed Electric Fields. *J. Food Process. Preserv.* **1993,** *17,* 47–73.

13. Chemat, F.; Hum, Z.; Khan, M. K. Applications of Ultrasound in Food Technology: Processing, Preservation and Extraction. *Ultrason. Sonochem.* **2010,** *18,* 813–835.

14. D'Amico, F. A. N.; Oliveira, F. I. P.; Guo, M. R. Inactivation of Microorganisms in Milk and Apple Cider Treated with Ultrasound. *J. Food Prot.* **2006,** *69*(3), 556–563.

15. Demirdoven, A.; Baysal, T. The Use of Ultrasound and Combined Technologies in Food Preservation. *Food Rev. Int.* **2009,** *25,* 1–11.

16. Devlieghere, F.; Vermeiren, L.; Debevere, J. New Preservation Technologies: Possibilities and Limitations. *Int. Dairy J.* **2004,** *14*(4), 273–285.

17. FDA. *Code of Federal Regulations,* 21CFR179.41; FDA: Washington, DC, 1996; Vol. 31, pp 1–11.

18. Fine, F.; Gervais, P. Efficiency of Pulsed UV Light for Microbial Decontamination of Food Powders. *J. Food Prot.* **2004,** *67,* 787–792.

19. Frappart, M.; Akoum, O.; Ding, L. H.; Jaffrin, M. Y. Treatment of Dairy Process Waters Modelled by Diluted Milk Using Dynamic Nanofiltration with a Rotating Disk Module. *J. Membr. Sci.* **2006,** *282,* 465–472.

20. Gervilla, R.; Ferragut, V.; Guamis, B. High Pressure Inactivation of Microorganisms Inoculated into Ovine Milk of Different Fat Contents. *J. Dairy Sci.* **2000,** *83,* 674–682.

21. Giri, S. K.; Mangaraj, S. Application of Pulsed Electric Field Technique in Food Processing—A Review. *Asian J. Dairy Food Res.* **2013**, *32*(1), 1–12.

22. Guerrero-Beltrán, J. A.; Barbosa-Cánovas, G. V.; Swanson, B. G. High Hydrostatic Pressure Processing of Fruit and Vegetable Products. *Food Rev. Int.* **2005**, *21*, 411–425.

23. Hendrickx, M.; Ludikhuyze, L.; Van den Broeck, I.; Weemaes, C. Effects of High Pressure on Enzymes Related to Food Quality. *Trends Food Sci. Technol.* **1998**, *9*, 197–203.

24. Hite, B. H. *The Effects of Pressure in the Preservation of Milk*; Agriculture Experiment Station Bulletin 58; Va. University: Washington, 1999; pp 15–35.

25. Jeyamkondan, S.; Jayas, D. S.; Holley, R. A. Pulsed Electric Field Processing of Foods: A Review. *J. Food Prot.* **1999**, *62*(9), 1088–1096.

26. Keenan, D. F.; Brunton, N. P.; Gormley, T. R.; Butler, F.; Tiwari, B. K.; Patras, A. Effect of Thermal and High Hydrostatic Pressure Processing on Antioxidant Activity and Color of Fruit Smoothies. *Innov. Food Sci. Emerg. Technol.* **2010**, *11*, 551–556.

27. Knorr, D. Effects of High-hydrostatic-pressure Processes on Food Safety and Quality. *Food Technol.* **1993**, *47*, 156–161.

28. Liyun, Z., Sun, D. Innovative Application of Power Ultrasound During Food Freezing Processes—A Review. *Trends Food Sci. Technol.* **2006**, *17*(1), 16–23.

29. Mackey, B. M.; Forestiere, K.; Isaacs, N. Factors Affecting the Resistance of *Listeria monocytogenes* to High Hydrostatic Pressure. *Food Biotechnol.* **1995**, *9*, 1–11.

30. Mason, T. J.; Paniwnyk, L.; Lorimer, J. P. The Uses of Ultrasound in Food Technology. *Ultrason. Sonochem.* **1996**, *3*, 253–260.

31. McClements, J. M. J.; Patterson, M. F.; Linton, M. The Effect of Stage of Growth and Growth Temperature on High Hydrostatic Pressure Inactivation of Some Psychotropic Bacteria in Milk. *J. Food Prot.* **2001**, *64*, 514–522.

32. Muthukumaran, S.; Kentish, S. E.; Ashokkumar, M.; Stevens, G. W. Mechanisms for the Ultrasonic Enhancement of Dairy Whey Ultrafiltration. *J. Membr. Sci.* **2005**, *258*, 106–114.

33. Qin, B. L.; Pothakamury, U. R.; Vega, H.; Martin, O.; Barbosa-Cánovas, G. V.; Swanson, B. G. Food Pasteurization Using High Intensity Pulsed Electric Fields. *J. Food Technol.* **1995**, *49*(12), 55–60.

34. Raso, J.; Pagan, R.; Condon, S.; Sala, F. J. Influence of Temperature and Pressure on the Lethality of Ultrasound. *Appl. Environ. Microbiol.* **1998**, *64*, 465–471.

35. Raso, J.; Palo, A.; Pagan, R.; Condon, S. Inactivation of *Bacillus subtilis* Spores by Combining Ultrasonic Waves Under Pressure and Mild Heat Treatment. *J. Appl. Microbiol.* **1998**, *85*, 849–854.

36. Rastogi, N. K.; Raghavarao, K. S. M. S.; Balasubramaniam, V. M.; Niranjan, K.; Knorr, D. Opportunities and Challenges in High Pressure Processing of Foods. *Crit. Rev. Food Sci. Nutr.* **2007**, *47*, 69–112.

37. Rowan, N. J.; MacGregor, S. J.; Anderson, J. G.; Fouracre, R. A.; McIlvaney, L.; Farish, O. Pulsed-light Inactivation of Food-related Microorganisms. *Appl. Environ. Microbiol.* **1999**, *65*, 1312–1315.

38. Salvia-Trujillo, L.; la Peña M. M.; Rojas-Graü, M. A.; Martín-Belloso, O. Microbial and Enzymatic Stability of Fruit Juice–Milk Beverages Treated by High Intensity Pulsed Electric Fields or Heat During Refrigerated Storage. *Food Control* **2011**, *22*, 1639–1646.

39. San Martín, M. F.; Barbosa-Cánovas, G. V.; Swanson, B. G. Food Processing by High Hydrostatic Pressure. *Crit. Rev. Food Sci. Nutr.* **2002,** *42,* 627–645.
40. Tao, Y., García, J. M.; Sun, D. W. Advances in Wine Ageing Technologies for Enhancing Wine Quality and Accelerating the Ageing Process. *Crit. Rev. Food Sci. Nutr.* **2011,** *54*(6), 817–835.
41. Vega Mercado, H.; Martin-Belloso, O.; Qin, B. L.; Gongora-Neito, M. M.; Barbosa-Cánovas, G. V.; Swanson, B. G. Non-thermal Food Preservation: Pulsed Electric Fields. *Trends Food Sci. Sci. Technol.* **1997,** *8*(5), 151–157.
42. Vilkhu, K.; Mawson, R.; Simons, L.; Bates, D. Applications and Opportunities for Ultrasound Assisted Extraction in the Food Industry: A Review. *Innov. Food Sci. Emerg. Technol.* **2008,** *9*(2), 161–169.
43. Villamiel, M.; van Hamersveld, E. H.; Jong, P. Review: Effect of Ultrasound Processing on the Quality of Dairy Products. *Milchwissenschafts* **1999,** *54,* 69–73.
44. Zenker, M.; Heinz, V.; Knorr, D. Application of Ultrasound Assisted Thermal Processing for Preservation and Quality Retention of Liquid Foods. *J. Food Prot.* **2003,** *66*(9), 1642–1649.

CHAPTER 2

HIGH-PRESSURE PROCESSING OF MILK AND MILK PRODUCTS

DEVBRAT YADAV[1,*], PRABIN SARKAR[1], PRASAD PATIL[2], and NICHAL MAYUR[3]

[1]*Dairy Chemistry Division, National Dairy Research Institute (NDRI), Karnal 132001, Haryana, India*

[2]*Dairy Microbiology Division, National Dairy Research Institute (NDRI), Karnal 132001, Haryana, India*

[3]*GCMMF Ltd., Anand 388001, Gujarat, India*

Corresponding author. E-mail: dev.007.yadav@gmail.com

CONTENTS

ABSTRACT

High-pressure (HP) treatment of milk and products is a novel processing technique during which the product is treated in a vessel of suitable strength at a HP. Under pressure, biomolecules obey the Le Chatelier–Braun principle (i.e., whenever a stress is applied to a system in equilibrium, the system will react so as to counteract the applied stress); thus, reactions that result in reduced volume will be promoted under HP. Therefore, HP is considered an interesting alternative for milk heat pasteurization and possibly sterilization, because under HP conditions, microorganisms (vegetative cells) and certain enzymes are inactivated and fresh flavor, color, taste, and vitamins are only minimally affected (Anema et al., 2005). The technique offers several advantages: (1) preserved products with characteristics similar to those present before processing, (2) homogeneity of treatment due to the fact that pressure is uniformly applied around and throughout the food product, (3) shelf lives similar to thermal pasteurization, while maintaining the natural food quality parameters (nutrients, flavor, and sensorial preservation). There is no doubt that HPP represents another interesting and promising dimension for dairy processing not only because it inactivates microorganisms but also because it provides opportunities for development of new "value-added" dairy products. The need for an alternative to thermal processing as the primary means of eliminating pathogenic and spoilage microorganisms is substantial.

2.1 INTRODUCTION

Compared to past few decades, consumers are getting more and more health conscious and aware about the nutritional values of processed food, which impelled scientific approaches to search novel processing alternatives for retaining maximum nutritional properties of food. Thermal processing is a traditional technology, which has been used from ancient times for preserving food (especially milk and milk products). No doubt, that the heat treatment kills or inactivates pathogenic microbes in food, but it has detrimental effects on quality attributes of food in terms of loss of nutrients, change in physicochemical properties, and hence change in sensorial characteristics. High-pressure processing (HPP) of food moieties proved to be a highly effective nonthermal preservation technique in retaining fresh-like characteristics of food. HPP significantly inactivates microorganisms at ambient temperature without addition of preservatives.[14]

Presently, HPP in dairy industry is a well-known treatment applied for denaturing microbial enzymes and rupturing plasma membrane of microbes or microbial cell death without assistance of chemical preservatives, thereby fulfilling the increasing demand of consumers for quality assurance of milk and milk products. However, depending upon the pressure level and other processing variable such as pH, temperature, ionic strength, and processing time, high-pressure (HP) treatment leads to change in functional properties of milk components. Effects of HPP on different milk constituents have been studied broadly by various researchers, which are with mechanism elaborated in this chapter.

This chapter provides a wide and detailed knowledge of the influence of HP on milk constituents, which ultimately affects various physical and chemical properties of milk product.

2.2 PRINCIPLES OF OPERATION

Pressure and temperature are the two important process variables, which directly controls the occurrence and pathway of many biochemical reactions or microbial growth. Use of high-temperature treatment solely degrades heat-sensitive components of milk significantly. Pressure and temperature combination as cold pasteurization or HPP at ambient temperature for a definite interval of time extensively reduces microbial count in milk and milk products with no damage to heat-sensitive components. According to Heremans[24] and Tauscher,[61] HPP is based on two principles:

- *Le-Chatelier principle*, which states that any process (phase transition, chemical reaction, or change in the configuration of molecules) along with a decrease in volume ($-\Delta V$) will be enhanced by an increase in pressure and vice versa.
- *Isostatic pressure principle,* according to which pressure uniformly and instantaneously transmits independent of size and geometry of food.

In simple words, on applying HP on a food system (in equilibrium), the reduction in volume will be promoted at ambient temperature, thereby causing inactivation of microorganism or enzymes and textural changes in food.[3]

2.3   EFFECTS OF HPP ON MICROBIAL GROWTH IN MILK

The concept of HPP is just similar to pasteurization of milk for inactivation of microbes. Pasteurization of milk includes heat treatment at specific temperature and time duration, whereas HPP is the combination of pressure, temperature, and time for inactivation of microbes present in milk. HP for short-time treatment is more effective than low pressure for long time combination for significantly reducing the microbial count. Pressure treatment more than 350 MPa for 10 min completely inactivates bacteria (except bacterial spores). A combination of 350 MPa pressure and 32 min processing time reported to increase the shelf life of milk to 18 days at 5°C.[40] Pressure treatment of 500 MPa for 15 min is considered equivalent to pasteurization in reducing the raw milk microbial population to an acceptable level.[8]

In dairy industry, HPP is generally applied for microbial inactivation or inhibition of microbial growth. But there are various types of microorganisms with different physiological features and may possess a different level of pressure resistance for inactivation. HP-induced inactivation of bacteria depends on the intensity of pressure applied.[20] Effects of HPP on microbes are: (a) 50 MPa—inhibition of protein synthesis and reduction in ribosomal number; (b) 100 MPa—reversible denaturation of proteins and enzymes; (c) 200 MPa—disruption of cell membrane and internal structure of cells; (d) ≥300 MPa—irreversible denaturation of proteins and enzymes, rupture of cell membrane, and ultimately bacterial cell death.

Pressure treatment of 200 MPa is generally regarded as threshold pressure level, on which disruption of bacterial plasma membrane occurs. Hite[26] observed six decimal reduction in microbial load of milk treated with 689 MPa for 10 min. As compared to Gram-negative bacteria, yeast, molds, and Gram-positive bacteria are quite resistant to pressure-induced inactivation. This effect is due to the presence of teichoic acid (a bacterial polysaccharide), which makes Gram-positive bacteria resistant against HP inactivation. As compared to lactobacilli, lactococci inoculated in reconstituted skim milk were reported to be more sensitive to pressure ranged from 100 to 300 MPa, thereby causing less acidification rates than untreated milk.[10]

Microbial spores are most pressure resistant state of bacteria, requires severe pressure treatments along with the use of heat treatment. Calcium-rich dipicolinic acid present in bacterial spores, protect them from heat and HP disruption. But they can be inactivated by hurdle technologies or by using heat treatment and HPP together.[14,60] Heat treatment of 90–105°C and 700 MPa pressure combination, significantly decreases spore counts of *Bacillus licheniformis* in food samples.[1]

2.4   EFFECTS OF HP HOMOGENIZATION ON MILK COMPONENTS

HP homogenization is the mechanical process, which leads to reduction in the size of fat globule by applying HP on raw milk. Homogenization temperature should be 45–50°C because lipases from the milk itself and microbes, remains inactive at this temperature, thereby ensuring prevention of fat oxidation.[49] Mechanical HP on raw milk results formation of small-sized fat globules, which tends to incorporate into the milk proteins (casein and denatured whey proteins).[58] Caseins are much perfect in performing this action as it covers the newly formed milk fat globule membrane (MFGM) at around 0.2 g casein per gram of fat and therefore known as natural emulsifier[11] (Fig. 2.1). This effect is enhanced by HPP as it leads to disintegration of casein micelles to small size, which eases the uniform distribution of casein over the surface of fat globule.[57]

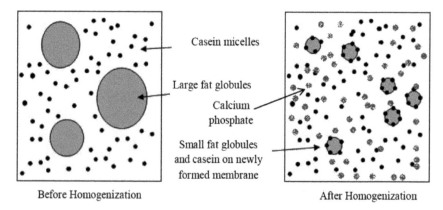

Before Homogenization   After Homogenization

FIGURE 2.1   Interaction of casein and fat globules during high-pressure homogenization.

2.5 EFFECTS OF HPP ON MILK PROTEINS

Structure of a protein is stabilized by various forces of interaction such as covalent bonds, electrostatic interactions, hydrogen bonds, and hydrophobic binding. Among all these forces of interactions, covalent bonds are the strongest force which majorly responsible for the basic structure of the protein. Covalent bonds are relatively stable against the denaturing or disintegrating action of HPP and hence cause no difference in primary structure of proteins after treatment.[45]

Under HPP, quaternary structure destabilizes very fast as compared to other form of proteins (tertiary, secondary, and primary). The destabilization of protein structures is reported to depend upon the extent of pressurization.[21] HPP treatment at >300 MPa decreases strength of hydrophobic and hydrogen bonds that lead to destabilization of tertiary structure. HPP treatment at ≥700 MPa decreases the strength of hydrophobic and ionic bonds that lead to destabilization of secondary structure. HPP treatment at ≤100 MPa decreases the strength of hydrophobic bonds that leads to destabilization of quaternary structure.

2.5.1 EFFECTS OF HPP ON CASEINS

The native form of milk casein is made up of primary structure, which is stabilized by strong covalent bonds and electrostatic interactions. Caseins are highly stable to very HP (even up to 300 MPa) due to its highly stabilized structure.[29] But on increasing pressure more than 300 MPa, defragmentation of casein micelles takes place. This effect occurs basically due two reasons:

- Solubilization of colloidal calcium phosphate (CCP)
- Disruption of hydrophobic and electrostatic interactions in casein micelle

Due to defragmentation, up to 50% size reduction was observed at >300 MPa by various researchers.[12,19,29,30,47] When bovine skim milk was treated with 100–600 MPa for 30 min, significant increase in the levels of α-S1 and β-caseins in soluble phase occurs. Order of casein dissociation depends on phosphate content, which leads to higher electrostatic repulsion under HPP.[31] Based on phosphate residues, fragmentation trend is as follows:

$$\kappa\text{-casein} > \beta\text{-casein} > \alpha\text{-S1casein} > \alpha\text{-S2 casein}$$

κ-casein dissociation chances are more as compared to other fragments, which leads to interaction with denatured β-lg and ultimately causes protein hydration and solubility increment.[37]

2.5.2 EFFECT OF HPP ON WHEY PROTEINS

β-lactoglobin (β-lg), α-lactalbumin (α-la), immunoglobulins (IGGs), and bovine serum albumin (BSA) are the main whey proteins, which are more prone to HP and heat denaturation.[21] Denatured whey proteins noticed to be responsible for the improvement in functional properties such as surface hydrophobicity, gelation, foaming, and emulsifying properties,[43] whereas, irreversibly denatured proteins are more readily consumed by probiotic bacteria and therefore causes increased microbial count. β-lg is most sensitive whey protein to HPP as it has only two disulfide bonds and one free –SH compared to α-la which possesses four disulfide linkages.[25] The effects of HPP on whey proteins have been studied in detail by Massoud et al.[42] According to them, HP homogenization at 100 MPa causes reversible denaturation compared to irreversible denaturation of whey proteins at >200 MPa.

Treatment of raw milk more than 100 MPa causes unfolding of β-lg and further interaction with κ-casein to form aggregates.[46] Application of 400 MPa pressure resulted in around 70–80% β-lg denaturation.[36,56] Approximately, 10% and 50% α-la denaturation takes place at 600 and 800 MPa, respectively. The sensitivity of whey proteins to HPP-induced denaturation is as follows:

$$\beta\text{-lg} > \alpha\text{-la} > \text{IGGs} > \text{BSA}$$

The presence of more number of disulfide bonds ensures the stability of whey proteins under HPP. The BSA is the most pressure stable polypeptide as it possesses a large number of disulfide bonds,[27] whereas IGGs are also resistant to pressure denaturation as compared to β-lg and α-la, but less resistant than BSA. Besides these facts denaturation also depends on other processing variables such as temperature, pH, and processing time.

2.6 EFFECT OF HPP ON CARBOHYDRATE OF MILK

Lactose present in milk and milk products may convert into lactulose and then degrade to form acids and other sugar moieties under the action of vigorous heating. No changes in these compounds are observed after applying pressure (100–400 MPa for 10–60 min at 25°C), indicating that no Mallard reaction products or by-products formation occurs in milk.[40]

2.7 EFFECT OF HPP ON MILK FAT

According to Buccheim et al.,[6] 100–400 MPa pressure treatment induces crystallization of fat with maximum intensity at 200 MPa. They studied the morphological difference in crystals of milk fat of HP-treated cream and conventional temperature-induced cream with the help of transmission electron microscopy and observed no differences in the structure of crystals. In fact, HP-treated cream had a higher solid fat content compared to untreated cream with maximum effect at 200 MPa.

The crystallization effect at HP treatment is due to the shifts of solid to liquid transition temperature of milk fat to a higher value. At 100 MPa, crystallization temperature increased by 16.3°C and melting temperature shifted to 15.5°C.[18] At very HP (>300 MPa), the crystallization effect may be hampered because of lower crystal growth of fat molecules having reduced motility at HP.[5,6] According to previous studies, HPP up to <400 MPa did not affect the diameter or size of milk fat globules but at pressure ranged from 400 to 800 MPa, the diameter of milk fat globule increases with no damage to MFGM and hence, no lipolysis takes place.[6,35]

2.8 EFFECT OF HPP ON BUFFERING CAPACITY OF MILK

The stability of milk in different processing parameters totally depends on the solubility of minerals (especially calcium), which can be easily understood as a concept of mineral balance. Mineral balance of milk is very important criteria of milk to be precipitated or coagulated. Calcium is important mineral present in milk, which is reported to found in two major forms as CCP and diffusible calcium. Colloidal calcium is that calcium, which is directly associated with casein molecules and integral part of CCP, whereas diffusible calcium is known as free ionic calcium (Ca^{2+}) or complexed with citrate or phosphate.[31] At HP, the colloidal calcium gets converted to diffusible phase. With the help of HPP treatment both native and heat-precipitated CCP can be converted into diffusible calcium phosphate, which leads to slight increase in pH.[33] HPP of milk increases diffusible calcium up to 200 MPa and on further increasing pressure (400–600 MPa), there is no effect reported in earlier literature.[12] Similarly, Lopez-Fandino et al.[38] observed that the diffusible phase calcium, magnesium, and phosphate increased on applying pressure treatment up to 300 MPa but decreases on further increasing pressure afterwards. This application of HPP is very good alternative to convert heat-precipitated milk to normal milk.[5,55] Generally, pH of milk increases on application of HP treatment and this change in pH is reversible.

Buffering capacity of milk is due to salts, organic acids, and basic and acidic amino acids. The buffering capacity of cow milk is more than buffalo milk.[53] HPP of milk causes significant vicissitudes in the physicochemical characteristics of casein micelles. Slight casein micelle demineralization is noted, which considerably affects the buffering capacity of milk.[19] The buffering peak noticed at pH 4.8–5.0 for untreated milk is shifted to pH 5.2–5.4 for treatments at 250, 450, or 600 MPa. Furthermore, the buffering capacity of HP-processed milk is lower than for untreated milk at pH around 5.0 but is greater between pH 6.0 and 5.2.[15] As discussed above, micellar calcium solubilization occurs at a higher pH value when milk is treated under pressure. When the pressure is released and the system returns to atmospheric pressure, casein micelles are differently organized and micellar calcium phosphate is not in its native form.[31]

2.9   EFFECT OF HPP ON ENZYMES

Pressure inactivation of enzymes generally depends on the structure and chemical nature of enzyme. Plasmin is the main native enzyme which shows 30% and 75% reduced activity at pressure 400 and 600 MPa at 20°C for 30 min.[28] A synergistic effect of temperature and HPP on plasmin inactivation was noticed at 300–600 MPa pressure range. Although an antagonistic effect is seen at a pressure above 600 MPa, due to structural changes, it stabilizes plasmin and plasminogen structure.[4] Inactivation of plasmin is reported to be enhanced by the presence of denatured β-lg,[57] due to the formation of thiol-disulfide bonds with unfolded β-lg. Lipoprotein lipase showed resistance against pressure-induced dissociation up to

400 MPa for 30 min at 3°C.[50] Alkaline phosphatase enzyme is the indicator enzyme, used for the assessment of sufficiency of pressure resistance even up to 500 MPa at 25–50°C for 10 min.[16] Due to this reason, xanthin oxidase suggested as indicator enzyme for adequacy of the pressure treatment, as exposure of more than 400 MPa pressure at 25°C causes inactivation following first-order kinetics.

2.10   EFFECT OF HPP ON TINY VOLATILE COMPOUNDS

Compared to thermal treatment (in which covalent as well as noncovalent bonds were reported to be affected), high pressurization at mild temperature only breaks relatively weak chemical bonds (hydrogen bonds, hydrophobic bonds, and ionic bonds). Therefore, tiny molecules such as vitamins, amino acids, and volatile flavoring compounds remain unaffected by HP treatment. Processing at 400 MPa at 25°C for 30 min, shows no significant loss of vitamin B1 and B6.[59]

2.11   EFFECT OF HPP ON PRESENCE OF WATER IN MILK

HP application leads to the depression in freezing point of water observed at 50, 100, or 210 MPa pressure, respectively.[34] Hence, this advanced technique allows subzero milk processing without ice crystal formation, thereby causing reduced microbial activity along with improved quality and extended shelf life of the product.

2.12   EFFECT OF HPP ON SENSORY ATTRIBUTES OF MILK AND MILK PRODUCTS

The appearance, taste and other sensory properties are considered as a measure of quality by consumers. The scattering of light by fat globules, proteins, and calcium phosphate interactions results in the white color of milk. High pressurization or HP-homogenized milk appears more whiter in color, due to small size casein micelles and fat globules and looks more appealing to consumers.[11] Hunter luminance value (L-value) is the measure of whiteness, which has been used to distinguish pressure-treated and untreated milk.[22] As pressure increases, the L-value decreases. Treatment of milk at 250–450 MPa pressure significantly decreases the *L*-value of milk, while at pressure ≥200 MPa slight reduction in *L*-value was noticed. Desobry-Banon et al.[12] treated skim milk at 600 MPa for 30 min and observed that the *L*-value decreased from 78 to 42 and skim becomes semitransparent. Besides these facts, HP-processed fermented milk products reported to possess better taste and flavor, due to release of diacetyl and acetaldehyde.[52,58] Entirely, the sensory properties of HP-processed milk and milk products are better than normally pasteurized milk.[52]

2.13   EFFECT OF HPP ON MILK PRODUCTS

2.13.1   CHEESE

2.13.1.1   RENNET COAGULATION PROPERTIES OF MILK

Rennet coagulation in milk is divided into two major steps as enzymatic hydrolysis and aggregation of enzymatically modified micelles.[12] According to Desobry-Banon et al.,[12] the HPP treatment at 400 MPa for 10 min of milk causes a reduction in the size of casein micelle causing an increase in the specific surface area/intra-particle collision and rennet coagulation time (RCT). Rennet or chymosin hydrolysis of κ-casein results formation of para κ-casein and caseinomacropeptides (CMP). CMP may be present in glycosylated or nonglycosylated forms, which promotes coagulation of milk.[41] HP treatment of milk improves rennet coagulation properties up to 200–250 MPa at 25°C and on further increasing pressure and temperature (to more than 300 MPa and 40°C), the RCT increases.[47]

The possible mechanisms of effect of HP treatment on rennet coagulation reported by different researchers are described in detail by Desobry-Banon et al.[12] and other investigators.[31,37] HPP induces disruption of micellar κ-casein and denaturation of β-lg, which interacts to form complex and hinders the action of rennet of chymosin on κ-casein. On the other hand, the HPP leads to the release of micellar negatively charged κ-casein thereby reducing electrostatic repulsion between casein molecules. This effect eases the coagulation of milk at a faster rate.[12] The RCT of HP-treated milk is also affected by the processing time and temperature. Pressure treatment of ≥200 MPa at 60°C or ≥300 MPa at 50°C hinders rennet coagulation of milk,[38] whereas, 400 MPa pressure for 10 min treatment leads to reduction in coagulation time significantly but longer processing times increased RCT considerably.[40]

2.13.1.2   INFLUENCE OF HPP ON YIELD OF CHEESE

Pressurization of milk at 300–400 MPa for 30 min increased cheese yield up to 20% compared to untreated milk. The reason behind this effect that the denatured whey protein incorporates in curd along with increased moisture retention in rennet-induced formation of cheese curd (Fig. 2.2).[2,39,40,47,51]

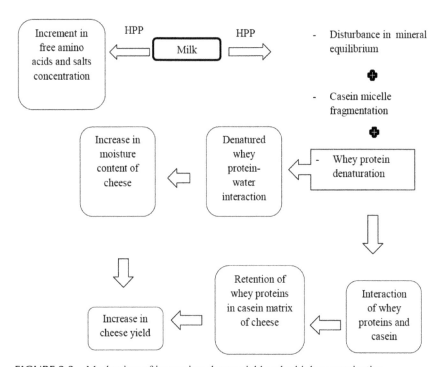

FIGURE 2.2   Mechanism of increasing cheese yield under high pressurization.

Reduction in water-holding capacity of rennet curd was noticed on decreasing pressure (200–400 MPa), temperature (3–21°C), and processing time (10–110 min).[51] A combination of 400 MPa and 40°C resulted increase in curd yield compared to 20°C.[39] Pressure treatment of 586 MPa for three 1 min cycle produced cheddar cheese with 7% increment in the yield.[13]

Sensory and microbiological analysis of pressure-treated milk revealed no differences in pasteurized and pressurized milk cheese.[13]

2.13.1.3   ACTION OF HPP ON CHEESE RIPENING

The cheese ripening is a very complex process of biochemical changes, which is required for the typical aroma, flavor, and texture of cheese. It is also a time-consuming and expensive process [depending on the variety of cheese, e.g., cheddar cheese (6–9 months), parmesan cheese (2 years)]. Extended ripening period involves increased cost due to (1) refrigerated storage, (2) loss in weight with time, (3) labor, (4) space, and (5) higher capital cost.

HP treatment (50 MPa for 3 days) accelerated cheese ripening with higher level of proteolysis of immature cheddar cheese. But the taste was comparable to that of matured commercial cheese.[48] HP treatment on camembert cheese (white mould-ripened cheese) resulted in increasing its degree of proteolysis depending on pressure and maturity stage of the cheese.[43] Pressure treatment of 50 MPa for 8 h considerably increased casenolytic activity of microbial enzymes and led to increase trichloro-acetic acid soluble nitrogen. Increased pH caused by HPP leads to a higher activity of metalloproteinase of *Penicillium camemberti*. Similarly, same treatment on smear-ripened cheese (such as Pere Joseph) showed a high level of proteolysis with increment in pH. *Brevibacterium linens* present in this cheese possesses extracellular proteases, which are responsible for degrading casein and release of small peptides and free amino acids in short interval of time.[44]

Exposure of 400 MPa for 5 min at 14°C enhances proteolysis in goat milk cheese reported to show same extent of proteolysis or ripening in 14 days as that of conventional 28 days.[54] HP-treated cheese made from goat milk had higher pH, salts, and moisture, favored the growth of lactic acid bacteria during ripening period, and developed strong flavor than pasteurized milk cheese. The sensory characteristics of HP-treated and pasteurized milk cheese before ripening phase observed to be different but at the end of ripening phase, there was no difference found.[9] Buffa et al.[7] found no significant differences between levels of lipolysis of HP-treated cheese and raw milk cheese. On the other hand, the free fatty acid content of HP-treated cheese was higher than pasteurized milk cheese. Pressure-resistant and heat-sensitive nature of lipase may be the reason behind this effect.

2.13.2 ACID COAGULATION OF MILK AND YOGURT MANUFACTURING PROPERTIES

HPP leads to casein disruption and release of free amino acids and denaturation of proteins, which are more easily utilized by probiotic bacteria to produce lactic acid in yogurt, indicating that HP treatment favors acid coagulation of milk.[27] On the other hand, compared to unpressurized milk the initiation of gelation occurs at higher pH in pressurized milk. Johnston et al.[32] observed that on inducing acid coagulation of HP-treated milk, the gel rigidity, strength, and resistance to syneresis increases, possibly due to increased protein hydration and cross-linking of network strands. The improvement in the mechanical strength of gel may also be due to the formation of aggregates of small-sized micelles with partially denatured β-lg on the surface of micelles.[23,42]

Acid coagulation of HP-processed reconstituted skim milk with glucono-δ-lactone showed that HP-treated milk coagulated at higher pH and formed a stronger gel than untreated sample.[12] Similarly, when the milk is exposed to 350–500 MPa pressure at 25°C or 55°C before fermentation, the coagulation occurred at higher pH and gel strength was comparable to yogurt made from pasteurized milk.[17]

2.14 SUMMARY

The application of HPP in dairy has attracted considerable research interest in recent decades. HP destroys vegetative microbes and bacterial spores without compromising quality attributes of dairy products compared to heat treatment. Pressurization up to 600 MPa destroys most of the microorganisms and increases shelf life of dairy milk and milk products. Casein micelles may disintegrate and whey proteins denature, depending on the level of pressure applied. HP also leads to increment in the level of soluble mineral. Change in the structure or configuration of milk proteins greatly affects chemical nature of products such as cheese produced from protein denaturation and disintegration. HPP also leads to enzymatic inactivation.

Alkaline phosphatase is pressure resistant and that is why, xanthin oxidase is recommended as indicator enzyme for determining the efficacy of pressurization treatment. Pressure treatment of 200 MPa causes fat crystallization due to shift in crystallization temperature to 16.3°C. In case of acid-coagulated or fermented milk products such as yogurt, HPP fastens acid coagulation and improves textural properties. The unique physical and sensory properties of pressure-treated dairy products offer new opportunities for new product development in dairy industry. This chapter provides a wide and detailed knowledge of the influence of HP on milk constituents, which ultimately affects various physical and chemical properties of milk product.

KEYWORDS

- high-pressure processing
- homogenization
- microbial growth
- microbial inactivation
- pressure-resistant bacteria
- sensory attributes
- viscosity
- water-holding capacity
- whey proteins
- yogurt texture

REFERENCES

1. Ahn, J.; Balasubramanium, V. M. Screening Food for Processing-resistant Bacterial Spores and Characterization of a Pressure and Heat Resistant *Bacillus licheniformis* spore. *J. Food Prot.* **2014,** *77*(6), 948–954.
2. Arias, M.; Lopez-Fandino, R.; Olano, A. Influence of pH on the Effects of High Pressure on Milk Proteins. *Milchwissenschaft* **2000,** *55*, 191–194.
3. Balci, A. T.; Wilbey, A. High Pressure Processing of Milk—The First 100 Years in the Development of a New Technology. *Int. J. Dairy Technol.* **1999,** *52*, 149–155.
4. Borda, D.; Smout, C.; Van Loey, A.; Hendrickx, M. High Pressure Thermal Inactivation of a Plasmin System. *J. Dairy Sci.* **2004,** *87*, 2351–2358.
5. Buchheim, W.; Schrader, K.; Morr, C. V.; Frede, E.; Schutt, M. Effects of High Pressure on the Protein, Lipid and Mineral Phase of Milk. *Int. Dairy Fed.* **1996,** *96*(2), 202–213.
6. Buchheim, W., Schutt, M., Frede, E. High Pressure Effects on Emulsified Fats. *Prog. Biotechnol.* **1996,** *13*, 331–336 (Proceedings of the International Conference on High Pressure Bioscience and Biotechnology).

7. Buffa, M.; Guamis, B.; Pavia, M.; Trujillo, A. J. Lipolysis in Cheese Made from Raw, Pasteurized or High Pressure Treated Goat Milk. *Int. Dairy J.* **2001,** *11*(3), 175–179.

8. Buffa, M.; Guamis, B.; Royo, C.; Trujillo, A. J. Microbiological Changes Throughout Ripening of Goat Cheese Made from Raw, Pasteurized and High-pressure-treated Milk. *Food Microbiol.* **2001,** *18*(1), 45–51.

9. Buffa, M.; Trujillo, A. J.; Pavia, M.; Guamis, B. Changes in Textural, Micro Structural and Color Characteristics During Ripening of Cheeses Made from Raw, Pasteurized and High Pressure Treated Goat Milk. *Int. Dairy J.* **2001,** *11*, 927–934.

10. Casal, V.; Gomez, R. Effect of High Pressure on the Viability and Enzymatic Activity of Mesophilic Lactic Acid Bacteria Isolated from Caprine Cheese. *J. Dairy Sci.* **1999,** *82*(6), 1092–1098.

11. Chandan, R. C.; Kilara, A. Dairy Ingredients for Food Processing: An Overview. In *Dairy Ingredients for Food Processing*; Chandan, R. C., Ed.; Blackwell Publishing Ltd.: Iowa, USA, 2011; pp 3–34 (Chapter 1).

12. Desobry-Banon, S.; Richard, F.; Hardy, J. Study of Acid and Rennet Coagulation of High Pressurized Milk. *J. Dairy Sci.* **1994,** *77*, 3267–3274.

13. Drake, M. A.; Harrison, S. L.; Asplund, M.; Barbosa-Canovas, G.; Swanson, B. G. High Pressure Treatment of Milk and Effects on Microbiological and Sensory Quality of Cheddar Cheese. *J. Food Sci.* **1997,** *62*, 843–860.

14. Espejoa, G. G. A.; Hernández-Herreroa, M. M.; Juana, B.; Trujilloa, A. J. Inactivation of *Bacillus* Spores Inoculation in Milk by Ultrahigh Pressure Homogenization. *Food Microbiol.* **2014,** *44*, 204–210.

15. Famelart, M. H.; Gaucheron, F.; Mariette, F.; Le Graet, Y.; Raulot, K.; Boyaval, E. Acidification of Pressure-treated Milk. *Int. Dairy J.* **1997,** *7*, 325–330.

16. Felipe, X.; Capellas, M.; Law, A. J. R. Comparison of the Effects of High-pressure Treatments and Heat Pasteurization on the Whey Proteins in Goat Milk. *J. Agric. Food Chem.* **1997,** *45*, 627–631.

17. Ferragut, V.; Martinez, V. M.; Trujillo, A. J.; Guamis, B. Properties of Yogurts Made from Whole Ewe's Milk Treated by High Hydrostatic Pressure. *Milchwissenschaft* **2000,** *55*, 267–269.

18. Frede, E.; Buchheim, W. The Influence of High Pressure Upon the Phase Transition Behavior of Milk-fat and Milk-fat Fractions. *Milchwissenschaft* **2000,** *55*, 683–686.

19. Gaucheron, F.; Famelart, M. H.; Mariette, F.; Raulot, K.; Michel, F.; Le Graet, Y. Combined Effects of Temperature and High Pressure Treatments on Physicochemical Characteristics of Skim Milk. *Food Chem.* **1997,** *59*, 439–447.

20. Ghalavand, R.; Nikmaram, N.; Kamani, M. H. The Effects of High Pressure Process on Food Microorganisms. *Int. J. Farming Allied Sci.* **2015,** *4*(6), 505–509.

21. Goyal, A.; Sharma, V.; Upadhyay, N.; Sihag, M.; Kaushik, R. High Pressure Processing and Its Impact on Milk Proteins: A Review. *J. Dairy Sci. Technol.* **2013,** *2*(1), 1–9.

22. Harate, F.; Luedecke, L.; Swanson, B.; Barbosa-Canvas, G. V. Low Fat Set Yoghurt Made from Milk Subjected to Combinations of High Pressure and Thermal Processing. *J. Dairy Sci.* **2003,** *86*, 1074–1082.

23. Harte, F.; Amonte, M.; Luedecke, L.; Swanson, B. G.; Barbosa-Canovas, G. V. Yield-stress and Microstructure of Set Yogurt Made from High Hydrostatic Pressure-treated Full Fat Milk. *J. Food Sci.* **2002,** *67*, 2245–2250.

24. Heremans, K. High Pressure Effects on Proteins and Other Biomolecules. *Annu. Rev. Biophys. Bioeng.* **1982**, *11*, 1–21.
25. Hinrichs, J.; Rademacher, B. Kinetics of Combined Thermal and Pressure-induced Whey Protein Denaturation in Bovine Skim Milk. *Int. Dairy J.* **2004**, *14*, 315–323.
26. Hite, B. H. *The Effect of Pressure in the Preservation of Milk*; Bulletin of West Virginia University of Agricultural and Forestry Experimental Station: Morgantown, WV, 1899; Vol. 58, pp 15–35.
27. Huppertz, T.; Fox, P. F.; Kelly, A. L. Influence of High Pressure Treatment on the Acidification of Bovine Milk by Lactic Acid Bacteria. *Milchwissenschaft* **2004**, *59*, 246–249.
28. Huppertz, T.; Fox, P. F.; Kelly, A. L. Plasmin Activity and Proteolysis in High Pressure-treated Bovine Milk. *Le Lait* **2004**, *84*, 297–304.
29. Huppertz, T.; Fox, P. F.; Kelly, A. L. Dissociation of Caseins in High Pressure-treated Bovine Milk. *Int. Dairy J.* **2004**, *14*, 675–680.
30. Huppertz, T.; Grosman, S.; Fox, P. F.; Kelly, A. L. Heat and Ethanol Stabilities of High Pressure-treated Bovine Milk. *Int. Dairy J.* **2004**, *14*, 124–133.
31. Huppertz, T.; Kelly, A. L.; Fox, P. F. Effects of High Pressure on Constituents and Properties of Milk. *Int. Dairy J.* **2002**, *12*(7), 561–572.
32. Johnston, D. E.; Austin, B. A.; Murphy, R. J. Properties of Acid Set Gels Prepared from High Pressure Treated Milk. *Milchwissenschaft* **1993**, *48*, 206–209.
33. Johnston, D. E.; Austin, B. A.; Murphy, P. M. Effects of High Hydrostatic Pressure on Milk. *Milchwissenschaft* **1992**, *47*, 760–763.
34. Kalichevesky, M. T.; Knorr, D.; Lillford, P. J. Potential Applications of High Pressure Effects on Ice Water Transition. *Trends Food Sci. Technol.* **1995**, *6*, 253–259.
35. Kanno, C.; Uchimura, T.; Hagiwara, T.; Ametani, M.; Azuma, N. Effect of Hydrostatic Pressure on the Physicochemical Properties of Bovine Milk Fat Globules and the Milk Fat Globule Membrane. In *High Pressure Food Science, Bioscience and Chemistry*; The Royal Society of Chemistry: Cambridge, 1992; pp 182–192.
36. Lopez-Fandino, R.; Carrascosa, A. V.; Olano, A. The Effects of High Pressure on Whey Protein Denaturation and Cheese-making Properties of Raw Milk. *J. Dairy Sci.* **1996**, *79*(6), 929–936.
37. Lopez-Fandino, R. High Pressure-induced Changes in Milk Proteins and Possible Applications in Dairy Technology. *Int. Dairy J.* **2006**, *16*, 1119–1131.
38. Lopez-Fandino, R.; Olano, A. Effects of High Pressures Combined with Moderate Temperatures on the Rennet Coagulation Properties of Milk. *Int. Dairy J.* **1998**, *8*, 623–627.
39. Lopez-Fandino, R.; Olano, A. Cheese-making Properties of Ovine and Caprine Milks Submitted to High Pressures. *Le Lait* **1998**, *78*, 341–350.
40. Lopez-Fandino, R.; Carrascosa, A. V.; Olano, A. The Effects of High Pressure on Whey Protein Denaturation and Cheese-making Properties of Raw Milk. *J. Dairy Sci.* **1996**, *79*(6), 929–936.
41. Lopez-Fandino, R.; Ramos, M.; Olano, A. Rennet Coagulation of Milk Subjected to High Pressure. *J. Agric. Food Chem.* **1997**, *45*, 3233–3237.
42. Massoud, R.; Belgheisi, S.; Massoud, A. Effect of High Pressure Homogenization on Improving the Quality of Milk and Sensory Properties of Yogurt: A Review. *Int. J. Chem. Eng. Appl.* **2016**, *7*, 66–70.

43. Messen, W.; Foubert, I.; Dewettinck, K.; Huyghebaert, A. Proteolysis of a High-pressure-treated Mould-ripen Cheese. *Milchwissenschaft* **2001,** *56*(4), 201–204.

44. Messen, W.; Foubert, I.; Dewettinck, K.; Huyghebaert, A. Proteolysis of a High-pressure-treated Smear-ripened Cheese. *Milchwissenschaft* **2000,** *55*(6), 328–332.

45. Mozhaev, V. V.; Heremans, K.; Frank, J.; Masson, P.; Balny, C. Exploiting the Effects of High Hydrostatic Pressure in Biotechnological Applications. *Trends Biotechnol.* **1994,** *12*, 493–501.

46. Needs, E. C.; Capellas, M.; Bland, A. P.; Manoj, P.; Macdougal, D.; Paul, G. Comparison of Heat and Pressure Treatments of Skim Milk, Fortified with Whey Protein Concentrate, for Set Yoghurt Preparation: Effects on Milk Proteins and Gel Structure. *J. Dairy Res.* **2000,** *67*(3), 329–348.

47. Needs, E. C.; Stenning, R. A.; Gill, A. L.; Ferragut, V.; Rich, G. T. High-pressure Treatment of Milk: Effects on Casein Micelle Structure and on Enzymatic Coagulation. *J. Dairy Res.* **2000,** *67*, 31–42.

48. O'reilly, C. C.; O'connor, P. M.; Murphy, P. M.; Kelly, A. L.; Beresford, T. P. The Effect of Exposure to Pressure of 50 MPa on Cheddar Cheese Ripening. *Innov. Food Sci. Emerg. Technol.* **2000,** *1*, 109–117.

49. Olorunnisomo, O. A.; Ososanya, T. O.; Adedeji, O. Y. Homogenization of Milk and Effect on Sensory and Physico-chemical Properties of Yoghurt. *Afr. J. Food Sci.* **2014,** *8*, 465–470.

50. Pandey, P. K.; Ramaswamy, H. S. Effect of High-pressure Treatment of Milk on Lipase and γ-glutamyl Transferase Activity. *J. Food Biochem.* **2004,** *28*, 449–462.

51. Pandey, P. K.; Ramaswamy, H. S.; St-Gelais, D. Water-holding Capacity and Gel Strength of Rennet Curd as Affected by the High Pressure Treatment of Milk. *Food Res. Int.* **2000,** *33*, 655–663.

52. Patrignani, F.; Iucci, L.; Lanciotti, R.; Vallicelli, M.; Maina, J. M.; Holzapfel, W. H.; Guerzoni, M. Effect of High-pressure Homogenization, Nonfat Milk Solids, and Milkfat on the Technological Performance of a Functional Strain for the Production of Probiotic Fermented Milks. *J. Dairy Sci.* **2007,** *90*, 4513–4523.

53. Salaun, F.; Mietton, B.; Gaucheron, F. Buffering Capacity of Dairy Products. *Int. Dairy J.* **2005,** *15*, 95–109.

54. Saldo, J.; Sendra, E.; Guamis, B. High Hydrostatic Pressure for Accelerating Ripening of Goat Milk Cheese: Proteolysis and Texture. *J. Food Sci.* **2000,** *65*(4), 636–640.

55. Schrader, K.; Buchheim, W.; Morr, C. V. High Pressure Effects on the Colloidal Calcium Phosphate and the Structural Integrity of Micellar Casein in Milk. Part 1. High Pressure Dissolution of Colloidal Calcium Phosphate in Heated Milk System. *Nahrung* **1995,** *41*, 133–138.

56. Scollard, P. G.; Beresford, T. P.; Needs, E. C.; Murphy, P. M.; Kelly, A. L. Plasmin Activity, β-lactoglobulin Denaturation and Proteolysis in High Pressure Treated Milk. *Int. Dairy J.* **2000,** *10*(12), 835–841.

57. Serra, M.; Trujillo, A. J.; Guamis, B.; Ferragut, V. Flavor Profiles and Survival of Starter Cultures of Yoghurt Produced from High-pressure Homogenized Milk. *Int. Dairy J.* **2009,** *19*, 100–106.

58. Sfakianakis, P.; Tzia, C. Conventional and Innovative Processing of Milk for Yogurt Manufacture, Development of Texture and Flavor: A Review. *Foods* **2014,** *3*, 176–193.

59. Sierra, I.; Vidal-Valverde, C.; Lopez-Fandino, R. Effect of High Pressure on the Vitamin B_1 and B_6 Content of Milk. *Milchwissenschaft* **2000,** *55*(7), 365–367.
60. Smelt, J. M. Recent Advances in the Microbiology of High Pressure Processing. *Trends Food Sci. Technol.* **1998,** *9*, 152–158.
61. Tauscher, B. Pasteurization of Food by Hydrostatic High Pressure: Chemical Aspects. *Z. Lebensm. Unters. Forsc.* **1995,** *200*(1), 3–13.

CHAPTER 3

APPLICATIONS OF SUPERCRITICAL FLUID EXTRACTION IN THE FOOD INDUSTRY

AKANKSHA WADEHRA[1,*], PRASAD S. PATIL[2],
SHAIK ABDUL HUSSAIN[1], ASHISH KUMAR SINGH[1],
SUDHIR KUMAR TOMAR[2], and RUPESH S. CHAVAN[3]

[1]*Dairy Technology Division, NDRI, Karnal 132001, Haryana, India*

[2]*Dairy Microbiology Division, NDRI, Karnal 132001, Haryana, India*

[3]*Department of Quality Assurance, Mother Dairy Junagadh, Junagadh 362001, Gujarat, India*

Corresponding author. E-mail: smartakanksha@gmail.com

CONTENTS

ABSTRACT

Supercritical fluid technology is applied in various fields, including the nonthermal cell inactivation, permeabilization, extraction of fermentation products, removal of biostatic agents and organic solvents from fermentation broth, disruption of yeasts and bacteria, destruction of industrial waste, and removal of chlorinated compounds from water and others. This chapter has focused on SCF special applications in the field of food industry. The application of SCF is simple, inexpensive, and noninjurious to the structure and function of enzymes and protein activities. The supercritical carbon dioxide is the most commonly used fluid. It has low critical temperature of 31.1°C, and the pressure of 7.3 MPa makes it an ideal medium for processing volatile products. The nontoxicity and nonflammability as well as the selectivity of the process and the ease of recovery are the most important features.

3.1   INTRODUCTION

There is an increasing public awareness of the health, environment, and safety hazards associated with the use of organic solvents in food processing and the possible solvent contamination of the final products. The high cost of organic solvents and the increasingly stringent environmental regulations together with the new requirements of the medical and food industries for ultrapure and high added value products have pointed out the need for the development of new and clean technologies for the processing of food products. Supercritical fluid (SCF) extraction using carbon dioxide as a solvent has provided an excellent alternative to the use of chemical solvents. Over the past three decades, supercritical carbon dioxide (SC-CO$_2$) is used for the extraction and isolation of valuable compounds from natural products.[13,28,44]

The supercritical fluid extraction (SCFE) technology has advanced tremendously since its inception and is a method of choice in many food processing industries. Over the last two decades, SCFE received a status of clean and environmentally friendly "green" processing technique. The most recent advances of SCFE applications are in food science, natural products and by-product recovery, pharmaceutical, and environmental sciences.[30] Applications of SCFE have been studied by many scientists to extract high-value compounds from food and natural products, as well as heavy metals recovery, enantiomeric resolution, or drug delivery systems and for the development of new separation techniques such as using SCFs to separate components of the extract, resulting in augmented quality and purity.[64] This makes SCFE a valuable technique for the extraction of natural products, including fats and oils, removal of caffeine from coffee,[74] and harmful components from nutraceutical products.[36] SCFE extracts the oil or desired elements from the subjected material in a shorter time compared to the conventional methods. SCF extracts are typically sterilized, contamination free, and the valuable components that remain in chemically natural state.[73] The SCFE technology is also investigated for the degumming and bleaching of soybean oil[39] and palm oil,[43,54] purification of used frying oil,[77] and fractionation of butter oil and beef tallow.[37]

3.2 SUPERCRITICAL FLUIDS

A supercritical solvent is one that at a certain temperature and pressure does not condense or evaporate, it exists as a fluid. This happens when a substance is high above its critical temperature and pressure, and passes it to a condition called SCF state. Under these conditions, the densities of the liquid and vapor are identical, and for this reason, the meniscus disappears. Under SCF conditions, substances exhibit physical properties that are intermediate between gases and liquids of the start material. Thus, the density of an SCF can be altered by the variation of the pressure applied on the fluid. When the fluid is compressed at high temperatures, it can have a density varying between those displayed by gases up to typical values of liquids. Likewise, an SCF maintained at a relatively high density has the ability to dissolve a variety of materials, exactly as conventional liquids do, but with the power of gases penetration. The variation of properties with conditions of state is monotonous, when crossing critical conditions, as indicated in Figure 3.1 by the SCF regions.

Table 3.1 shows characteristic values for the gaseous, liquid, and supercritical state. In the supercritical state, liquid-like densities are approached, while viscosity is near that of normal gases and diffusivity is about two orders of magnitude higher than in typical liquids.[7]

In processes with SCFs, the driving potential for mass and heat transfer is determined by the difference from the equilibrium state. The equilibrium state provides information about:

- The capacity of a supercritical (gaseous) solvent, which is the amount of a substance dissolved by the gaseous solvent at thermodynamic equilibrium.
- The amount of solvent, which dissolves in the liquid or solid phase, and the equilibrium composition of these phases.
- The selectivity of a solvent, which is the ability of a solvent to selectively dissolve one or more compounds, expressed by the separation factor, α.
- The dependence of these solvent properties on conditions of state (P, T).
- The extent of the two-phase area as limiting condition for a two-phase process like gas extraction.

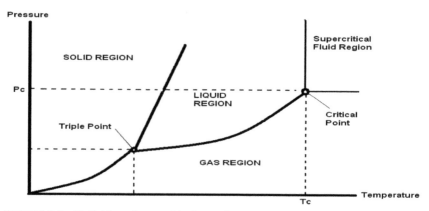

FIGURE 3.1   Definition of supercritical state for a pure component.

TABLE 3.1   Comparison of Gas, Supercritical Fluid, and Liquid.

State	Density	Viscosity	Diffusivity
	kg/m³	µPa·s	mm²/s
Gas	1	10	1–10
Supercritical fluid	100–1000	50–100	0.01–0.1
Liquid	1000	500–1000	0.001

Source: Edit Székely. "What is a supercritical fluid?". Budapest University of Technology and Economics. Retrieved 2014-06-26. http://sfe.kkft.bme.hu/en/current-research.html

The separation factor α is defined by

$$\alpha = \frac{\dfrac{y_i}{y_j}}{\dfrac{x_i}{x_j}} \tag{3.1}$$

where x_i and x_j are the equilibrium concentrations of component i and j, respectively, in the condensed phase, in mole or mass fractions; and y_i and y_j, the corresponding equilibrium concentrations of the same components using equivalent units.

The solubility in the subcritical (liquid) fluid (solvent) increases at constant pressure up to temperatures slightly below the T_C of the solvent. A further increase in temperature leads, at "low" pressures, to a decrease of the dissolved amount of low-volatility substance (triacylglyceride, caffeine, or naphthalene) in the subcritical liquid solvent and at "high" pressures to an increase. "High" and "low" pressures refer to a "medium" pressure level, which for most systems (including the abovementioned) is about 10 MPa. Analogous solubility behavior can be found in systems of an SCF with a solid substance or a low-volatility liquid.

3.2.1 ADVANTAGES OF SC-CO₂ AS A SOLVENT

The application of carbon dioxide as an SCF has been extensively studied over the past three decades, especially in food processing and also for natural oils and bioactive compounds.[27] Carbon dioxide can easily penetrate through the solid matrix and dissolve the desired extract due to its dual gaseous and liquid-like properties. Heat-labile products can be extracted by using SC-CO₂ extraction without changing the bioactivity of natural molecules. Extraction selectivity can be achieved by changing the temperature, pressure, and cosolvent; the extracted material is easily recovered by simply depressurizing, allowing the SC-CO₂ to return to gaseous state and evaporate leaving little or no traces of solvent.[46] The solvent power of SC-CO₂ is good since it dissolves nonpolar to slightly polar compounds. The addition of small quantities of polar organic solvent as modifiers can improve the extraction of polar compound by increasing the solubility of the analyte in CO_2, or by reducing its interaction with the sample matrix or both.[67] Other characteristics of carbon dioxide are, namely, easily recyclable, nonflammable nature, recognized as a natural substance, recovery from the final product is easier, and a convenient critical temperature (31.04°C), which enables extractions to be carried out at low (40–50°C) temperatures thereby protecting the thermolabile compounds.

Extraction of natural raw material with SC-CO_2 allows obtaining extracts whose flavor and taste are perfectly respected and reproducible. The SCF's ability to vaporize nonvolatile components (at moderate temperatures) reduces the energy spent, when comparing to distillation; hence, it is a cost-effective operation. Moreover, increased pressure in the vessel helps to eliminate the entry of oxygen which reduces the chances of oxidation in fat-rich products.[4] CO_2 used is largely a by-product of industrial processes of brewing, and its use as a supercritical solvent does not cause any extra emissions and cost.

Examples of the substances used thus far as supercritical solvents and their critical temperatures and pressures are given in Table 3.2. Among them, CO_2 is the most common SCF solvent, and has been extensively studied for its potential applications in many different fields, including the food processing industries.[78] As an example, cholesterol is more

TABLE 3.2 Supercritical Solvents and Their Respective Critical Temperature and Pressure.

Gas	Critical temperature (K)	Critical pressure (MPa)
Acetylene	308.70	6.24
Argon	150.66	4.86
Carbon dioxide	304.17	7.38
Ethane	305.34	4.87
Ethylene	282.35	5.04
Hydrogen	33.25	1.29
Methane	190.55	4.59
Neon	44.40	2.65
Nitrogen	126.24	3.39
Nitrous oxide	309.15	7.28
Oxygen	154.58	5.04
Propane	369.85	4.24
Xenon	289.70	5.87

soluble in supercritical ethane than in SC-CO_2.[10] As ethane is much more costly than CO_2, the use of CO_2/ethane and CO_2/propane mixtures can be a good alternative for removal of cholesterol from food by compromising between higher ethane cost and better cholesterol removal efficiency, so SCFE can reduce both the extraction and separation costs. During usage of CO_2 for supercritical extraction, following points are to be considered:[7,13]

- It can dissolve nonpolar or slightly polar compound, free fatty acids, glycerides. Water exhibits low solubility while proteins, polysaccharides, sugars, and mineral salts are insoluble.
- Molecular weight of the compound: low molecular weight compounds are highly soluble.
- Affinity with oxygenated organic compounds of medium molecular weight is higher.

The CO_2 technology has some disadvantages associated with it, that is, high capital cost; space needed to store the CO_2 cylinders; and requirement to design and build a prototype that satisfy the temperature, pressure, humidity, and agitation requirements for the CO_2 sterilization.

3.3 SCFE PROCESS

SCFE process involves four primary steps, namely, extraction, expansion, separation, and solvent conditioning; the four corresponding critical components needed are a high-pressure extractor, a pressure reduction valve, a low-pressure separator, and a pump for intensifying the pressure of the recycled solvent (Fig. 3.2). Other ancillary equipment includes heat exchangers, condenser, storage vessels, fluid makeup source, etc. The feed, generally ground solid, is charged into the extractor and CO_2 is fed to the extractor through a high-pressure pump (100–350 bars). The extract-laden CO_2 is sent to a separator (120–50 bars) via a pressure reduction valve. At reduced temperature and pressure conditions, the extract precipitates in the separator, while CO_2, free of any extract, is recycled to the extractor.

SCFE for solid feed is a semibatch process in which carbon dioxide flows in a continuous mode, whereas the solid feed is charged in the extractor basket in batches. For making the process semicontinuous at the commercial scale, multiple extraction vessels are sequentially used, such that when one vessel is on loading or unloading, the other vessels are kept in an uninterrupted extraction mode, as shown in Figure 3.3. A cosolvent is often pumped and mixed with the high-pressure CO_2 for enhancing the solvent power or selectivity of separation of the specified components.

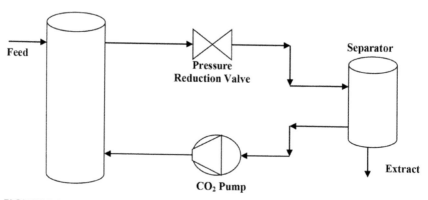

FIGURE 3.2 Schematic diagram of SCFE process.

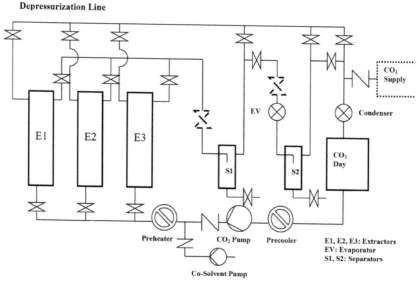

FIGURE 3.3 Schematic diagram of an SCFE plant on commercial scale.

Separation is often carried out in stages by maintaining different conditions in two or three separators for fractionation of the extract, depending on the solubility of the components and the desired specifications of the products. Similarly, by varying the pressure, it is possible to alter the solvent power of the extractant, the effect of which is equivalent to change the polarity of an extraction solvent. Thus, a production plant can have flexible operating conditions for multiple natural products; it is also possible to obtain different product profiles from a single botanical material by merely using a single solvent, namely, SC-CO_2.

3.3.1   ADVANTAGES OF SCFE

- The solubility of a solute in an SCF approaches that of the solubility in a liquid; the major advantage of SCFE is the improved selectivity of solvent rather than the degree of solubility.
- SCFs have relatively low viscosities and high diffusivities both of which result in increased mass transfer of solute into the fluid. However, despite this, the rate of mass transfer of solute into the SCF is usually the rate determining step in an SCFE process.
- There is no requirement for a distillation stage to separate the preferential solvent and solute.
- SCFs are volatile and recycling of solvent takes place at low temperatures (rather than at the high temperatures required in distillation) and therefore the solute, or extract, is not subjected to thermal damage; this is of particular relevance to temperature-sensitive foods.
- This process is inherently safer because of the use of nontoxic and noninflammable solvents.

3.3.2   DISADVANTAGES OF SCFE

- The need for high-pressure equipment fetches high initial capital cost.
- SCFE is a batch process but with some modifications it can be converted into a continuous process.

3.4   APPLICATIONS OF SCFS

3.4.1   FOOD PROCESSING

3.4.1.1   REMOVAL OF FAT FROM FOODS

Edible oils and their components are subjected to SCF processing since early 1970s. Several principles can be utilized to affect the extraction of lipid, namely, optimizing the solubility of lipids in SC-CO_2, enriching and fractionating with respect to a particular target lipid analyte, and appreciating the role of extraction kinetics in recovering lipids from sample matrices. Considerable data are now available on the solubility of seed oils, pure triglycerides, and lipids in SC-CO_2 over a range of pressures and temperatures. Although triacylglycerides are only fairly soluble in SC-CO_2, the advantages of organic solvent-free processing have stimulated research and development in various areas. One of these is the removal of fat from food. The process of extraction of fat from potato chips has the advantage of producing fat-free or fat-reduced potato chips. According to the expected taste, the amount of remaining fat in the potato chips can easily be controlled.

Supercritical extraction can be used as an alternative extraction method to remove oil content from high-protein diet source for producing healthy food with low calorie and fat content. The optimum conditions for peanut extraction with yield of 48.53% peanut oil are at 35 MPa, 60°C, and 15 g/min of CO_2 flow rate, while for black-eyed pea with yield of 5.4% black-eyed pea oil are at 25 MPa, 60°C, and 10 g/min of CO_2 flow rate.

Fish is the main natural source of omega-3 polyunsaturated fatty acids which are of a great importance in the food industries. Comparing to conventional fish oil extraction processes such as cold extraction, wet reduction, or enzymatic extraction, SFCE with CO_2 at 25 MPa and 313K can be used to reduce fish oil oxidation, especially when it is rich in omega-3 such as salmon oil. Furthermore, particularly in the meat processing industry, a significant amount of wastewater is produced. The treatment of food processing wastewater to separate the water (and nonionically charged soluble components) from the solids (including anionically charged soluble components) and reduce the overall waste disposal requirements are an important function within any meat processing operation which can be achieved by implementation of SFCE.

3.4.1.2   ENRICHMENT OF VITAMINS

Fat-soluble vitamins are organic molecules that are nutritionally essential to the human body. Lack of vitamins can lead to serious diseases such as night blindness (vitamin A), rickets and weakening of bones (vitamin D), rupturing of blood cells and cancer (vitamin E), and blood coagulation diseases (vitamin K). New methods are constantly being developed regarding the extraction and enrichment of fat-soluble vitamins from natural sources such as plants, oilseeds, and vegetables. Fat-soluble vitamins are sensitive toward exposure to extreme pH, oxygen, light, and heat, which complicate both sample preparation and subsequent separation.

Using SCFE-CO_2, vitamin E can be extracted from wheat germ having optimized processing conditions of the extraction of natural vitamin E in wheat germ by SCFE-CO_2 were extracting pressure 5000 PSI, extracting temperature 316K, flow rate of CO_2 1.7 mL/min to have a yield of 2307 mg/100g vitamin E. In another approach, a process based on SCFE) at pilot-scale plant has been optimized to obtain fractions highly enriched in vitamin E from microalga *Spirulina platensis*. The estimated model demonstrated that the extraction temperature, the quadratic term of temperature, the extracting pressure, and the interaction pressure × temperature had a significant effect on the final concentration of vitamin E. Yield of 29.4 mg/g was achieved which signified a concentration of 12 times the initial concentration of tocopherol in the raw material.

Vitamin D_2, or ergocalciferol, is a fungal source suitable as a dietary supplement for vegans. Dried brewer's yeast was used as a raw material for the enrichment of D_2 from the precursor ergosterol. Dried yeast was reconstituted in water and exposed to ultraviolet light (UV) with continuous stirring for an optimized period of time to convert ergosterol to ergocalciferol. Freeze-dried D_2-enriched yeast was mixed with rice hulls and wheat germ oil to be subjected for SCFE, with parameters of extraction temperature of 40°C, a solvent to feed of 75 (g CO_2/g material), and a flow rate of 75 g/min in a 4 L extraction vessel. 100 g of material was extracted and separated in a two-stage separator (separator 1, 1500 psi, 40°C; separator 2, 750 psi, 40°C) which to a D_2 enhancement from nondetectable levels to 3.8 mg/g (152,000 IU/g). The percentage of D_2 extracted from the yeast at the extraction pressure of 3000, 6000, and 9000 psi was 55%,

35%, and 31% of the starting concentration, respectively; hence, SCFE extraction is more efficient at lower extraction pressures.

SCFE offers several advantages for the enrichment of tocochromanols over conventional techniques such as vacuum distillation, in particular a lower operating temperature. As a starting material, one can use various edible oils or their distillates. Most promising as feed materials are crude palm oil (CPO) and soybean oil deodorizer distillate (SODD). CPO contains several tocotrienols and tocopherols at a total concentration of approximately 500 ppm. SODD may contain (after several conventional concentration steps) about 50% tocopherols. Both materials can be used for the production of enriched fractions of tocochromanols. Distribution coefficient is smaller than 1, whereas that for the carotenes is much smaller than 1, meaning that these components stay in the liquid oil phase. Thus, tocochromanols can be extracted as the top-phase product in a separation column, whereas carotenes remain in the bottom-phase product together with triglycerides.

Concentration profiles in a pilot-scale plant are described in detail by Gast and Brunner,[23] fitted with an extraction column of 6 m height, inner diameter of 17.5 mm, and equipped with Sulzer EX packing, which can be operated either in a true stripping mode, with the solvent introduced at the bottom and the feed at the top of the column or in reflux mode. Extraction of SODD as feed material was conducted at temperatures (353K and 363K), pressures (23 and 26 MPa), solvent-to-feed (S/F) ratio (33 and 171), and the reflux ratio (1 and 38). After extraction process, the top layer made of squalene was enriched from 3.1% (w/w) in the feed to 18.8% (w/w) and the bottom phase was completely enriched with sterols. Upon increased concentration, the products become too viscous and clog the packing mash and the column; the sterols and tocochromanols were not separated. In order to simulate a taller extraction column, the top-phase product of the previous run was collected and introduced as feed for a second time which made it possible to separate squalene and tocochromanols, which helped to increase the tocochromanol concentration from 48.3 to 94.4% (w/w)[23] and an experimental parameters of pressure (23MPa), temperature (353 K), S/F ratio (110), and a reflux ratio (4.6). A further purification of these compounds is possible, for example, with adsorptive or chromatographic techniques, again using SCFs.

3.4.1.3 REMOVAL OF ALCOHOL FROM WINE AND BEER

Reduction of ethanol content in alcoholic drinks can be accomplished by restricting fermentation or by removing ethanol after the complete fermentation process. In general, better taste is obtained in the second case. Vacuum distillation, membrane techniques, and CO_2 SCFE can be used in the dealcoholization process.

In an attempt to remove ethanol from aqueous solution with about 10% (w/w) ethanol, SC-CO_2 in a stripping column was deployed.[8] Rate of ethanol removal is temperature dependent, for example, to reduce alcohol content to less than 0.5% (w/w), it can take 2.5 h at 45°C. Recovery of aroma compounds can be achieved by a side column in which a separation from ethanol can be carried out.

SCC technique offers minimal thermal damage ensuring aroma conservation, low entrainment, and low liquid residence with high efficiency. The removal of wine alcohol by spinning cone column (SCC) involves two-step process. The delicate wine aroma compounds (nearly 1% of total wine) are recovered from the first pass of wine at low temperature (around 26–28°C) and vacuum conditions (0.04 atm). Dearomatized wine is passed second time aiming at alcohol removal at slightly higher temperatures (around 38°C) and vacuum conditions. Ethanol concentration of a wine can be reduced from 15% to less than 1% v/v by using SCC.

A similar process can be used for production of absolute alcohol using complete miscible ethanol and CO_2.[24] Ethanol can be concentrated above azeotropic composition whenever the pressure in the ternary mixture, CO_2 + ethanol + water, is below the critical pressure of the binary mixture, CO_2 + ethanol.[9,22] The separation factor of ethanol–water has been related as a function of the concentration of ethanol in the solvent-free liquid phase.[9] The separation factor decreased from around 30 at infinite dilution of ethanol in water to approximately 1.25 at infinite dilution of water in ethanol. No azeotrope was formed at the conditions investigated. Separation factors are larger compared to data at atmospheric conditions.

In another attempt, countercurrent multistage extraction was carried out in an extraction column (6 m height; 25 mm inner diameter), equipped with of Sulzer EX packing and a set temperature of 333K and a pressure of 10 MPa. Solvent and extract were separated by pressure reduction down to 5 MPa and washing the extract phase with liquid CO_2 in countercurrent flow. With a feed of 94% (w/w) ethanol, an extract with 99.5% (w/w) ethanol was produced at a reflux ratio of 4 and an S/F ratio of 60 (using 9 kg CO_2/h and 150 g feed/h).

Apart from removal of alcohol, CO_2 under high pressure is also studied for flavor enhancement of light wines. Liquid carbon dioxide is a selective solvent for typical food flavor constituents such as esters, aldehydes, ketones, and alcohols and can be used to extract wine flavors from a "Flavortech" SCC distillate. The extractions can be performed in a batch process at 3–5°C with liquid carbon dioxide at its vapor pressure. The white wine extracted in liquid carbon dioxide has an increased level of major volatile wine flavor compounds.

3.4.1.4 ENCAPSULATION OF LIQUIDS

Encapsulation is a process to entrap active agents within a carrier material and it is a useful tool to improve delivery of bioactive molecules and living cells into foods. Materials used for design of protective shell of encapsulates must be food-grade, biodegradable, and able to form a barrier between the internal phase and its surroundings. Among all materials, the most widely used for encapsulation in food applications are polysaccharides. Proteins and lipids are also appropriate for encapsulation. Spray-drying is the most extensively applied encapsulation technique in the food industry because it is flexible and continuous, but more important an economical operation. Most of encapsulates are spray-dried ones, rest of them are prepared by spray chilling, freeze-drying, melt extrusion, and melt.

A liquid product can be entrapped by adsorption onto solid particles (liquid at the outside of solid particles), by agglomeration (liquid in the free volumes between the solid particles), or by impregnation (liquid within the pore system of the solid particles). Microspheres or larger capsules can be formed, totally encapsulating the liquid. The solid material provides a coating for the liquid inside. Such particulate products can be achieved by means of SCF processing also. An example is the so-called concentrated powder form process, wherein CO_2 is mixed (dissolved) in the liquid feed by static mixing. The CO_2–liquid feed mixture is then sprayed into a spray chamber at ambient conditions together with the substrate material. CO_2 is suddenly released from the liquid, and the liquid forms small droplets. During the spraying process, solid substrate and liquid droplets are intensively mixed and combined to a solid particulate product of the type described above. The product is finally removed from the chamber as a free-flowing powder and separated from the outgoing gas stream by a cyclone. With this type of process, a wide variety of solid substrates can be applied to uptake liquids of different kind and up to about 90%.

As an example, encapsulation or adsorption of tocopherol acetate on silica gel in which about 50% of tocopherol acetate can be incorporated without apparent change of morphology and flow properties of the powder. The amount which can be adsorbed at high pressures is comparable to that of normal pressure. Autoclave was used to saturate the SC-CO$_2$ current with tocopherol acetate, and the density of the solvent was changed in the nozzle where the loaded SC-CO$_2$ phase was fed to the adsorber. This adsorption at high pressures makes possible the direct product formation in the SCF, with the advantageous effect that the supercritical solvent can easily be recycled without substantial compression.

3.4.1.5   FRACTIONATION OF ESSENTIAL OILS

The supercritical extraction of the flavoring/aromatic compounds in vegetable, fruits, and flowers are a promising field for the industrial application. Indeed, there is considerable interest in replacing the steam distillation and solvent extraction processes traditionally used to obtain these products. The characteristic smell of plant materials is usually the result of the complex interactions occurring among hundreds of compounds. The presence of thermolabile compounds, the possibility of hydrolysis, and hydrosolubilization are serious obstacles in the reproduction of natural fragrances. CO$_2$ is the supercritical solvent of choice in the extraction of fragrance compounds, since it is nontoxic and allows supercritical operation at relatively low pressures and near room temperature. SC-CO$_2$ behaves similar to a lipophilic solvent but, compared to liquid solvents, it has the advantage that its selectivity or solvent power is adjustable and can be set to values ranging from gas-like to liquid-like.[49]

SCFE of flavors and fragrances has been reviewed by Kerrola[34] and Reverchon;[60] and the different substrates subjected to extraction are:

Angelica (root);[52] anise (seed);[53] basil (leaf);[63] cardamom (seed);[26] chamomile (flower);[63] cinnamon (leaf); citron (peel); coriander (fruit);[33,35] dragonhead (leaf);[29] *Evodia rutaecarpa* (herb);[40] frankincense (herb);[40] geranium (leaf);[41] ginger (rhizome);[2] hops (fruit);[38,75] iris (rhizome);[5] jasmine (flower);[58,65] lavender (flower);[1] *Lavandula* (flower); kumquat (peel); marjoram (leaf);[62] massoi (bark);[15] mirrh (herb);[40] onion (bulb);[72] oregano (leaf);[53] Pampelmousse (peel); paprika (fruit); peppermint (leaf);[61] rose (flower);[60] rosemary (leaf);[45,62] sage (leaf);[64] savory (leaf);[29] spearmint (leaf);[3] strawberry (berry);[56] thyme (leaf); *Tsuga canadiensis* (leaf); and vanilla (pod).[51]

3.4.1.6   EXTRACTION OF CAROTENE AND LUTEIN

The use of artificial dyes is a common practice in the modern food industry, but there is a growing concern about their actual or potential effect on human health. This concern has led to an increasing interest and utilization of natural products as alternative food colorants. Carotenoids are one of the major groups of natural pigments that find widespread utilization in the food industry, in particular ß-carotene. They represent a group of bioactive compounds that are responsible for the yellow–red pigmentation of many plant organs and plant-derived raw and processed foods, known to improve food healthiness and stability because of their provitamin A activity, antioxidant power, and boost of cell-mediated and humoral immune response.[12,59,69,70] Tomato, green leafy vegetables (spinach, broccoli, endive, lettuce, etc.), cereals, and red pepper are the main sources of lycopene, lutein, zeaxanthin, and β-cryptoxanthin, respectively, while α- and β-carotene are taken up mainly by the consumption of carrot, sweet potato, and pumpkin.

Traditional natural sources of carotenoids are the fruit of *Bixa orellana*, paprika, carrots; however, a large portion of commercially available β-carotene is also synthetically produced. Concerns regarding the effect of artificial dyes on human health suggest an expanding economic market potential for natural pigments, and a need for additional research into alternative natural substances for the production of carotenoids. A key step in the production of alternative sources of carotenoids is the development of new extraction techniques, which can minimize molecular alteration of the carotenoids during the extraction process.

Leaf protein concentrates (LPC) have not only high protein content but also an appropriate amino acid composition for use as a source of good quality protein for enhancing human nutrition. Therefore, much research has been carried out on the production techniques and nutritional properties of LPC.[16,18,32] The high pigment content in LPC has been one of the major obstacles for its utilization in the human diet. Carotenoids are a group of pigments that can easily degrade from exposure to heat and light. For example, the high temperatures (130°C) applied during the extraction of annatto causes isomerization and degradation of the bixin molecule, thus affecting its pigmentary properties.[57] In another study by Favati et al.,[18] SCF carbon dioxide was used to extract carotene and lutein from LPC in which extractions were performed using pressures of 10–70 MPa at 40°C and CO_2 flow rates were 5–6 L/min. Over 90% of the carotene contained in LPC was removed at extraction pressures in excess of 30 MPa. Removal of lutein was achieved at higher extraction pressures (70 MPa) and gas volumes to attain a 70% recovery level.

SC-CO_2 extraction of carotenoids from pumpkin have gained great attention and studies have revealed that pumpkin SC-CO_2 extracts are rich with α- and β-carotene as well as other bioactive compounds. The best operative conditions for pumpkin carotenoids extraction by SC-CO_2 as reported by scientists are 35 MPa and 50–70°C, for pressure and temperature, respectively, with highest yield by the combination of water (10%) and olive oil (10%) or ethanol (10%) and olive oil (10%), as entrainers, at 50°C and 80°C, respectively. This process offered the possibility of obtaining a selective extraction of natural colorants, free of solvent residues, which can be used as food dyes.

3.4.1.7 DECAFFEINATION OF COFFEE

Coffee is one of the most popular beverages consumed throughout the world. Caffeine is of major importance with respect to the physiological properties of coffee and also in determining the bitter character. Caffeine is one of the most widely used psychoactive substances in the world. There have been several reports on the psychological effects of caffeine consumption[25,31,76] and its effect in conjunction with the consumption of alcohol, smoking,[17,55,66] and drugs.[11,42] In human beings, a fatal dose of caffeine is estimated to be 10 g or about 50–100 cups of average coffee (depending on strength and volume). Runge first discovered and isolated caffeine in the late nineteenth century, and Roselius developed the first decaffeinating process for coffee in Germany in 1900.[68] The most selective process for removing just caffeine and not the other flavor precursors from coffee is carbon dioxide.[71] Carbon dioxide does not affect the carbohydrates (sugars, starch) and peptides (proteins) that during roasting are converted into the various compounds responsible for the flavor and aroma of brewed coffee. There are several patents available on carbon dioxide decaffein-ation of coffee. Caffeine can be extracted from green beans using SCFs, for example, carbon dioxide, nitrogen, nitrous oxide, methane, ethylene, and propane as the extractants. Efficiency of carbon dioxide can be improved by treating it with dimethyl sulfoxide before decaffeination of either wet or dry green coffee seeds. Caffeine can also be extracted from roasted coffee using liquid carbon dioxide. In general, the process of decaffeination using carbon dioxide can be described as follows:

- It begins with mixing green coffee beans with water to bring the moisture content to about 50%.
- The beans swell, pores open, and caffeine is converted into a mobile form that can diffuse out of the bean.
- The swollen beans are loaded into a thick-walled stainless steel extractor that is then sealed. Liquid carbon dioxide is pumped in at the operating pressure of about 300 atm and is heated to about 150°F. After the system is filled, carbon dioxide is recircu-lated between the extractor and scrubber. Carbon dioxide flowing through the extractor dissolves caffeine.
- In the scrubber, caffeine is removed from carbon dioxide using water. The scrubber water preferentially carries away the caffeine. The recirculation is continued for 8–12 h, depending on the initial caffeine content of the green beans. The target caffeine content for the decaffeinated beans is usually 0.08%.
- At the end of recirculation, carbon dioxide is pumped from the system to a strong tank and is recycled to the next batch.
- The beans are unloaded from the extractor for drying which reduces the moisture to its original level and vaporizes the carbon dioxide.

The advantages of this process are: (1) there is no harmful residue, (2) the product is of a superior quality in terms of flavor and appearance, and (3) losses of coffee solubles (other than caffeine) are quite low, less than other decaffeination methods, resulting in greater product yield.[50]

3.4.1.8   EXTRACTION OF ANTHOCYANIN

Phenolic compounds from blueberry (*Vaccinium myrtillus* L.) residues can be extracted by solvent extraction and SCFE. SCFE can be performed using

CO_2 as solvent at temperature of 40°C and pressure of 15, 20, 25, and 30 MPa at solvent flow rate of 3 L/min which is proved to be more beneficial than conventional Soxhlet extraction in recovery of anthocyanins. Similarly, SCFE can be employed for the extraction of bioactive compounds from grape (*Vitis labrusca* B.) peel. The optimum extraction conditions for extraction of bioactive compounds are temperature of 45–46°C, pressure at 160–165 kg/cm², and ethanol of 6–7%. Anthocyanin from Indian blackberry (*Syzygium curnini* L.), commonly known as jamun, with the maximum yield can be performed at pressure of 162 bar, temperature of 50°C, and cosolvent flow rate of 2 g/min. Extractions of anthocyanins from peels of eggplant (*Solanum melongena* L.) using GRAS solvents and SC-CO_2 (SC-CO_2) can also be performed with the optimized conditions of 10 g of peels at 60°C, 10 MPa, 1.5 h extracting time, and 2 L/min of CO_2.

3.4.2 MILK AND MILK PRODUCTS

SC-CO_2 extraction technique is also implemented for fractionation of anhydrous milk fat into fractions with specific properties as per end use using batch or continuous systems (Fig. 3.4). The melting properties of the fractions obtained by SCFE are not as pronounced as with melt crystallization. Nevertheless, niche applications could be developed if fractions rich in short-chain fatty acids or a solid fraction in unsaturated fatty acids are required. Low melting fraction having a melting point of <15°C usually have a strong butter flavor and can be incorporated into milk powder to improve functionality; it can be incorporated into confectionery products and used to manufacture spreadable butter at refrigerator temperatures.

Milk fat fractions having a melting point of 15–30°C can be used as shortening to provide a crusty, flaky texture to croissants and pastries and in making cakes and biscuits such as shortbread. High melting fraction (melting point 30°C) when used in chocolate manufacturing as a substitute for cocoa butter, apart from cost, is also reported to inhibit chocolate bloom. In milk chocolates, hard fraction can be incorporated to improve the buttery flavor and smooth texture. Hard fraction can also improve the whipping properties of cream which is desirable in ice cream manufacturing. Milk fractions are also mixed with vegetable fats to increase their functionality, reduce cost as in case of cocoa butter, improve flavor profile, and increase the nutritional quality.

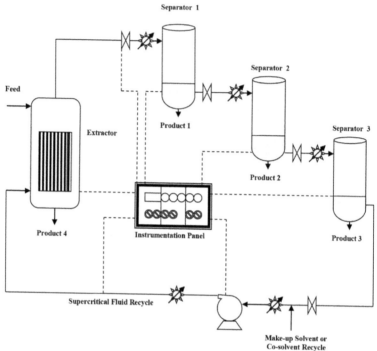

FIGURE 3.4   Flow diagram for fractionation of milk fat.

Supercritical extraction is also used for manufacturing of dairy ingredients rich in phospholipids by using other conventional techniques. One such approach to manufacture milk fat globule membrane (MFGM)-rich dairy ingredient using whey buttermilk is by concentrating using ultrafiltration (10×) and then subsequently filtered (5×) (10 kDa molecular mass cutoff membrane) at 25°C followed by spray-drying of final retentate. The spray-dried whey buttermilk powder can then be subsequently submitted to supercritical extraction (350 bar, 50°C) using carbon dioxide which can produce a phospholipids-rich dairy powder containing 73% protein and 21% lipids, of which 61% were phospholipids. This ingredient could be used as an emulsifier in different food systems.

In spite of being highly nutritious, the consumption of milk is hindered because of its high cholesterol content, which is responsible for numerous cardiac diseases. SC-CO_2 using ethanol as cosolvent was employed to extract cholesterol from whole milk powder (WMP). Scientist have also attempted to quantify the cholesterol content of WMP using HPLC and the impact of the extraction conditions on the fat content (FC), solubility index (SI), and lightness (L*). They optimized the process parameters using response surface methodology. About 46% reduction in cholesterol was achieved at the optimized conditions of 48°C, 17 MPa, and 31 mL cosolvent; flow rate of expanded CO_2 and static time of extraction were 6 L/min and 80 min, respectively. The treated WMP retained its FC, SI, and L* at moderate limits of 183.67 g/kg, 96.3%, and 96.90, respectively. This study demonstrated the feasibility of ethanol-modified SCFE of cholesterol from WMP with negligible changes in its physicochemical properties.[14]

The cholesterol content of butter oil with supercritical ethane and CO_2 was reduced at 40°C, 55°C, and 70°C and pressures ranging from 8.5 to 24.1 MPa, using a semicontinuous flow and high-pressure apparatus. To enhance the cholesterol removal process selectivity, the extraction operation was subsequently coupled to adsorption on alumina. The solubility of cholesterol and butter oil in supercritical ethane was found to be much higher than those in SC-CO_2. The combined extraction/adsorption process resulted in butter oil fractions with a cholesterol content limited to about 3% of that in the original oil. Selective blending of these low cholesterol fractions can provide yet further possibilities in the formulation of high-value products.

A summary of the main products containing cholesterol and their extraction with SCFs is presented in Table 3.3. These results clearly indicate the great potential of SCFE in the recovery of meat products with acceptable cholesterol and fat contents.

TABLE 3.3 Supercritical Fluid Extraction of Cholesterol.

Product	Pressure (MPa)	Temperature (°C)	Cholesterol (mg/g)		Yield (%)
			Before	After	
Beef patties (cooked)[19]	17.2–55.1	40–50	1.94	0.12	93.8
Dehydrated beef	23.4–38.6	45–55	1.56	0.19	87.8
Dried chicken meat[20]	30.6–37.6	45–55	4.96	0.54	90.0
Dried egg yolk[21]	16.5–37.8	40–55	18.52	6.34	65.8
Dried egg yolk[6]	24.1–37.8	45–55	18.94	0.38	98.0
Milk fat[47]	10.1–36.4	40–70	2.50	0.21	91.5
Milk fat*[48]	8.0–24.0	40–70	2.50	0.20	93.4
Pork (cooked)	7.3–34.4	50–150	0.80	0.22	70.1

*Using supercritical ethane as solvent.

Apart from fractionation and extraction of fats from milk and milk products, SCFE can also be used for sterilization of milk. Sterilizing effect of SC-CO_2 on various microorganisms at 20.3 MPa and 35°C (water content of 70–90) is reported but dried cells were not sterilized when treated under the same conditions. The inactivation effect of the native microorganisms in raw milk and raw cream is nearly the same. Fat does not influence the inactivation. Milk enzymes including phosphohexose isomerase, gamma-glutamyltransferase, and alkaline phosphatase also are reported to be inactivated as a result of SC treatment.

3.5   SUMMARY

Increasing environmental concerns, government measures, and population growth drive the search for green processes to replace the conventional ones. SCFE has received growing interest as a green alternative technology in the food industry. This technique is now a real option for product development, mainly those that will be used for human consumption such as new foods, food ingredients/additives, or pharmaceutical products. Moreover, SCFE has also demonstrated some advantages in the environmental field such as, to reduce solvent waste, to get new useful compounds from industrial by-products, and to allow quantification and/or removal of toxic compounds from the environment.

KEYWORDS

- anthocyanin
- carotene
- critical pressure
- critical temperature
- spinning cone column
- supercritical carbon dioxide extraction
- supercritical fluid state

REFERENCES

1. Adaşoğlu, N.; Dinçer, S.; Bolat, E. Supercritical-fluid Extraction of Essential Oil from Turkish Lavender Flowers. *J. Supercrit. Fluids* **1994,** *7*(2), 93–99.
2. Bartley, J. P.; Foley, P. Supercritical Fluid Extraction of Australian Grown Ginger (*Zingiber officinale*). *J. Sci. Food Agric.* **1994,** *66*(3), 365–371.
3. Barton, P.; Hughes, R. E.; Hussein, M. M. Supercritical Carbon Dioxide Extraction of Peppermint and Spearmint. *J. Supercrit. Fluids* **1992,** *5*(3), 157–162.
4. Berger, T. A.; Fogleman, K.; Staats, T.; Bente, P.; Crocket, I.; Farrell, W.; Osonubi, M. The Development of a Semi-preparatory Scale Supercritical-fluid Chromatograph for High-throughput Purification of 'Combi-chem' Libraries. *J. Biochem. Biophys. Methods* **2000,** *43*(1), 87–111.
5. Bicchi, C.; Rubiolo, P.; Rovida, C. Analysis of Constituents of Iris Rhizomes. Part II. Simultaneous SFE of Irones and Iridals from *Iris pallida* L. Rhizomes. *Flavor Fragr. J.* **1993,** *8*(5), 261–267.
6. Bohac, C. E.; Rhee, K. S.; Cross, H. R.; Ono, K. Assessment of Methodologies for Calorimetric Cholesterol Assay of Meats. *J. Food Sci.* **1988,** *53*(6), 1642–1644.
7. Brunner, G. Stofftrennungmit überkritischenGasen (Gasext-raktion). *Chemie Ingenieur Technik* **1987,** *59*, 12–22.
8. Brunner, G.; Kreim, K. ZurTrennung von Ethanol–Wasser Gemischenmittels Gasextraktion. *Chemie Ingenieur Technik* **1985,** *57*, 550–551.
9. Budich, M.; Brunner, G. Supercritical Fluid Extraction of Ethanol from Aqueous Solutions. *J. Supercrit. Fluids* **2003,** *25*, 45–55.
10. Casas, L.; Mantell, C.; Rodríguez, M.; Torres, A.; Macías, F. A.; de la Ossa, E. M. Extraction of Natural Compounds with Biological Activity from Sunflower Leaves Using Supercritical Carbon Dioxide. *Chem. Eng. J.* **2009,** *152*(2), 301–306.
11. Conway, T. L.; Vickers, R. R.; Ward, H. W.; Rahe, R. H. Occupational Stress and Variation in Cigarette, Coffee, and Alcohol Consumption. *J. Health Soc. Behav.* **1981,** *2*, 155–165.
12. Cutler, R. G. Carotenoids and Retinol: Their Possible Importance in Determining Longevity of Primate Species. *Proc. Natl. Acad. Sci.* **1984,** *81*(23), 7627–7631.
13. del Valle, J. M.; Aguilera, J. M. Articles-high Pressure CO_2 Extraction. Fundamentals and Applications in the Food Industry. *Food Sci. Technol. Int. Frederick* **1999,** *5*(1), 1–24.
14. Dey Paul, I.; Jayakumar, C.; Mishra, H. N. Optimization of Process Parameters for Supercritical Fluid Extraction of Cholesterol from Whole Milk Powder Using Ethanol as Co-solvent. *J. Sci. Food Agric.* **2016,** *96*(15), 4885–4895.
15. Dick, R.; Quirin, K. W.; Cresp, P. Massoïa Lactone: Extraction et Concentration Parle CO_2 Supercritique. *Parfums Cosmétiques Arômes* **1992,** *103*, 91–93.
16. Edwards, R. H.; Miller, R. E.; De Fremery, D.; Knuckles, B. E.; Bickoff, E. M.; Kohler, G. O. Pilot Plant Production of an Edible White Fraction Leaf Protein Concentrate from Alfalfa. *J. Agric. Food Chem.* **1975,** *23*(4), 620–626.
17. Engs, R. C. The Drug-use Patterns of Helping-profession Students in Brisbane, Australia. *Drug Alcohol Depend.* **1980,** *6*(4), 231–246.
18. Favati, F.; King, J. W.; Friedrich, J. P.; Eskins, K. Supercritical CO_2 Extraction of Carotene and Lutein from Leaf Protein Concentrates. *J. Food Sci.* **1988,** *53*(5), 1532–1536.

19. Fenton, M.; Sim, J. S. Determination of Egg Yolk Cholesterol Content by On-column Capillary Gas Chromatography. *J. Chromatogr. A* **1991**, *540*, 323–329.

20. Froning, G. W.; Fieman, F.; Wehling, R. L.; Cuppett, S. L.; Niemann, L. Supercritical Carbon Dioxide Extraction of Lipids and Cholesterol from Dehydrated Chicken Meat. *Poult. Sci.* **1994**, *73*(4), 571–575.

21. Froning, G. W.; Wehling, R. L.; Cuppett, S. L.; Pierce, M. M.; Niemann, L.; Siekman, D. K. Extraction of Cholesterol and Other Lipids from Dried Egg Yolk Using Supercritical Carbon Dioxide. *J. Food Sci.* **1990**, *55*(1), 95–98.

22. Furuta, S.; Ikawa, N.; Fukuzato, R.; Imanshi, N. Extraction of Ethanol from Aqueous Solutions Using Supercritical Carbon Dioxide. *Kagaku Kogaku Ronbunshu* **1989**, *15*, 519–525.

23. Gast, K.; Jungfer, M.; Saure, C.; Brunner, G. Purification of Tocochromanols from Edible Oil. *J. Supercrit. Fluids* **2005**, *34*(1), 17–25.

24. Gilbert, M. L.; Paulaitis, M. E. Gas–Liquid Equilibrium for Ethanol–water–carbon Dioxide Mixtures at Elevated Pressures. *J. Chem. Eng. Data* **1986**, *31*, 296–298.

25. Gilliland, K.; Andress, D. Ad Lib Caffeine Consumption, Symptoms of Caffeinism, and Academic Performance. *Am. J. Psychiatry* **1981**, *138*(4), 512–514.

26. Gopalakrishnan, N.; Narayanan, C. S. Supercritical Carbon Dioxide Extraction of Cardamom. *J. Agric. Food Chem.* **1991**, *39*(11), 1976–1978.

27. Gracia, I.; García, M. T.; Rodríguez, J. F.; Fernández, M. P.; De Lucas, A. Modelling of the Phase Behaviour for Vegetable Oils at Supercritical Conditions. *J. Supercrit. Fluids* **2009**, *48*(3), 189–194.

28. Hartono, R.; Mansoori, G. A.; Suwono, A. Prediction of Solubility of Biomolecules in Supercritical Solvents. *Chem. Eng. Sci.* **2001**, *56*(24), 6949–6958.

29. Hawthorne, S. B.; Rickkola, M. L.; Screnius, K.; Holm, Y.; Hiltunen, R.; Hartonen, K. Comparison of Hydrodistillation and Supercritical Fluid Extraction for the Determination of Essential Oils in Aromatic Plants. *J. Chromatogr. A* **1993**, *634*(2), 297–308.

30. Herrero, M.; Mendiola, J. A.; Cifuentes, A.; Ibanez, E. Supercritical Fluid Extraction: Recent Advances and Application. *J. Chromatogr. A* **2010**, *1217*(16), 2495–2511.

31. Hollands, M. A.; Arch, J. R.; Cawthorne, M. A. A Simple Apparatus for Comparative Measurements of Energy Expenditure in Human Subjects: The Thermic Effect of Caffeine. *Am. J. Clin. Nutr.* **1981**, *34*(10), 2291–2294.

32. Huang, K. H.; Tao, M. C.; Boulet, M.; Riel, R. R.; Julien, J. P.; Brisson, G. J. A Process for the Preparation of Leaf Protein Concentrates Based on the Treatment of Leaf Juices with Polar Solvents. *Can. Inst. Food Technol. J.* **1971**, *4*(3), 85–90.

33. Kallio, H.; Kerrola, K. Application of Liquid Carbon Dioxide to the Extraction of Essential Oil of Coriander (*Coriandrum sativum L.*) Fruits. *Zeitschriftfür Lebensmittel-Untersuchung Forschung* **1992**, *195*(6), 545–549.

34. Kerrola, K. Literature Review: Isolation of Essential Oils and Flavor Compounds by Dense Carbon Dioxide. *Food Rev. Int.* **1995**, *11*(4), 547–573.

35. Kerrola, K., Kallio, H. Volatile Compounds and Odor Characteristics of Carbon Dioxide Extracts of Coriander (*Coriandrum sativum L.*) Fruits. *J. Agric. Food Chem.* **1993**, *41*(5), 785–790.

36. Korhonen, H. Technology Options for New Nutritional Concepts. *Int. J. Dairy Technol.* **2002**, *55*(2), 79–88.

37. Kwon, Y. A.; Chao, R. R. Fractionation and Cholesterol Reduction of Beef Tallow by Supercritical CO_2 Extraction. *Food Biotechnol.* **1995**, *4*, 234–242.

38. Langezaal, C. R.; Chandra, A.; Katsiotis, S. T.; Scheffer, J. J.; De Haan, A. B. Analysis of Supercritical Carbon Dioxide Extracts from Cones and Leaves of *Humulus lupulus L* Cultivar. *J. Sci. Food Agric.* **1990**, *53*(4), 455–463.

39. List, G. R.; King, J. W.; Johnson, J. H.; Warner, K.; Mounts, T. L. Supercritical CO_2 Degumming and Physical Refining of Soybean Oil. *J. Am. Oil Chem. Soc.* **1993**, *70*(5), 473–476.

40. Ma, X.; Yu, X.; Zheng, Z.; Mao, J. Analytical Supercritical Fluid Extraction of Chinese Herbal Medicines. *Chromatographia* **1991**, *32*(1–2), 40–44.

41. Machado, A. S. R.; de Azevedo, E. G.; Sardinha, R. M.; da Ponte, M. N. High Pressure CO_2 Extraction from Geranium Plants. *J. Essent. Oil Res.* **1993**, *5*(2), 185–189.

42. Macklin, A. W.; Szot, R. J. Eighteen Month Oral Study of Aspirin, Phenacetin and Caffeine in C57BL/6 Mice. *Drug Chem. Toxicol.* **1980**, *3*(2), 135–163.

43. Manan, Z. A.; Siang, L. C.; Mustapa, A. N. Development of a New Process for Palm Oil Refining Based on Supercritical Fluid Extraction Technology. *Ind. Eng. Chem. Res.* **2009**, *48*(11), 5420–5426.

44. Mansoori, G. A.; Schulz, K.; Martinelli, E. E. Bioseparation Using Supercritical Fluid Extraction/Retrograde Condensation. *Nat. Biotechnol.* **1988**, *6*(4), 393–396.

45. Mendes, R. L.; Coelho, J. P.; Fernandes, H. L.; Marrucho, I. J.; Cabral, J. M.; Novais, J. L. M.; Palavra, A. A. F. Applications of Supercritical CO_2 Extraction to Microalgae and Plants. *J. Chem. Technol. Biotechnol.* **1995**, *62*(1), 53–59.

46. Mohamed, R. S.; Mansoori, G. A. The Use of Supercritical Fluid Extraction Technology in Food Processing. *Food Technol. Mag.* **2002**, *20*, 134–139.

47. Mohamed, R. S.; Neves, G.; Kieckbusch, T. G. Reduction in Cholesterol and Fractionation of Butter Oil Using Supercritical CO_2 with Adsorption on Alumina. *Int. J. Food Sci. Technol.* **1998**, *33*(5), 445–454.

48. Mohamed, R. S.; Saldaña, M. D.; Socantaype, F. H.; Kieckbusch, T. G. Reduction in the Cholesterol Content of Butter Oil Using Supercritical Ethane Extraction and Adsorption on Alumina. *J. Supercrit. Fluids* **2000**, *16*(3), 225–233.

49. Molero Gómez, A.; Pereyra López, C.; Martinez De La Ossa, E. Recovery of Grape Seed Oil by Liquid and Supercritical Carbon Dioxide Extraction: A Comparison with Conventional Solvent Extraction. *Chem. Eng. J. Biochem. Eng. J.* **1996**, *61*(3), 227–231.

50. Muhlnickel, T. The Ins and Outs of CO_2 Decaffeination. *Tea Coffee Trade J.* **1992**, *164*(9), 35–36.

51. Nguyen, K.; Barton, P.; Spencer, J. S. Supercritical Carbon Dioxide Extraction of Vanilla. *J. Supercrit. Fluids* **1991**, *4*(1), 40–46.

52. Nykänen, I.; Nykänen, L.; Alkio, M. Composition of Angelica Root Oils Obtained by Supercritical CO_2 Extraction and Steam Distillation. *J. Essent. Oil Res.* **1991**, *3*(4), 229–236.

53. Ondarza, M.; Sanchez, A. Steam Distillation and Supercritical Fluid Extraction of Some Mexican Spices. *Chromatographia* **1990**, *30*(1–2), 16–18.

54. Ooi, C. K.; Bhaskar, A.; Yener, M. S.; Tuan, D. Q.; Hsu, J.; Rizvi, S. S. H. Continuous Supercritical Carbon Dioxide Processing of Palm Oil. *J. Am. Oil Chem. Soc.* **1996**, *73*(2), 233–237.

55. Ossip, D. J.; Epstein, L. H.; McKnight, D. Modeling, Coffee Drinking, and Smoking. *Psychol. Rep.* **1980,** *47*(2), 408–410.

56. Polesello, S.; Lovati, F.; Rizzolo, A.; Rovida, C. Supercritical Fluid Extraction as a Preparative Tool for Strawberry Aroma Analysis. *J. High Resolut. Chromatogr.* **1993,** *16*(9), 555–559.

57. Preston, H. D.; Rickard, M. D. Extraction and Chemistry of Annatto. *Food Chem.* **1980,** *5*(1), 47–56.

58. Rao, G. R.; Srinivas, P.; Sastry, S. V. G. K.; Mukhopadhyay, M. Modeling Solute-co-solvent Interactions for Supercritical-fluid Extraction of Fragrances. *J. Supercrit. Fluids* **1992,** *5*(1), 19–23.

59. Rettura, G.; Stratford, F.; Levenson, S. M.; Seifter, E. Prophylactic and Therapeutic Actions of Supplemental β-carotene in Mice Inoculated with C3HBA Adenocarcinoma Cells: Lack of Therapeutic Action of Supplemental Ascorbic Acid. *J. Natl. Cancer Inst.* **1982,** *69*(1), 73–77.

60. Reverchon, E. Supercritical Fluid Extraction and Fractionation of Essential Oils and Related Products. *J. Supercrit. Fluids* **1997,** *10*(1), 1–37.

61. Reverchon, E.; Osseo, L. S.; Gorgoglione, D. Supercritical CO_2 Extraction of Basil Oil: Characterization of Products and Process Modeling. *J. Supercrit. Fluids* **1994,** *7*(3), 185–190.

62. Reverchon, E.; Senatore, F. Isolation of Rosemary Oil: Comparison Between Hydrodistillation and Supercritical CO_2 Extraction. *Flavor Fragr. J.* **1992,** *7*(4), 227–230.

63. Reverchon, E.; Senatore, F. Supercritical Carbon Dioxide Extraction of Chamomile Essential Oil and Its Analysis by Gas Chromatography-mass Spectrometry. *J. Agric. Food Chem.* **1994,** *42*(1), 154–158.

64. Reverchon, E.; Taddeo, R. Extraction of Sage Oil by Supercritical CO_2: Influence of Some Process Parameters. *J. Supercrit. Fluids* **1995,** *8*(4), 302–309.

65. Reverchon, E.; Taddeo, R.; Porta, G. D. Extraction of Sage Oil by Supercritical CO_2: Influence of Some Process Parameters. *J. Supercrit. Fluids* **1995,** *8*(4), 302–309.

66. Reynolds, V.; Jenner, D. A.; Palmer, C. D.; Harrison, G. A. Catecholamine Excretion Rates in Relation to Life-styles in the Male Population of Otmoor, Oxfordshire. *Ann. Hum. Biol.* **1981,** *8*(3), 197–209.

67. Rial-Otero, R.; Gaspar, E. M.; Moura, I.; Capelo, J. L. Gas Chromatography Mass Spectrometry Determination of Acaricides from Honey After a New Fast Ultrasonic-based Solid Phase Micro-extraction Sample Treatment. *Talanta* **2007,** *71*(5), 1906–1914.

68. Schoenholt, D. N. Decaffeinated-not Decapitated. *Tea Coffee Trade J.* **1993,** *165*(2), 32–34.

69. Seifter, E.; Rettura, G.; Padawer, J.; Levenson, S. M. Moloney Murine Sarcoma Virus Tumors in CBA/J Mice: Chemopreventive and Chemotherapeutic Actions of Supplemental β-carotene. *J. Natl. Cancer Inst.* **1982,** *68*(5), 835–840.

70. Seifter, E.; Rettura, G.; Padawer, J.; Stratford, F.; Goodwin, P.; Levenson, S. M. Regression of C3HBA Mouse Tumor Due to X-ray Therapy Combined with Supplemental β-carotene or Vitamin A. *J. Natl. Cancer Inst.* **1983,** *71*(2), 409–417.

71. Sims, M. Decaffeinating with Carbon Dioxide. *Tea Coffee Trade J.* **1990,** *162*(9), 8–10.

72. Sinha, N. K.; Guyer, D. E.; Gage, D. A. Lira, C. T. Supercritical Carbon Dioxide Extraction of Onion Flavors and Their Analysis by Gas Chromatography–Mass Spectrometry. *J. Agric. Food Chem.* **1992**, *40*(5), 842–845.
73. Staby, A.; Mollerup, J. Separation of Constituents of Fish Oil Using Supercritical Fluids: A Review of Experimental Solubility, Extraction and Chromatographic Data. *Fluid Phase Equilibria* **1993**, *91*(2), 349–386.
74. Tello, J.; Viguera, M.; Calvo, L. Extraction of Caffeine from Robusta Coffee (*Coffeacanephora var.* Robusta) Husks Using Supercritical Carbon Dioxide. *J. Supercrit. Fluids* **2011**, *59*, 53–60.
75. Verschuere, M.; Sandra, P.; David, F. Fractionation by SFE and Microcolumn Analysis of the Essential Oil and the Bitter Principles of Hops. *J. Chromatogr. Sci.* **1992**, *30*(10), 388–391.
76. White, B. C.; Lincoln, C. A.; Pearce, N. W.; Reeb, R.; Vaida, C. Anxiety and Muscle Tension as Consequences of Caffeine Withdrawal. *Science* **1980**, *209*(4464), 1547–1548.
77. Yoon, J.; Han, B. S.; Kang, Y. C.; Kim, K. H.; Jung, M. Y.; Kwon, Y. A. Purification of Used Frying Oil by Supercritical Carbon Dioxide Extraction. *Food Chem.* **2000**, *71*(2), 275–279.
78. Zaidul, I. S. M.; Norulaini, N. A. N.; Mohd Omar, A. K.; Sato, Y.; Smith, R. L. Separation of Palm Kernel Oil from Palm Kernel with Supercritical Carbon Dioxide Using Pressure Swing Technique. *J. Food Eng.* **2007**, *81*, 419–428.
79. Edit Székely. "What is a supercritical fluid?". Budapest University of Technology and Economics. Retrieved 2014-06-26.

PART II
Novel Technological Interventions in Dairy Science

CHAPTER 4

PRESERVATION OF MILK USING BOTANICAL PRESERVATIVES (ESSENTIAL OILS)

ASHWINI S. MUTTAGI[1], RUPESH S. CHAVAN[2,*], and SHRADDHA B. BHATT[3]

[1]*DRDC-Foods, Dabur India Limited, Plot No 22, Site 4, Sahibabad, Ghaziabad 201301, Uttar Pradesh, India*

[2]*Department of Quality Assurance, Mother Dairy Junagadh, Junagadh 362001, Gujarat, India*

[3]*Department of Biotechnology, Junagadh Agricultural University, Junagadh 362001, Gujarat, India*

[*]Corresponding author. E-mail: rschavanb_tech@rediffmail.com

CONTENTS

ABSTRACT

Essential oils are natural aromatic compounds found in the roots, rhizomes, wood bark, leaves, stems, fruit, flowers and seeds, and other parts of plants. Essential oils contain many complex chemical compounds having anti-inflammatory, antibacterial, antimicrobial, and antiviral properties. Essential oils are highly concentrated and a small amount is very potent and consists of esters, aldehydes, ketones, and alcohols. Preserving milk by adding essential oils is a nontraditional technique in order to satisfy customers need, and the chapter deals in details with application of essential oils in milk and milk products and the different techniques used for masking the strong odor of the oils. Quantification of the minimum and noninhibitory concentration is also discussed in brief along with the mechanism of antimicrobial action.

4.1   INTRODUCTION

Milk is the primary food with perishable nature. A number of nontraditional preservation techniques are being developed to satisfy consumers' need and demand. Preservation of the milk can be done using plants and spices because their components/extracts contain antimicrobial as well as bactericidal effect, which can help in controlling microbial growth as well as inhibit their activity. The application of botanical preservative is beneficial as it does not alter taste and flavor of the milk; with proper and uniform distribution of preservative, using nanodispersion and microencapsulation.

Ensuring food safety and at the same time meeting the demands for retention of nutrition and quality attributes have resulted in increased interest in alternative preservation techniques for inactivating microorganisms and enzymes in food. Keeping quality in mind, important factors included are flavor, odor, color, and nutritional value. This increasing demand has resulted in new dimensions for the use of preservatives derived from natural sources such as plants. Extensive research has indicated that usage of plants in milk also can be done in order to preserve it. Antimicrobial compounds present in the foods can extend shelf life of unprocessed or processed foods by reducing microbial growth rate or viability.[9] Edible, medicinal, and herbal plants and their derived essential oils (EOs) and isolated compounds contain a large number of secondary metabolites that are known to retard or inhibit the growth of bacteria, yeast, and mold.[13] Natural antimicrobials in preservation can be used alone or can be used in combination to increase the efficiency of preservation. Sources of antimicrobial agents from plants are commonly found in EO fraction of leaves, flower or bud, fruit, or other parts of the plant.

The main problem with milk after pasteurization is recontamination: during thermal treatment psychotropic bacteria are killed, but if storage is not proper then recontamination takes place.[24] During storage after the expiration date, the growth and enzyme activities of psychotropic bacteria within the milk is higher.[25] The spoilage of pasteurized milk may result in microbial and chemical changes in the milk.[56] Due to extracellular enzymes, especially protease which degrades protein and lipase which degrades lipids, the addition of antibacterial and aromatic compounds can help in reduction in spoilage of the milk. Milk subjected for preservation can be either raw or pasteurized. The addition of extract after pasteurization, instead of before, has certain advantages such as:

- The total bacterial counts after pasteurization are lower than that before.
- The pathogenic bacteria are killed after pasteurization and are not killed before pasteurization.
- Materials extracted are used for inhibiting psychrotrophic bacteria, which causes the spoilage of the milk at storage.[38]

The synergistic effect can also be brought into existence like combination of various EOs depending on experimental base, usage of nonthermal preservation with antimicrobials, creation of hurdle technology. Pina-Pérez et al.[54] showed that the synergistic effect attributed to hurdle technology using pulse electric field (PEF): 10 kV/cm, 3000 μs, and addition of cinnamon (5% w/v) represented 52% of the total inactivation achieved in pasteurized skim milk samples. Whole milk is more easily preserved due to antimicrobial lipids in the milk than skim milk. Lipids are one of the nonspecific protective factors present in the milk, which function at mucosal surfaces. In addition to lipids, nonspecific protective factors in the milk include lactoferrin, lactoperoxidase, lysozyme, receptor oligosaccharides, and antimicrobial peptides.[10]

This chapter explores the possibility of using botanical preservatives to increase the shelf life of the milk, thus avoiding its spoilage.

4.2   ESSENTIAL OILS

EOs are well known for their potential antimicrobial activity to control foodborne pathogenic and spoilage microflora. In this regard, applications of plant essential oils (PEOs), being potent antimicrobials, can be a good strategy to control or inhibit such foodborne pathogenic and spoilage bacteria in the milk and dairy products with better consumer acceptability. EOs are volatile, hydrophobic, and plant extracts even are referred to as volatile oils or plant oils. EOs are normally present in special cells or group of cells, found in leaves and stems, and commonly concentrated in one particular region such as leaves, bark, or fruit. These oils are usually extracted using steam distillation and contain a range of oxygenated and nonoxygenated terpene hydrocarbons. However, nowadays, usage of other techniques for extraction of EOs is practiced which includes conventional methods such as hydrodistillation (HD), turbohydrodistillation (THD), ultrasound-assisted extraction (US-SD), microwave steam distillation (MSD), microwave hydrodiffusion and gravity (MHG), and microwave steam diffusion (MWSD). Among all these methods, MHG method gave good quantitative yield without loss of quality, with reduced extraction time, and gave no differences in EO yield and sensorial perception. Many EOs also have relatively rapid antimicrobial action, with significant cell death occurring at concentrations equivalent to or greater than the minimum bactericidal or fungicidal concentrations. Reports of 90% reduction in aflatoxin production at a 5–10 mg/mL concentration of turmeric, an effect attributed to the antioxidant curcumin in turmeric have been reported. In contrast, bacteriostatic agents inhibit the growth but do not kill. The majority of EOs is broad spectrum in activity, meaning that they are active against a wide range of bacteria and fungi. EOs includes various biocompounds such as:

- Aloe vera has glucomannans, uronic acid, glycoproteins, anthraquinone, saccharides, and phenolic compounds.[48]
- Black pepper contains adipic acid, piperine, and oleorecin.[19]
- Cardamom contains aromatic cardamom.[46]
- Cinnamon contains cinnamic acid, cinnamon oil, ethanol, methylene chloride, eugenol, and benzyl benzoate.[63]
- Citronella contains citronella oil and kingisidic acid.
- Clove contains clove oil, isobiflorin, biflorin, eugenol, phenolic compound, and ferulic acid.

- Cumin contains glycosides of 2-C-methyl-D-erythritol and cumin oil.
- Galangal contains galangal oil (cineol, α-pinene, eugenol, camphor, methyl cinnamate, and sesquiterpenes).
- Galingale contains aromatic galingale.
- Laurel like that has aromatic laurel.
- Nutmeg contains diphenylpropanoids and ethyl acetate.
- The dried leaves of green tea contain polyphenol, phenolic acid, catechins, caffeine, flavonoid, epicatechin, and ascorbic acid.[4]
- The fresh leaves of bamboo contain cinnamic derivatives.[64]
- Turmeric has turmeric oil.[42]
- Wild ginger roots contain wild ginger oil.
- Zingiber contains zingiber oil, acyclic oxygenated and monoterpenes.

These compounds in the plants and spices have been reported to cause alterations. Curcumin has the ability to bind casein micelles after heat-induced surface changes. Curcumin binds to the hydrophobic moieties of the casein proteins, with a 10 nm blueshift in its fluorescence emission peak, and causes quenching of the intrinsic fluorescence spectra of the proteins. The increased capacity of the milk proteins to bind curcumin can be attributed to denaturation of whey protein.

4.3   APPLICATIONS OF EOs IN FLAVORED MILK

Conventional milk processing operations (e.g., pasteurization, sterilization, concentration, drying, etc.) adversely affect the overall quality of the milk, like destroying some vitamins and denaturation of some protein. Therefore, in recent years, considerable effort has been made to find natural antimicrobials that can inhibit bacterial and fungal growth in foods in order to improve quality and shelf life. The EOs are used in the milk in such a way that these do not alter the flavor and acceptability of milk. The herbal milk represents great in-between meal and medicine. The flavored herbal milk contains calcium, phosphorus, iron, and other essential nutrients, which makes it a potential food supplement for adults and children. It can be flavored with different herbal plants to change the medicinal properties. EOs usually added to liquid milk are obtained from various plants and spices such as cinnamon, citronella (sweet); ginger, radish, turmeric, galingale, zingiber (roots); wild ginger, nutmeg, cardamom, cumin, pepper (seeds); garlic, clove, javanoni, galangal (bubs); green tea, laurel like (dried leaves); bamboo leaf, banana leaf, guava leaf, avocado leaf, betel vine, celery, garlic leaf, aloe vera (fresh leaves).[38]

Chandler et al.[18] studied the effect of 27 EOs on microbial quality of supplemented milk. The total bacterial counts of the supplemented milk after 5 days of the expiration date were in the range of 6.0 × 10^2–8.4 × 10^4 cfu/mL (whole milk) and 8 × 10^2 to 9.4 × 10^5 cfu/mL (skim milk), while the total bacterial counts of the milk without supplements (control) were 8.5 × 10^5 cfu/mL (whole milk) and 9.7 × 10^6 cfu/mL (skim milk), respectively. Similarly, the protease activities of the supplemented milk were in the range of 0.20–0.40 U/mL (whole milk) and 0.30–0.50 U/mL (skim milk), while that of without supplements (control) were 0.50 U/mL (whole milk) and 0.60 U/mL (skim milk), respectively. The lower bacterial growth of acceptable whole milk as compared to skim milk may be attributed to increased inhibiting level of psychrotrophic bacteria.

Studies on flavored milk, enriched with plant-derived EO component are very limited. Samaddar et al.[58] used *trans*-cinnamaldehyde and eugenol for preparation of EO-enriched flavored milk and evaluated its sensory quality and biological activities. EO-enriched flavored milk was prepared according to the flow diagram shown in Figure 4.1. After storage for 7 days at refrigeration temperature (4–7°C) the control, *trans*-cinnamaldehyde and eugenol-enriched flavored milk had a total viable count (TVC) (log10 cfu/mL) of 5.15, 4.37, and 4.39, respectively. During storage, no significant increase in bacterial count was observed due to the antimicrobial property of EO components which correlated with the report of Alves et al.[3] who indicated that anthraquinones isolated from the exudate of aloe vera have shown wide antimicrobial activity. In a similar fashion, cardamom- and curcumin-flavored milk can be prepared with acceptable sensory qualities in bottles with an estimated shelf life of 6 months at room temperature (Fig. 4.2).

Ultrahigh temperature processing (UHT)-treated curcumin, ginger, and tulsi-enriched flavored milk with a shelf life of 6 months can be manufactured and packed under aseptic conditions (Fig. 4.3). Palthur et al.[53] studied the partial substitution of tulsi powder with milk powder to develop sterilized tulsi-flavored milk with a shelf life of 30 days at refrigerated temperature (5°C). The product was highly acceptable on organoleptic parameters. Antioxidant and iron chelating activity of tulsi [*Ocimum*

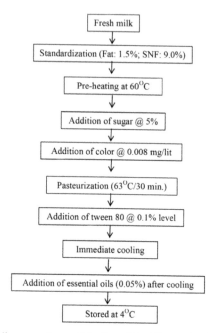

FIGURE 4.1   Flow diagram of essential oil-enriched flavored milk.

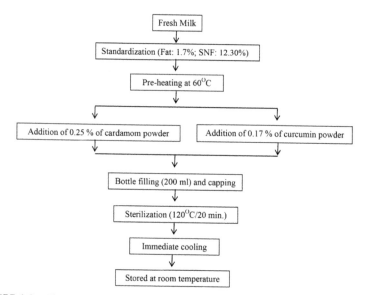

FIGURE 4.2   Flow diagram of cardamom- and curcumin-flavored milk.

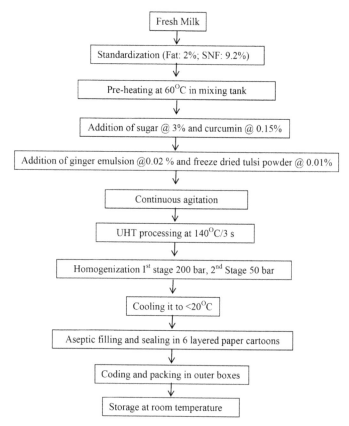

FIGURE 4.3 Flow diagram of UHT-treated curcumin-, ginger-, and tulsi-enriched flavored milk.

tenuiflorum, also known as *Ocimum sanctum*, holy basil, or *tulasi*, or *tulsi* (also sometimes spelled *thulasi*)] offered the capability to reduce the microbial count including yeast and mold, coliform and *Escherichia coli*, thereby making the product safer for consumption.

4.4 APPLICATIONS OF EOS IN FERMENTED DAIRY PRODUCTS

Thabet et al.[65] in an attempt to extent the shelf life of concentrated yogurt added plant-derived EOs cinnamon, cumin, and mint oils at a concentration of 0.3%, 0.5%, and 0.8%, respectively. Total therapeutic bacterial count, *Streptococcus thermophilus* and *Lactobacillus delbrueckii* ssp. *bulgaricus* in the treated yogurt increased and reached a maximum after 8 days of storage whereafter it was decreased until the end of the storage period at $6 \pm 1°C$. Coliform and *staphylococcus* bacteria were not detected, while yeasts and molds were detected at insignificant in some treated yogurt. They reported that yogurt containing 0.3% cinnamon, cumin, or mint oils were most acceptable with a good body and texture as similar to the untreated one with an increased shelf life of 24 days, with higher level of total volatile free fatty acid and therapeutic bacteria counts and low level of total viable, molds and yeast count.

Abou et al.[1] studied the effect of addition of thyme oil, black cumin oil, fennel oil, mint oil, and chamomile oil on the properties of the concentrated yogurt (prepared from goat's milk). Goat's milk fat was replaced by using 4% of corn or sunflower oils and using skim milk powder standardized to 14% total solids (TS). The control sample was prepared from whole fresh goat's milk which was standardized to 4% milk fat and 14% TS. One ml of thyme, black cumin, fennel, mint, and chamomile oils separately were added to each of the goat milk samples (control and fat replaced). All the milk samples were heated at 95°C/15 min; homogenized at 250 kg cm^{-2} and cooled to 40°C then inoculated with 3% ABT culture *(Lactobacillus acidophilus; Bifidobacterium bifidum, and Streptococcus salivarius subsp. thermophilus)* and incubated at 40°C until the final pH 4.8 was achieved. The fermented milk samples were then cooled to 10°C and stored over-night, mixed, and put into sterilized cloth bags. The bags were hanged in refrigerated room at 6–8°C for 12 h to allow whey drainage followed by addition of 0.5% NaCl. Labneh (concentrated yogurt) was filled into plastic containers and stored at 6–8°C for 30 days. Labneh containing chamomile or thyme oils were the preferable treatments as they had higher levels of acetaldehyde, diacetyl, and therapeutic bacteria counts while lowest level of total viable, psychrotrophic, molds and yeasts count.

Noël et al.[49] studied the effects of two EOs from *Xylopia aethiopica* (Dunal) (A. Rich) and *Pimenta racemosa* (Mill.) (J. W. Moore) on the physicochemical, microbiological, and organoleptic characteristics of the fermented milk. EOs to be added in the fermented milk were extracted and detected for their minimum inhibitory concentration (MIC) using the serial dilution method and Muller–Hinton broth (MHB) medium and the reference strains used were *Staphylococcus aureus* ATCC 25923 and *E. coli* ATCC 25922. During 2 weeks of storage period, it was reported that *X. aethiopica* had a lowest MIC. After end of storage period, a decrease of mesophilic flora and lactic bacteria was observed and a disappearance was noticed for coliforms, thermostolerant coliforms, and *S. aureus*. These results revealed a significant contribution of EOs in inhibition of microbial flora. Indeed, oil extracted from *P. racemosa* and *X. aethiopica* were bacteriostatic and fungistatic.[33,36] Moreover, antimicrobial activity of *X. aethiopica* was more pronounced which can be explained by the fact that its EO contains more bacteriostatic substances than *P. racemosa*.

4.5   APPLICATIONS OF EOS IN CHEESE AND CHEESE-BASED PRODUCTS

The use of aromatic plants and their extracts with antimicrobial properties may be compromised in the case of cheese, as some type of fungal starter is needed during its production. Lower concentration of EO would be recommended for addition in cheese as it is reported by several researchers that EO at high concentration, required to be effective in cheese quality, could raise concerns regarding changes in the organoleptic properties. Foda et al.,[28] prepared white cheese with different concentrations of spearmint EO (0.5–2.5 mL/kg retentate) and stored at 7°C ± 2 for 5 weeks. The chemical composition and ripening index of spearmint white cheese revealed that lower concentrations of spearmint EO increased titratable acidity values significantly, while ripening index was increased significantly by increasing the concentration of EO. Lower concentration of EO also was reported to be more acceptable among the panelists. Similarly, curd-type cheese was prepared by using some EOs as natural antimicrobial agents from cinnamon; cassia oil had the highest antibacterial effect and *Sambucus nigra* had the lowest antimicrobial activity against all tested microorganisms. Highest antibacterial activity was reported against Gram-negative bacteria compared to Gram-positive bacteria tested; *Listeria monocytogenes* was sensitive, while *E. coli* was the most resistant microorganism against the EO.

Moro et al.,[45] using solvent-free microwave extraction technology extracted EOs and studied its effect on *Penicillium verrucosum* which is considered as a common cheese spoiler. The most effective EOs against *P. verrucosum* were obtained from *Anethum graveolens*, *Hyssopus officinalis*, and *Chamaemelum nobile*, yielding 50% inhibition of fungal growth at concentration values lower than 0.02 µL mL^{-1}. Thus, the use of EO as natural covers on cheese can inhibit the growth of some mycotoxicogenic fungal spoilers and which was confirmed by the presence of volatile compounds including α- and β-phellandrene.

Olmedoa et al.[50] evaluated the cream cheese (CC) and cream cheese with oregano (CO) and rosemary (CR) EOs during storage as stability indicators by estimating peroxide (PV) and anisidine (AV) values, descriptive analysis, and fermentation parameters. The samples CO and CR showed higher stability during storage and exhibited lower PV (11.70 and 12.32 meq O_2/kg, respectively) than CC at the end of 35 days of storage. Similarly, intensity of rancid flavor was much higher in CC having a rating of 26.27 with respect to 20.22 in CO and 20.67 in CR. On storage day 35, CO samples had the highest pH (4.68) and the lowest acidity (1.24 mg lactic acid/100 g) and TVC (2.35 cfu/g). Thus, it was concluded that EOs extracted from oregano and rosemary had a protective effect against lipid oxidation and fermentation in flavored cheese prepared with CC base.

Govaris et al.[30] studied the antibacterial activity of oregano and thyme added at doses of 0.1 or 0.2 and 0.1 mL/100 g, respectively, to feta cheese inoculated with *E. coli* O157:H7 or *L. monocytogenes* during storage under modified atmosphere packaging (MAP) of 50% CO_2 and 50% N_2 at 4°C. In feta cheese treated with oregano EO at dose of 0.1 mL/100 g, *E. coli* O157:H7 or *L. monocytogenes* survived up to 22 and 18 days, respectively, whereas at the dose of 0.2 mL/100 g up to 16 or 14 days, respectively and the efficacy of thyme EO was not significantly different. Although both EOs exhibited equal antibacterial activity against both pathogens, the populations of *L. monocytogenes* was decreased faster than those of *E. coli* O157:H7 during the refrigerated storage, indicating a stronger antibacterial activity of both EOs against the former pathogen, which can be attributed to the presence of phenols including carvacrol and thymol.

Similarly, Argentinean oregano EOs (EOs: Compacto, Cordobes, Criollo, and Mendocino) and thymol were used to manufacture flavored organic cottage cheese. During a 30-day storage under thermal conditions, it was found that the stored organic cheese flavored with Cordobes EO and thymol presented lower conjugated dienes (15.94 and 15.53, respectively), whereas the control sample exhibited the maximum value (17.54). Due to oxidative deterioration, unsaturated fatty acids decreased significantly in cottage cheese samples without EO but cheese flavored with Compacto, Cordobes, and Criollo EOs showed lower saturated/unsaturated fatty acid ratios than the control (1.67, 1.62, and 1.68, respectively) which can be due to reduced production of organic acids during storage.[5]

Agaricus bohusii Bon is an edible and prized mushroom especially common in Serbia and Southern Europe and is well known for its antifungal preserving properties. Reis et al.[57] evaluated antifungal properties of *A. bohusii* extract on *P. verrucosum* var. *cyclopium* in CC. The *p*-hydroxybenzoic and *p*-coumaric acids were the phenolic acids present in the extract and γ-L-glutaminyl-4-hydroxybenzene and cinnamic acid were found in larger amounts which are responsible for strong antioxidant capacity. Malic, oxalic, and fumaric acids were the organic acids identified and quantified in *A. bohusii*. Methanol extract successfully inhibited the development of *P. verrucosum* var. *cyclopium* in CC, tested at room temperature after 7 days of inoculation.

It is evident from the reports for different types of cheese added with EOs that there is a great possibility to use EOs as biopreservatives. But the challenges for future are to determine the conditions, which avoid the loss of lactic bacteria; to preserve organoleptic characteristic of cheese after incorporation of EO; and preserve the effects of EOs on foodborne pathogens.

4.6 QUANTIFICATION OF THE MINIMUM AND NONINHIBITORY CONCENTRATION

The use of antimicrobials as preservatives in food systems can be constrained when the effective antimicrobial doses exceed organoleptic acceptability levels. This is particularly true in case of EOs as their impact on sensory attributes is more. The most commonly used methods in the laboratory for their simplicity and quickness are: agar diffusion disk technique which is used to obtain qualitative results; and dilution in broth and agar which is used to obtain quantitative results. Tests of antimicrobial activity can be classified as diffusion, dilution, or bioautographic methods.

Agar diffusion disk method due to the simplicity and quick interpretation of results is a widely used method for determination of the antimicrobial activity of EOs. In this method, a fixed amount of an antimicrobial substance or similar on a substrate (usually paper disks) is applied onto the agar surface which is already inoculated with the test microorganism. After incubation, a halo around the disk appears which is then measured for its diameter for estimating the sensitivity of the microorganism under test. Apart from the sensitivity of the microorganism, diameter is also affected by the concentration of the microorganism, thickness of the agar, pH, and the composition of the culture medium, the ability of the tested product to diffuse in that medium, the temperature, the incubation atmosphere, the speed of bacterial growth, the amount of inoculum, and the phase of the bacterial growth. The agars for estimating the efficacy of EOs for their bactericidal activity commonly used are Mueller–Hinton agar, tryptone soya agar, nutritive agar, and brain heart infusion agar.[15]

Another evaluation method based on usage of agar is the *agar dilution method* in which the EOs are added in a medium with agar. The EOs to be tested are added when the medium is still liquid. In order to achieve the desired dilution range, several plates are prepared, each one containing a fixed concentration of EO, followed by inoculation with the microorganism to be tested once the medium is solid. There are several comprehensive studies that have screened a large number of EOs and/or component using this assay, but it is not ideal for water-insoluble compounds such as EOs.[8,21]

The quantification of the in vitro activity of the EOs is commonly carried using different dilution methods which are based on the determination of the growth of the microorganism when present with increasing concentrations of the EO which is diluted in the culture medium (broth). *Macrodilution* is a technique, in which assays are carried out using many tubes containing broth culture media with a fixed range of EO, but this method requires a great amount of time as well as sophisticated equipments, hence it is not practically used. In *microdilution method,* micropipettes and micro-well plates are used to determine MIC of plant extracts. The standard culture medium used to evaluate antibiotics by means of the broth dilution method is MHB although for the evaluation of MIC of plants extracts other culture media are used such as nutritive broth (NB), tryptone soya broth (TSB), and sometimes adding supplements such as serum.[16,44,62,66]

Two terms are often used during expressing the inhibitory action of EO, namely: noninhibitory concentration (NIC) and MIC. NIC is defined as the concentration above which the inhibitor begins to have negative effect on growth, and MIC is the concentration above which no growth is observed when compared to the control.[17] Therefore, these concentrations are quantified with the aim of defining the boundaries of sensory acceptability and antimicrobial efficacy of antimicrobials[41] and are included in quantitative approaches evaluating the efficacy of antimicrobial activity based on plant origin, that is, EOs and their components. The MIC and NIC values are affected by experimental conditions which include, incubation temperature, type of organism under examination, and inoculum size.[41]

Studies for identification of MIC can be divided into groups such as diffusion, dilutions, impedance, and optical density (or absorbance) methods. Lambert and Pearson[17] developed a fully quantitative approach known as Lambert–Pearson model (LPM) to evaluate the dose responses of microorganisms against inhibitors present in EOs. This modelling approach can be examined for optical density as below:

$$MIC = P_1 \cdot \exp\left(\frac{1}{P_2}\right) \tag{4.1}$$

$$NIC = P_1 \cdot \exp\left[\frac{1-e}{P_2}\right] \tag{4.2}$$

where P_1 = concentration at maximum slope; and P_2 = slope parameter

When a broth dilution is used, the sampling of each dilution to quantify the surviving organisms allows NIC, which is referred as minimum concentration of oil required to kill 99.9% of cells originally inoculated into the assay for bacteria, fungi. It has been shown that the MIC of lemon EO for *Geotrichum candidum* in skimmed milk was >4 μL/mL revealing a strong protection of the milk against the effect of the EO. MIC values for common EOs are shown in Table 4.1.

TABLE 4.1 MIC Values for Common Essential Oils.

Common name of oil	Plant source	Organism	MIC, approximate range (μL mL^{-1})
Thyme	*Thymus vulgaris*[27]	*E. coli*	0.05–0.15
		S. aureus	0.02–0.13
Oregano	*Origanum vulgare*	*E. coli*	0.05
		S. aureus	0.125
Clove	*Syzygium aromaticum*	*E. coli*	0.125
		S. aureus	0.25
Marjoram	*Origanum marjorana*	*E. coli*	0.26
		S. aureus	0.53
Turmeric	*Curcuma longa*	*Aspergillus*	0.2
α-Terpineol	*Sardinian thymus*[23]	*E. coli*	0.450–0.9
		Salmonella typhimurium	0.225
		S. aureus	0.9

TABLE 4.1 *(Continued)*

Common name of oil	Plant source	Organism	MIC, approximate range (μL mL^{-1})
		Listeria monocytogenes	>0.9
		Bacillus cereus	0.9
Carvacrol	Sardinian Thymus[23]	E. coli	0.225–5
		S. typhimurium	0.225–0.25
		S. aureus	0.175–0.450
		L. monocytogenes	0.375–5
Citral	S. thymus[23]	Bacillus cereus	0.1875–0.9
Citral[39]		E. coli	0.5
		S. typhimurium	0.5
		S. aureus	0.5
		L. monocytogenes	0.5
Eugenol[39]		E. coli	1.0
		S. typhimurium	0.5
		L. monocytogenes	>1.0
Thymol	S. thymus[23]	E. coli	0.225–0.45
		S. typhimurium	0.056
		S. aureus	0.140–0.225
		L. monocytogenes	0.450
		B. cerus	0.450

4.7 MECHANISM OF ANTIMICROBIAL ACTION

The EOs are substances with antimicrobial effects and are found in plants, herbs, and spices and are phenolic compounds, terpenes, aliphatic alcohols, aldehydes, ketones, acids, and isoflavonoids.[13] As EOs contain large number of different groups of chemical compounds hence, their antibacterial activity cannot be attributed to one specific mechanism as there are several targets in the cell. Simple and complex derivatives of phenol are reported to be the main antimicrobial compounds present in EOs derived from spices. It has also been reported that some nonphenolic compounds of EOs are more effective against Gram-negative bacteria, and many Gram-positive fungi, for example, allyl isothiocyanate. Plant EOs are generally more inhibitory against Gram-positive bacteria than Gram-negative bacteria,[31,32] but there are some agents which are effective against both groups such as clove, cinnamon, thyme.[39]

The possible effect of phenolic compound can be concentration dependent,[35] for example, at low concentration phenols affect enzyme activity particularly those associated with energy production, while at high concentration they cause protein denaturation. The antimicrobial activity of phenolic compounds may be due to their ability to alter microbial cell permeability, thereby permitting the loss of macromolecules from the interior (e.g., ribose, Na-glutamate) and also could interfere with membrane function such as electron transport, nutrient uptake, protein, nucleic acid synthesis, enzyme activity,[6,7] and interact with membrane proteins, causing deformation in structure and functionality.

The mode of action of EOs is multiple and they act at different sites in cell which causes deterioration of cell wall, damage to cytoplasmic membrane, damage to membrane proteins, leakage of cell contents, coagulation of cytoplasm, depletion of proton motive active sites, inactivation of essential enzymes, and disturbance of genetic material functionality[13] as well as morphological structure. Majority of the EOs and their derivatives are hydrophobic in nature, which makes it possible to partition in the lipids of the bacterial cell membrane and mitochondria, disturbing the structures and rendering them more permeable and destruction of target microorganisms. EOs and their components have the capability to impair the integrity of cell membrane of microorganisms, which eventually leads to loss of cell homeostasis, leakage of intracellular constituents, and eventually cell death.[67] Lethality of the EOs is affected by two factors, namely, dosage concentration and exposure time. It may be noted that higher concentrations of EOs are able to cause severe effects within a short time and the reverse is true for lower concentrations.

Carvacrol and thymol are able to disintegrate the outer membrane of Gram-negative bacteria, releasing lipopolysaccharides (LPS) and increasing the permeability of the cytoplasmic membrane to adenine triphosphate (ATP) which causes death of microorganism. EO-containing p-cymene is hydrophobic in nature which causes swelling in greater extent as compared to carvacrol of the cytoplasmic membrane resulting into death. But the effectiveness of p-cymene as antibacterial substance is greater when used in combination with other EO then used alone. Carvone [2-methyl-5-(1-methylethenyl)-2-cyclohexen-1-one] is able to dissipate the pH gradient and membrane potential of cells. Decrease in count of

E. coli, S. thermophilus, and *Lactococcus lactis* with increased carvone concentration have proven that carvone is able to disturb the metabolic energy status of cells thereby by causing death of the cell. Cinnamaldehyde (3-phenyl-2-propenal) is able to inhibit growth of *E. coli* O157:H7 and *Salmonella typhimurium* at similar concentrations to carvacrol and thymol without disintegrating the outer membrane or depleting the intracellular ATP pool.[52] The antibacterial effect is due to presence of carbonyl group which is believed to bind to proteins thereby preventing the action of amino acid decarboxylases in *Enterobacter aerogenes*.

Terpenes containing EO are able to alter the physical properties of membranes, which result in insertion between the fatty acyl chains of lipid bilayer, disturbing the van der Waal interaction between acyl chains, disrupting lipid packaging, and decreasing lipid order. Increased levels of terpene molecule in the bilayer of the cell results in increased lipid volume, which causes membranes to swell and thicken. The expansion of the membrane is associated with decreased membrane integrity and loss of intracellular compounds including hydrogen, potassium, and sodium which leads into a decreased membrane potential thereby causing disintegration of the cells. At higher dosage level of EO or extended exposure, high level of membrane damage can occur. The extent of damage may be quantified by loss of intracellular materials which are able to absorb wavelengths of either 260 or 280 nm; 260 nm is for DNA, whereas 280 nm is for protein. The presence of high levels of DNA or proteins in cell-free preparations indicates major membrane damage to the cell. Apart from antimicrobial effect, some organisms are tolerated toward antimicrobial action, but the true resistance is not apparent.

Various mechanisms are been suggested for microorganisms to become resistant for EOs, namely, alteration of drug target site, inactivation of antimicrobial agents, reduced permeability and the upregulation of efflux, and genetic mutations within organism or the acquisition of external genetic material.[69]

4.8 FACTORS AFFECTING ANTIMICROBIAL ACTIVITY OF ESSENTIAL OILS

Antimicrobial efficacy of EOs is affected by different factors, which include growth of resistant microorganisms, situations that can destabilize the biological activity of antimicrobial agents, binding to food components such as fat particles or protein surfaces, inactivation by other additives, poor solubility and uneven distribution in the food matrix, and pH effects on stability and activity of antimicrobial agents. Tserennadmid et al.[68] reported that marjoram EO had no effect on the growth rate and lag phase of *G. candidum* at pH 5, but at all other pH values, the lag phase was lengthened and the growth rate decreased when tested on malt extract medium where milk was investigated.

Factors also include history such as botanical source, time of harvesting, stage of development, and method of extraction.[34] The composition, structure as well as functional groups of the oils play an important role in determining their antimicrobial activity. Usually, compounds with more phenolic groups are most effective. For optimized antimicrobial effect on the milk and milk products by application of EOs factors which need be considered are: identifying the target microorganisms of concern, use of two or more EOs in combination to provide a synergistic effect, matching the activity of the compounds to the composition, processing and storage conditions of the final product, and more importantly the effects on organoleptic properties.

It has been reported by scientists that when the EOs are applied in a food system, the efficacy of EOs is reduced due to interaction with food components. When EOs are used in combination, the technique can minimize application concentrations required, thus reducing any adverse effect on organoleptic property.[32] The composition of food also affects the antimicrobial activity of EO, for example, activity of thyme is increased in high protein concentration and pH values of 5.[31] Accordingly, the challenge for practical application of EOs is to develop optimized low-dose combinations which can be delivered in a variety of ways to match product profiles as well as maintaining product safety and shelf life, thus minimizing the undesirable flavor and sensory changes associated with the addition of high concentration of EOs.

4.9   MICRO-ENCAPSULATION AND NANODISPERSIONS FOR USAGE OF ESSENTIAL OILS IN MILK AND MILK PRODUCTS

The strong odor of EO may limit its use to be directly added into the food. In order to prevent the modification of organoleptic properties of food product, it is recommended to add a small concentration of EOs to the product.[29] The EOs are lipophilic, immiscible with water, and at the same time, they are sensitive toward the chemical modification under effect of some external factors such as: temperature, light, presence of oxygen, etc. In order to diminish these drawbacks, the EOs are encapsulated through different techniques as follows: emulsification, coacervation, internal ionotropic gelation, molecular inclusion, spray-drying, spray–freeze-drying, fluid bed coating, cocrystallization, extrusion, etc.

In order to retain the volatile substances present in EOs, proper coating is done, usually by using biopolymers. These coating materials must have the property to retain and protect the encapsulated volatiles from their loss and chemical degradation during manufacture, storage, and handling and should be in such manner that when introduced in final product it must release them during desirable step of manufacture or consumption.[40] Each encapsulant material has its unique emulsifying and film-forming properties and therefore, rigorous selection procedure must be adopted before selecting and implementing the coating material best suited for an application. Carbohydrate-based matrixes that are starch, maltodextrins (MD), gum Arabic, and other gums, as well in protein-based matrixes[14] are commonly used as flavor encapsulates.

Oregano (*Origanum vulgare* L.) and aroma extracts of citronella (*Cymbopogon nardus* G.) and sweet marjoram (*Majorana hortensis* L.) were subjected to spray drying using skimmed milk powder (SMP) and whey protein concentrate (WPC) as encapsulates. After spray drying, efficiency of microencapsulation expressed as a percentage of flavoring entrapped into the microcapsules varied from 54.3% (marjoram in WPC) to 80.2% (oregano in SMP). Consequently, the changes in the composition of individual flavor compounds during encapsulation were considerably smaller for oregano EO as compared with citronella and marjoram aroma extracts. Particle size varied from 6 to 280 μm for SMP and from 2 to 556 μm for WPC microencapsulated products.

Nanoencapsulation has been proposed as one such technology which is having a great potential to protect the EOs for its antibacterial properties. Bioactive EO components can be encapsulated into nanosized emulsified droplets. Nanoencapsulated bioactive oils emulsions can be prepared by high-amplitude ultrasonic homogenization with food grade ingredients. The incorporation of bioactive oils into nanoencapsulation emulsion can enhance its antimicrobial activity, which is significantly affected by the physicochemical properties of the delivery systems such as emulsifier composition and the mean droplet size.

If EOs are to be considered for antibacterial effect, then use should be in such way that it does not change the accepted flavor of the original food.[47] For this purpose, use of nanodispersions is carried out. Nanodispersion enables even distribution of EO in the milk at concentrations above the solubility limit which can enhance the antimicrobial efficacy in spite of interfering constituents present in milk. Thus, nanodelivery systems can help to reduce the amount of EOs to be added for antimicrobial effect without making the milk products turbid (Fig. 4.4).[59,60] Major challenge in the usage of EOs as antimicrobial substances is their hydrophobic nature, which can be countered by dissolving them in a solvent or solvent mixture having a decreased polarity or dispersing them in emulsion droplets or biopolymer particles. The milk proteins especially whey proteins are best utilized emulsifiers and their emulsifying capacity can be improved by conjugation with more hydrophilic oligosaccharides and polysaccharides.[37,43] Conjugates of whey protein isolate (WPI) and MD have already been applied to manufacture emulsions of tomato volatiles and conjugated linoleic acid with an aim to minimize the degradation and improve its dispersibility.[20,22]

In aqueous dispersions, Brownian motion is the mechanism dominating the dynamics of nanoparticles, which prevents gravitational sedimentation or creaming of dispersed particles. Smaller particles scatter visible light less effectively and can eventually enable transparent dispersions at a sufficiently small dimension. The solubility of compounds with limited solubility increases when dispersed in a smaller structure, which, along with increased surface area for bacterial contact and even distribution in food matrices, improves the antimicrobial effectiveness.

An emulsion-evaporation process can be used to encapsulate EOs.[59] An oil phase [with 20% (wt/v) EO in hexane] can be emulsified at 10% (v/v) into an aqueous phase with 11.1% (wt/v) conjugates, followed by spray drying. Conjugates can be prepared by spray drying a base solution containing WPI and MD 180 in a ratio of 1:2. The dispersion with the smallest diameter by using MD40 (WPI:MD ratio of 1:2) can be prepared which is transparent and able to retain maximum antimicrobial activity.[60] The dispersion can then be subjected to spray-drying (inlet temperature of 150°C, feed rate of 0.11 mL/s, 600 kPa compressed air pressure, a 35 m³/h airflow rate, outlet temperature of 80–90°C) followed by hot air drying at 90°C for 2 h for conjugation (Mallard reaction) which are stored at −18°C.

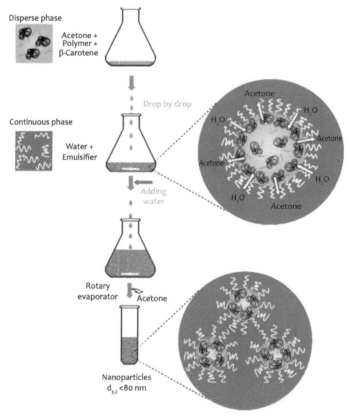

FIGURE 4.4 Flow diagram of preparation of essential oils nanodispersions.

High-performance liquid chromatography can be used to estimate the amount embedded during spray drying. The particle size distribution can be used to calculate volume-length mean particle diameter $(d_{4,3})$ using following equation where n_i is the number of particles corresponding to diameter d_i:

$$d_{4,3} = \frac{\sum_{i=1} n_i d_i^4}{\sum_{i=1} n_i d_i^3} \tag{4.3}$$

Emulsifying properties of proteins can be increased in conjugation with carbohydrates as the protein molecules anchor the oil droplets during emulsion formation, while the hydrophilic carbohydrate moieties position themselves into the aqueous phase, which forms a steric stabilizing layer that prevents oil droplets from coalescing.[2,37] Glycation of whey protein with MD prevents aggregation, enabling transparent dispersions, particularly at pH 5.0, where unconjugated whey proteins (β-lactoglobulin, α-lactalbumin, bovine serum albumin) would aggregate easily.[12]

Shah et al.[59] studied antimicrobial activities of free and nanodispersed thymol used at 4.5 g/L in 2% reduced-fat milk against *E. coli* O157H7 at 35°C and *L. monocytogenes* at 32°C. Activities of free and nondispersed thymol were within 2 to 3 log (cfu/mL) against *E. coli* and *L. monocytogenes*.

Inhibition activity of nanodispersed and free eugenol was studied in the milk with three fat levels [full fat (~4%), reduced fat (2%), and skim (<0.5%)] against *E. coli* O157:H7 at concentrations of 3.5, 4.5, and 5.5 g/L, respectively. In skim milk, inhibition was observed at 3.5 g/L eugenol, with nanodispersion showing better activity, while inactivation was observed at two higher eugenol levels. As fat levels increased, a higher eugenol level was needed to inhibit the growth and treatment with 4.5 g/L was ineffective in inhibiting *E. coli* O157:H7 in full fat milk. The best improvement in antilisterial properties after nanodispersion was observed at 5.5 g/L eugenol in full fat milk. The eugenol concentrations required to inhibit bacteria in the milk were much higher, verifying the interference of protein and fat on eugenol activity. The improved antimicrobial properties likely resulted from improved dispersibility of eugenol in the liquid dispersion and can be released locally to keep the concentration sufficiently high to inhibit the bacteria.[55,60]

Donsi et al.[26] encapsulated terpenes mixture and D-limonene into nanometric delivery systems in order to enhance their antimicrobial activity while minimizing the impact on the quality attributes The nanoemulsions were prepared by high-pressure homogenization at 300 MPa and antimicrobial efficacy was determined by the MIC and minimum bactericidal concentration for *L. delbrueckii*, *Saccharomyces cerevisiae*, and *E. coli*. Gas chromatography–mass spectrometry (GC–MS) analysis revealed that high-intensity processing for nanoemulsion production may affect the chemical stability of several active compounds.

4.10   TOXICITY OF ESSENTIAL OILS

Despite of their image as "natural," EOs are far from nontoxic. The majority of EOs in high doses will cause toxic effects. Poisoning is largely occurred in children due to ingestion which directly causes nervous system depression. Irritant reactions EOs are dose-dependent causing redness, itching, a burning sensation, or blistering.[51,70] A factor in the development of EO allergy is the use of aged or oxidized oil, whereby exposure of the oil to air has led to the formation of oxidation products which have higher allergenic potential.[11]

4.11 SUMMARY

Milk can be preserved using botanical preservatives, but to obtain good results use of proper combinations of EOs or hurdle technology combination with natural preservatives is necessary. Using botanical preservatives, one takes into consideration various factors especially flavor of the milk, proper emulsion of EO, and proper encapsulation of volatile compounds. Nanodispersions help in proper dispersion of EOs without any loss in volatile compounds and without disturbing the flavor of the milk.

KEYWORDS

- essential oils
- gas chromatography–mass spectrometry
- microwave steam distillation
- minimum inhibitory concentration
- Mueller–Hinton broth
- noninhibitory concentration
- nutritive broth
- pulse electric field
- ultrasound assisted extraction

REFERENCES

1. Abou Ayana, I. A. A.; Gamal El Deen, A. A. Improvement of the Properties of Goat's Milk Labneh Using Some Aromatic and Vegetable Oils. *Int. J. Dairy Sci.* **2011,** *6,* 112–123.

2. Akhtar, M.; Dickinson, E. Whey Protein-maltodextrin Conjugates as Emulsifying Agents: An Alternative to Gum Arabic. *Food Hydrocoll.* **2007**, *21*, 607–616.
3. Alves, D. S.; Perez-Fons, L.; Estepa, A.; Micol, V. Membrane-related Effects Underlying the Biological Activity of the Anthraquinones, Emodin and Barbaloin. *Biochem. Pharmacol.* **2004**, *68*, 549–561.
4. An, B. J.; Kwak, J. H.; Son, J. H.; Park, J. M.; Lee, J. Y.; Jo, C.; Byun, M. W. Biological and Anti-microbial Activity of Irradiated Green Tea Polyphenols. *Food Chem.* **2004**, *88*, 549–555.
5. Asensio, C. M.; Grosso, N, R.; Juliani, H. R. Quality Preservation of Organic Cottage Cheese Using Oregano Essential Oils. *LWT—Food Sci. Technol.* **2015**, *60*(2), 664–671.
6. Bajpai, V. K.; Rahman, A.; Dung, N. T.; Huh, M. K.; Kang, S. C. In Vitro Inhibition of Food Spoilage and Foodborne Pathogenic Bacteria by Essential Oil and Leaf Extracts of *Magnolia liliflora* Desr. *J. Food Sci.* **2008**, *74*, 314–320.
7. Bajpai, V. K.; Shukla, S.; Kang, S. C. Chemical Composition and Antifungal Activity of Essential Oil and Various Extract of *Silenearmeria L. Bioresour. Technol.* **2008**, *99*, 8903–8908.
8. Bauer, A. W.; Kirby, W. M.; Sherris, J. C.; Turck, M. Antibiotic Susceptibility by Standardized Single Disk Method. *Am. J. Clin. Pathol.* **1966**, *45*(4), 493–496.
9. Beuchat, L. R.; Golden, D. A. Antimicrobials Occurring Naturally in Foods. *Food Technol.* **1989**, *43*(1), 134–142.
10. Blewett, H. J.; Cicalo, M. C.; Holland, C. D.; Field, C. J. The Immunological Components of Human Milk. *Adv. Food Nutr. Res.* **2008**, *54*, 45–80.
11. Brared, C. J.; Forsstrom, P.; Wennberg, A. M.; Karlberg, A. T.; Matura, M. Air Oxidation Increases Skin Irritation from Fragrance Terpenes. *Contact Dermatitis* **2009**, *60*, 32–40.
12. Bryant, C. M.; McClements, D. J. Molecular Basis of Protein Functionality with Special Consideration of Cold-set Gels Derived from Heat-denatured Whey. *Trends Food Sci. Sci. Technol.* **1998**, *9*,143–151.
13. Burt S. A.; Rrinders, R. D. Antibacterial Activity if Selected Plant Essential Oils Against *Eshcherchia coli* and *Staphylococcus aureus* O157:H7. *Lett. Appl. Microbiol.* **2003**, *36*, 162–167.
14. Bylaite, E.; Venskutonis, P. R.; Mażdżieriene, R. Properties of Caraway (*Carum carvi L.*) Essential Oil Encapsulated into Milk Protein-based Matrices. *Eur. Food Res. Technol.* **2001**, *212*, 661–670.
15. Calvo, M. A.; Angulo, E.; Costa-Batllori, P.; Shiva, C.; Adelantado, C.; Vicente, A. Natural Plant Extracts and Organic Acids: Synergism and Implication on Piglet's Intestinal Microbiota. *Biotechnology* **2006**, *5*(2), 137–142.
16. Calvo, M. A.; Asensio, J. J. Métodos para el análisis y control de la actividad antimicrobiana de productos útiles en alimentación animal. *Anaporc* **1999**, *191*, 97–102.
17. Carson, C. F.; Hammer, K. A.; Riley T. V. Broth Microdilution Method for Detection of Susceptibity of *Eshcherchia coli* and *Staphylococcus aureus* to the Essential Oil of *Melaeuca alternifoloia* (Tea Tree Oil). *Microbios* **1995**, *82*, 181–185.
18. Chandler, R. E.; Ng, S. Y.; Hull, R. R. Bacterial Spoilage of Specialty Pasteurized Milk Products. *CSIRO Food Res. Q.* **1990**, *50*, 11–14.

19. Chitwood, R. L.; Pangborn, R. M.; Jennings, W. GCMS and Sensory Analysis of Volatiles from Three Cultivars of Capsicum. *Food Chem.* **2003**, *11*, 201–216.

20. Choi, K. O.; Ryu, J.; Kwak, H. S.; Ko. S. Spray-dried Conjugated Linoleic Acid Encapsulated with Maillard Reaction Products of Whey Proteins and Maltodextrin. *Food Sci. Biotechnol.* **2010**, *19*, 957–965.

21. Chorianopoulos, N.; Evergetis, E.; Mallouchos, A.; Kalpoutzakis, E. G. J.; Haroutounian S. A. Characterization of Essential Oil Volatiles of *Satureja thymbra* and *Satureja parnassics*: Influence of Harvesting Time and Microbial Activity. *J. Agric. Food Chem.* **2006**, *54*, 3139–3145.

22. Christiansen, K. F.; Olsen, E.; Vegarud, G.; Langsrud, T.; Lea, P.; Haugen, J. E.; Egelandsdal, B. Flavor Release of the Tomato Flavor Enhancer, 2-Isobutylthiazole, from Whey Protein Stabilized Model Dressings. *Food Sci. Technol. Int.* **2011**, *17*, 143–154.

23. Cosentino, S.; Tuberoso, C. I. G.; Pisano, B.; Satta, M.; Mascia, V.; Arzedi, E.; Palmas, F. In Vitro Antimicrobial Activity and Chemical Composition of Sardinian Thymus Essential Oils. *Lett. Appl. Microbiol.* **1999**, *29*, 130–135.

24. Craven, H. M.; Macauley, B. J. Microorganism in Pasteurized Milks After Refrigerated Storage 3. Effect of Milk Processor. *Aust. J. Dairy Technol.* **1992**, *47*, 50–55.

25. Deeth, H. C.; Khusniati, T.; Datta, N.; Wallace, R. B. Spoilage Patterns of Skim and Whole Milks. *J. Dairy Res.* **2002**, *69*, 227–241.

26. Donsì, F.; Annunziata, M.; Sessa, M.; Ferrari, G. Nanoencapsulation of Essential Oils to Enhance Their Antimicrobial Activity in Foods: A Review. *LWT-Food Sci. Technol.* **2011**, *44*, 1908–1914.

27. Faleiro, M. L.; Miguel, M. G.; Ladeiro, F.; Venancio, F.; Tavares, R.; Brito, J. C.; Figueiredo, A. C.; Barroso, J. G.; Pedro, L. G. Antimicrobial Activity of Essential Oils Isolated from Portuguese Endemic Species of Thymus. *Lett. Appl. Microbiol.* **2003**, *36*(1), 35–40.

28. Foda, M. I.; El-Sayed1, M. A.; Hassan, A. A.; Rasmy, N. M.; El-Moghazy, M. M. Effect of Spearmint Essential Oil on Chemical Composition and Sensory Properties of White Cheese. *J. Am. Sci.* **2010**, *6*(5), 272–279.

29. Gouin, S. Microencapsulation: Industrial Appraisal of Existing Technologies and Trends. *Trends Food Sci. Sci. Technol.* **2004**, *15*, 330–347.

30. Govaris, A.; Botsoglou, E.; Sergelidis, D.; Chatzopoulou, P. S. Antibacterial Activity of Oregano and Thyme Essential Oils Against *Listeria monocytogenes* and *Escherichia coli* O157:H7 in Feta Cheese Packaged Under Modified Atmosphere. *LWT-Food Sci. Technol.* **2011**, *44*(4), 1240–1244.

31. Guiterrez, J.; Bourke, P.; Lonchamp, J.; Barry–Ryan, C. Impact of Plant Essential Oil Onmicrobiological, Organoleptic and Quality Markers of Minimally Processed Vegetables. *Innov. Food Sci. Innov. Emerg. Technol.* **2009**, *10*, 195–202.

32. Guiterrez, J.; Rodriguez, G.; Barry-Ryan, C.; Bourke, P. Efficacy of Plant Essential Oils Against Foodborne Pathogens and Spoilage Bacteria Associated with Ready to Eat Vegetables: Antimicrobials and Sensory Screening. *J. Food Prot.* **2008**, *71*, 1846–1854.

33. Guylène, S. A.; Jacqueline, A.; Paul, B.; Joëlle, L. Antibacterial and Antifungal Activities of the Essential Oils of *Pimenta racemosa* var. *racemosa* P. Miller (J. W. Moore) (*Myrtaceae*). *J. Essent. Oil Res.* **1998**, *10*, 161–164.

34. Janssen, A.; Scheffer, M. M.; Baerhein-Svendsen, A. Antimicrobial Activity of Essential Oils: A 1976–1986 Literature Review on Aspects of Test Methods. *Plant Med.* **1986,** *53,* 395–398.

35. Juven, B. J.; Kanner, J.; Schved, F.; Weisslowicz, H. Factors That Interact with the Antibacterial Action of Thyme Essential Oils and Its Active Constituents. *J. Appl. Bacteriol.* **1994,** *76,* 626–631.

36. Karioti, A.; Hadjipavlou-Litina, D.; Mensah, M. L. K.; Theophilus, C.; Skaltsa, F. H. Composition and Antioxidant Activity of the Essential Oils of *Xylopia aethiopica* (Dun) A. Rich. (*Annonaceae)* Leaves, Stem Bark, Root Bark, and Fresh and Dried Fruits, Growing in Ghana. *J. Agric. Food Chem.* **2004,** *52*(26), 8094–8098.

37. Kato, A. Industrial Applications of Maillard-type Protein-polysaccharide Conjugates. *Food Sci. Technol. Res.* **2002,** *8,* 193–199.

38. Khusniatil, T.; Widyastuti, Y. The Preservation of Milk with the Addition of Antibacterial and Aromatic Supplements Produced in Indonesia. *Biotropia* **2008,** *15,* 50–64.

39. Kim, J.; Marshall, M. R.; Wei, C. I. Antibacterial Activity of Some Essential Oil Components Against Five Foodborne Pathogens. *J. Agric. Food Chem.* **1995,** *43,* 2839–2845.

40. Kim, Y. D.; Morr, C. V. Microencapsulation Properties of Gum Arabic and Several Food Proteins: Spray-dried Orange Oil Emulsion Particles. *J. Agric. Food Chem.* **1996,** *44,* 1314–1320.

41. Lambert, R. J. W.; Pearson, J. Susceptibility Testing: Accurate and Reproducible Minimum Inhibitory Concentration (MIC) and Noninhibitory (NIC) Values. *J. Appl. Microbiol.* **2000,** *91,* 453–462.

42. Lantz, R. C.; Chen, G. J.; Solyom, A. M.; Jolad, S. D.; Timmermann, B. N. The Effect of Turmeric Extracts on Inflammatory Mediator Production. *Phytomedicine* **2005,** *12,* 445–452.

43. Ly, M. H.; Aguedoc, M.; Goudota, S.; Lea, M. L.; Cayotd, P.; Teixeirac, J. A.; Leb, T. M.; Belina, J. M.; Wachéa, Y. Interactions Between Bacterial Surfaces and Milk Proteins, Impact on Food Emulsions Stability. *Food Hydrocoll.* **2008,** *22,* 742–751.

44. Mann, C. M.; Markham, J. L. A New Method for Determining the MIC of Essential Oils. *J. Appl. Bacteriol.* **1998,** *84,* 538–544.

45. Moro, A.; Librán, C. M.; Berruga, M. I.; Zalacain, A.; Carmona, M. Mycotoxicogenic Fungal Inhibition by Innovative Cheese Cover with Aromatic Plants. *J. Sci. Food Agric.* **2013,** *93*(5), 1112–1118.

46. Mummenhoff, K.; Hurka. H. Isoelectric Focusing Analysis of RUBISCO in Lepidium (*Brassicaceae*), Sections Lepia, Lepiocardamon and Cardamon. *Biochem. Syst. Ecol.* **1991,** *19,* 47–52.

47. Nazer, A. I.; Kobilinsky, A.; Tholozana, J. L.; Dubois-Brissonneta, F. Combinations of Food Antimicrobials at Low Levels to Inhibit the Growth of *Salmonella sv. Typhimurium*: A Synergistic Effect? *Food Microbiol.* **2005,** *22,* 391–398.

48. Ni, Y.; Turner, D.; Yates, K. M.; Tizard, I. Isolation and Characterization of Structural Components of Aloe Vera L. Leaf Pulp. *Int. Immunopharmacol.* **2004,** *4,* 1745–1755.

49. Noël, T. S.; Kifouli, A.; Boniface, Y.; Edwige, D.; Farid, D.; Fatiou, T. Antimicrobial and Physico-chemical Effects of Essential Oils on Fermented Milk During Preservation. *J. Appl. Biosci.* **2016,** *99,* 9467–9475.

50. Olmedoa, R. H.; Nepoteb, V.; Grossoa, N. R. Preservation of Sensory and Chemical Properties in Flavored Cheese Prepared with Cream Cheese Base Using Oregano and Rosemary Essential Oils. *LWT-Food Sci. Technol.* **2013**, *53*(2), 409–417.

51. Orafidiya, L.; Agbani, E.; Oyedele, A., Babalola, O. O.; Onayemi, O. Preliminary Clinical Test on Topical Preparations of *Ocimum gratissimim* Linn Leaf Essential Oil for the Treatment on Acne Vulgaris. *Clin. Drug Investig.* **2002**, *22*, 313–319.

52. Oussalah, M.; Caillet, S.; Saucier, L.; Lacroix, M. Inhibitory Effects of Selected Plant Essential Oils on the Growth of Four Pathogenic Bacteria: *E. coli* O157:H7, *Salmonella Typhimurium, Staphylococcus aureus* and *Listeria monocytogenes. Food Control* **2006**, *18*(5), 414–420.

53. Palthur, S.; Devanna, N.; Anuradha, C. M. Antioxidant and Organoleptic Properties of Tulsi-flavored Herbal Milk. *Int. J. Plant Anim. Environ. Sci.* **2014**, *4*(4), 35–40.

54. Pina-Pérez, M. C.; Martínez-López, A.; Rodrigo, D. Cinnamon Antimicrobial Effect Against *Salmonella typhimurium* Cells Treated by Pulsed Electric Fields (PEF) in Pasteurized Skim Milk Beverage. *Food Res. Int.* **2012**, *48*, 777–783.

55. Qiumin, M.; Davidson, P. M.; Zhong, Q. Antimicrobial Properties of Lauric Arginate Alone or in Combination with Essential Oils in Tryptic Soy Broth and 2% Reduced Fat Milk. *Int. J. Food Microbiol.* **2013**, *166*, 77–84.

56. Reinheimer, J. A.; Suarez, V. B.; Haye, M. A. Microbial and Chemical Changes in Refrigerated Pasteurized Milk in the Santa Fe Area (Argentina). *Aust. J. Dairy Technol.* **1993**, *48*, 5–9.

57. Reis, F. S.; Stojkovi, D.; Sokovi, M.; Glamo, J.; Ana, L.; Barros, L.; Ferreira, I. C. F. R. Chemical Characterization of *Agaricus bohusii*, Antioxidant Potential and Antifungal Preserving Properties When Incorporated in Cream Cheese. *Food Res. Int.* **2012**, *48*(2), 620–626.

58. Samaddar, M.; Ram, C.; Sen, M. Assessment of Storage Stability of Essential Oil-enriched Flavored Milk. *Indian J. Dairy Sci.* **2015**, *68*(4), 357–363.

59. Shah, B.; Davidson, P. M.; Zhong, Q. Nanocapsular Dispersion of Thymol for Enhanced Dispersibility and Increased Antimicrobial Effectiveness Against *Escherichia coli* O157:H7 and *Listeria monocytogenes* in Model Food Systems. *Am. Soc. Microbiol.* **2012**, *78*, 8448–8453.

60. Shah, B.; Ikeda, S.; Davidson, P. M.; Zhong, Q. Nanodispersing Thymol in Whey Protein Isolate-maltodextrin Conjugate Capsules Produced Using the Emulsion–evaporation Technique. *J. Food Eng.* **2012**, *113*, 79–86.

61. Sikkema, J.; de Bontt, J. A. M.; Poolman, B. Interaction of Cyclic Hydrocarbons with Biological Membranes. *J. Biol. Chem.* **1994**, *269*, 8022–8028.

62. Smith-Palmer, A.; Stewart, J.; Fyfe, L. Antimicrobial Properties of Plant Essential Oils and Essences Against Five Important Foodborne Pathogens. *Lett. Appl. Microbiol.* **1998**, *26*(2), 118–122.

63. Tabak, M.; Armon, R.; Neeman, I. Cinnamon Extracts' Inhibitory Effect on *Helicobacter pylori. J. Ethnopharmacol.* **1999**, *67*, 269–277.

64. Tachibana, S.; Ohkubo, K.; Towers, G. H. N. Cinnamic Acid Derivatives in Cell Walls of Bamboo and Bamboo Grass. *Phytochemistry* **1992**, *31*, 3207–3209.

65. Thabet, H. M.; Nogaim, Q. A.; Qasha, A. S.; Abdoalaziz, O.; Alnsheme, N. Evaluation of the Effects of Some Plant Derived Essential Oils on Shelf Life Extension of Labneh. *Merit Res. J. Food Sci. Technol.* **2014**, *2*(1), 8–14.

66. Toit, E. A.; Rautenbach, M. Sensitive Standardised Micro-gel Well Diffusion Assay for the Determination of Antimicrobial Activity. *J. Microbiol. Methods* **2000,** *42,* 159–165.

67. Trombetta, D.; Castelli, F.; Sarpietro M. G., Venuti, V.; Cristani, M.; Daniele, C.; Saija, A; Mazzanti, G.; Bisignano, G. Mechanism of Antibacterial Action of Three Monoterpenes. *Antimicrob. Agents Chemother.* **2005,** *49,* 2474–2478.

68. Tserennadmid, R.; Takó, M.; Galgóczy, L.; Papp, T.; Pesti, M.; Vágvölgyi, C.; Almássy, K.; Krisch, J. Anti-yeast Activities of Some Essential Oils in Growth Medium, Fruit Juices and Milk. *Int. J. Food Microbiol.* **2010,** *144,* 480–486.

69. Ultee, A.; Kets, E. P. W.; Alberda M., Hoekstra, F. A.; Smid, E. J. Addition of Food Borne Pathogen *Bacillus cereus* to Carvacrol. *Arch. Microbiol.* **2000,** *174,* 233–238.

70. Weber, F. J.; de Bontt, J. A. M. Adaption Mechanisms of Microorganisms to the Toxic Effects of Organic Solvents on Membranes. *Biochim. Biophys. Acta* **1996,** *1286,* 225–45.

CHAPTER 5

ROLE OF HUMAN COLONIC MICROBIOTA IN DISEASES AND HEALTH

AKRUTI JOSHI[1,*], VIJENDRA MISHRA[2], SHRADDHA BHATT[3], and RUPESH S. CHAVAN[4]

[1]*Shree P. M. Patel Institute of P.G. Studies and Research, Anand 388001, Gujarat, India*

[2]*National Institute of Food Technology Entrepreneurship and Management, HSIIDC, Kundli 131028, Sonepat, India*

[3]*Department of Biotechnology, Junagadh Agricultural University, Junagadh 362001, Gujarat, India*

[4]*Department of Quality Assurance, Mother Dairy at Junagadh, Junagadh 362001, Gujarat, India*

*Corresponding author. E-mail: akruti.joshi@gmail.com

CONTENTS

ABSTRACT

Milk and milk products are associated with human health across the lifespan. A major advantage of dairy foods is that consumers are already familiar with them and many believe that dairy products are healthy, natural products. Milk and dairy products constitute one of the four major food groups that make up a balanced diet. Milk is one of the major sources of conjugated linoleic acid in the diet, although it is a minor component of milk fat. In addition to bioavailable calcium, milk also provides phosphorus, zinc, and fluoride, which are essential for optimal bone health and prevent osteoporotic fractures. Milk is an excellent natural source of nutrients; research also shows that milk-derived components have many beneficial physiological properties. The importance of microbiota in human health, its different functions, and various milk and milk products which contribute to the microbiota are discussed in the present chapter.

5.1   INTRODUCTION

During the last decades, it has become clear that the human body lives in close harmony with a complex ecosystem that is composed of more than 1000 different bacterial species inhabiting the oral cavity, upper respiratory tract, gastrointestinal tract (GIT), vagina, and skin. This collection, known as the microbiota, is acquired soon after birth and persists throughout life. Together, these microbes play an important role in the physiology of their host, by digestion and assimilation of nutrients, protection against pathogen colonization, modulation of immune responses, regulation of fat storage, and stimulation of intestinal angiogenesis. The number of microbes residing in our gut exceeds the number of our own body cells by a factor of ten and thousands.[6] The large bowel is the main area of permanent microbial colonization of the human GIT and several hundred bacterial strains and species have been isolated from this complex ecosystem where viable counts in feces typically reach 10^{11}–10^{12}/g, with anaerobic bacteria predominating.

The present chapter highlights the importance of microbiota in human health and its different functions. Factors affecting the microbiota are also discussed in detail, including factors such as: host genetics, birth delivery mode, feeding, gestational age, geographical impacts, and also influence of ageing, diet, antibiotics, stress, prebiotics, probiotics, and synbiotics.

5.2   DEVELOPMENT AND ESTABLISHMENT OF THE MICROBIOTA

We are born sterile but immediately after birth a microbial community starts to establish. The first inocula come from the mother, by the birth canal, and/or a fecal–oral route, and from the surrounding area. The microbial composition and colonization pattern varies widely from baby to baby. During the first months of life, the community is highly variable and dominated with a facultative anaerobic microbiota. When space and nutrients are not limiting, bacteria with high division rates are more prevalent. However, over time, when the system becomes more densely populated and nutrient supply is limited, a more specialized microbiota takes over. The microbiota of the children becomes progressively more complex and at the age of two the microbial community is stabilized and an adult-like flora is established with a large predominance of anaerobic microbes.[36] Without outer environmental stresses the dominant microbiota is considered to be remarkably stable over time.[35] However, by studying subgroups of the microbiota, temporal variation was observed in the *Lactobacilli* population and *Enterococcus* population, whereas *Bacteroides* spp. and *Bifidobacteria* remained relatively stable.[39]

5.3   HUMAN GASTROINTESTINAL MICROFLORA

The stomach microflora constitutes total count of 10^1 to 10^3 cfu/mL, comprising Gram-positive facultative anaerobes that include *Lactobacillus, Streptococcus, Enterobacterium,* and yeasts. The jejunum/ileum microflora constitutes of 10^4 to 10^7 cfu/mL comprising facultative or strict anaerobes, including *Bifidobacteria, Streptococcus, Enterobacteria, Clostridium,* yeasts, and *Bacteroides*, whereas the colonic microflora constitutes the total count of 10^{11} to 10^{12} cfu/mL including strict anaerobes Gram-positive or Gram-negative that includes *Peptostreptococcus* (strict anaerobes), *Eubacteria, Bacteroides, Bifidobacteria, Clostridium, Lactobacillus, Streptococcus, Enterobacteria,* and yeasts.

5.4   FUNCTIONS OF THE MICROBIOTA

The commensal microbiota has important and specific functions for the host. These can be divided into three main groups: trophic, protective, and metabolic. Trophic functions include control of proliferation, differentiation of epithelial cells, and immune system maturation. The protective role of the microbiota is occupying intestinal surfaces and creating a stability of the system and a milieu that prevents invasion from exogenous microbes, for example, by production of antimicrobial compounds. Metabolic functions include breakdown of nondigestible compounds by anaerobic fermentation, such as resistant starch and plant polysaccharides, which generate short-chain fatty acids (SCFA) that serve as growth signals and fuel for the intestinal epithelium as well as for other microbes. In particular, butyrate has gained large attention for its health-promoting properties.[57] Another example of metabolic function provided by the microbiota is the production of some vitamins such as vitamin K and B_{12}, synthesis of amino acids, and digestion of proteins.

5.5   FACTORS INFLUENCING THE MICROBIAL STRUCTURE

The dominant microbial communities of the large bowel have shown a remarkable stability over time during adulthood, yet the microbial community profiles are unique for each individual.[20] However, there are numerous external factors that have potential to influence the microbial composition in the gut.[37]

5.5.1   HOST GENETICS

The structure of the intestinal microbiota has also shown to be dependent on host genetics. The culturable fraction of the fecal microbiota between mono- and dizygotic twins found higher similarities within monozygotic pairs. This difference was confirmed in young children, living in the same household.

5.5.2   BIRTH DELIVERY MODE

The mode of delivery shapes the acquisition and the structure of the initial microbiota in newborns.[19] The birth delivery mode also influences the composition of the microbiota, with slower diversification and lack of anaerobic species, such as *Clostridia*, in infants delivered by caesarean section.[26] In another study, a lower prevalence of *Clostridial* species in the microbiota of a caesarean section delivered child was prolonged over a period of several years.[49] In vaginally delivered neonates, bacteria begin to appear in the stool during day 1. These are mainly anaerobic but are followed within the first 5 days by *Bifidobacterium* spp. By day 10, most healthy full-term neonates display colonization by a heterogeneous microbiota. In contrast, neonates delivered by caesarean section, microbiota is characterized by lack of anaerobic bacteria, since they are generally colonized only by microaerophilic microorganisms, facultative anaerobes, and sporulated forms such as *Clostridium*.[47]

The intestinal microbiota after caesarean delivery is characterized by an absence of *Bifidobacteria* species. Vaginally delivered neonates, even if they show individual microbial profiles, are characterized by predominant groups such as *Bifidobacterium longum* and *Bifidobacterium catenulatum*.[11,23,24] By comparison of Greek healthy neonates, the caesarean delivered infants were less colonized with *lactobacilli* and *Bifidobacteria* at day 4 compared with the vaginal delivered infants.

5.5.3   FEEDING

Historically, breastfed infants are thought to have relatively higher proportions of *Bifidobacteria* than formula-fed babies of the same age, who possess a more complex composition. Breastfed babies traditionally have colonic populations that are dominated by *Bifidobacteria* and lactic acid bacteria, with very few *bacteroides, Blostridia, and coliforms*. More variation occurs in the microflora of formula-fed babies that tend to contain large number of *bacteroides, Clostridia*, and enteric bacteria. With the introduction of solid foods and by about 2 years after birth, infants start to adopt microflora profiles in proportions that approximate to those seen in adults. *Bacteroides* and other Gram-negative bacteria begin to dominate and a wider range of genera is seen. The populations then seem to be relatively stable (>99% anaerobic), aside for perturbations by diet and habit, until advanced ages when a significant decline in *Bifidobacteria*, plus increases in *Elostridia* and *Enterobacteriaceae* are reported. Once this climax microflora is achieved, we generally maintain the same pattern of bacterial genera for the rest of our lives, although it will vary tremendously at strain level throughout this time. The climax microflora is dominated by nonspore-forming, strictly anaerobic bacteria. Of these, *bacteroides* and *fusobacteria* represent the predominant Gram-negative genera, while *Bifidobacteria* tend to be the principal Gram-positive populations, with fewer numbers of *Clostridia, lactobacilli,* and Gram-positive cocci.

5.5.4 GESTATIONAL AGE

These processes of colonization and bacterial succession at delivery refer to infants born at term, whereas the situation in preterm infants is different. *Lactobacilli* numbers are lower in preterm children, and this reduction has is due to the antibiotic treatment time spend in the incubator and delayed colonization.[47]

5.5.5 GEOGRAPHICAL IMPACTS

Geographical region can have an impact on the gut microbial composition. The microbiota of Estonian infants differed from that of Swedish infants.[10,13] In one study, local variations in the microbiota of the western European population were correlated to age, that is, adult compared with elderly.[41] For example, country and age interactions were observed for German and Italians with inverse interactions for the predominating *Clostridium coccoides* and *Bacteroides–Prevotella* groups. Contrary to this observation, no significant differences in the microbial composition were observed in individuals from five European countries.[35] Swedish infants delivered vaginally and also breastfed and Pakistani infants delivered through both vaginally and caesarean deliveries with incomplete breast feeding were colonized with *enterobacteria* then the Swedish infants. *Escherichia coli* and *Klebsiella* were more frequently isolated from Swedish infants and *E. coli* was dominated in both Pakistani groups with colonization of *Proteus, Klebsiella, Enterobacter,* or *Citrobacter.*[1] By comparison of three tribal groups (Maguzawa, Hausa, and Fulani), groups with distinctive dietary patterns in a rural area of northern Nigeria, the number of *bacteroides* and *Clostridia* were lower in Maguzawa than in the other dietary groups. Comparison of breastfed infants, weaned children, and adults were examined in rural Nigeria and urban United Kingdom, the breastfed infants had a similar anaerobic flora dominated by *Bifidobacteria* but *bacteroides* were isolated less in quarter of either community. Weaned children in both communities had greater number of *bacteroides* and *clostridia* than breastfed infants; higher numbers of *bacteroides* and *Clostridia* were present in UK adults but not in Nigerian adults.[54]

In comparison to 45 healthy infants in Stockholm, the Ethiopian infants had more *enterococcus* and less *Staphylococcus* and *bacteroides* during the first 2 weeks of life. Changes in the bacterial flora occur during childhood and adolescence characterized by reduction in *lactobacillus* and *Bifidobacterium* species and an increase in *bacteroides, Eubacterium rectale,* and *Faecalibacterium prausnitzii* peaked during late childhood from 130 healthy children and adolescents in the age group 2–17 years and from 30 healthy adults residing in Southern India.[7] By comparing the fecal microbiota of European children (EU) and children from a rural African village of Burkina Faso (BF) high in fiber content, a significant enrichment in *bacteroides,* and depletion in *Firmicutes* was found with more SCFA in BF than in EU, whereas *enterobacteriaceae* were lower in BF than in EU. Higher proportion of *lactobacilli* was observed in rural younger elderly (70 years) than in urban elderly (98 years) and proportion of *E. coli* lower in rural than in urban elderly (109 years). The counts of anaerobes, that is, *Bifidobacteria* and *bacteroides* remain constant from day 7 to 12 months but no consistent pattern was found to be in facultative bacteria, that is, *E. coli, Klebsiella*, other *enterobacteria,* and *enterococcus*.[1,2]

Total number of anaerobic flora of 15 healthy elderly persons, aged 84 years, were smaller in rural (Yuzurihara) than urban people, 68 years of age, in Tokyo. Large *Bifidobacterial* counts were found in rural population than in urban as compared with the *clostridia*, which was found to be more in urban region than in rural population. The *enterobacteria* and *Streptococci* could be identified even in meconium samples passed in the 4–7 h of life in a study carried out by Mata and Urutia in 1971 on Guatemalan infants. The assessment of the faecal microbiota of infants from five European countries, Sweden, Scotland, Germany, Italy, and Spain, with different lifestyle characteristics showed differences in microbial populations.

5.5.6 INFLUENCE OF AGEING

Due to an altered physiology in elderly, with a decreased intestinal motility, reduced secretion of gastric acid, and a change of dietary habits and lifestyle, the microbiota in elderly persons differs from younger adults, although with great individual variations. In elderly persons, it is common to have a reduction in number and diversity of *bacteroides* and *Bifidobacteria*, and a reduced production of SCFA and amylolytic activity. In addition, increased number of facultative anaerobes, *Fusobacteria, Clostridia,* and *Eubacteria* but decreased numbers of *Bifidobacteria* have been reported in elderly people.[58] The abundance and diversity of *Bifidobacteria* is consistently reported to decrease in elderly individuals.[41,58] The gut microbiotas were compared among in children, young adults, healthy older people, and elderly patients with *Clostridium difficile*-associated diarrhea.[29] The main difference between adults and children is higher number of *enterobacteria* in the latter group, as determined by viable counting and relative 16S ribosomal ribonucleic acid (rRNA) abundance measurements. Furthermore, a greater proportion of faecal rRNA in children is hybridized by the three probes (*Bifidobacteria, enterobacteria,* and *bacteroides*), showing a less developed gut microbiota. Species diversity is also significantly lower in the *C. difficile*-associated diarrhea group, which is characterized by high numbers of *enterobacteria* and low levels of *Bifidobacteria* and *bacteroides*.

To determine if there is any change with age in the distribution of *Bifidobacterium* and *enterobacteriaceae* species in human intestinal microflora, strains were isolated from a total of 54 samples of human feces (15 children, 3–15 years old; 17 adults, 30–46 years old; and 22 elderly, 69–89 years old). After 2 years of age, when the microbiota is practically identical to that of adults, *bacteroides, bifidobacterium, eubacterium, clostridium, peptococcus, peptostreptococcus,* and *ruminococcus* are the predominant species. Finally, *bifidobacterium* counts progressively decreased with ageing.[27] The *bacteroides* spp. and *lactobacilli* increased with age, while *enterococci* and *E.coli* decreased, and *bifidobacteria* were found to be relatively stable.[44]

5.5.7 INFLUENCE OF DIET

Diet is often of interest for research studies due to its potential for modulation of the intestinal microbiota of the host, either in a beneficial or detrimental way. Dietary habits impact the microbial composition in the gut, in particular during the first years of life. However, broadly defined diets such as "Western diet" comprised a high intake of fat and animal proteins but low fiber, and a "Japanese diet," with less fat and more vegetables, have only revealed moderate differences in the gut microbiota involving a few bacterial genera. In another study, a shift in the bacterial SCFA profile was observed between individuals that had an uncooked extreme vegan diet for 1 month compared to individuals that had a westernized diet. However, this shift was not reflected in the bacterial community composition.[28,46] However, recently it was shown that diet could promote a division-wide change of the microbiota.

The colonic microflora derive substrates for growth from the diet (e.g., nondigestible oligosaccharides, dietary fiber, undigested protein reaching the colon) and from endogenous sources such as mucin, the main glyco-protein constituent of the mucus which lines the walls of the GIT. The two main fermentative substrates of dietary origin are nondigestible carbohy-drates (e.g., resistant starch, nonstarch polysaccharides and fibers of plant origin and nondigestible oligosaccharides) and protein which escapes digestion in the small intestine. Of these, carbohydrate fermentation is more energetically favorable, leading to a gradient of substrate utilization spatially through the colon. The proximal colon is a saccharolytic environ-ment with the majority of carbohydrate entering the colon being fermented in this region. As digesta moves through toward the distal colon, carbo-hydrate availability decreases and protein and amino acids become a more dominant metabolic energy source for bacteria in the distal colon. Resistant starch is readily fermented by a wide range of colonic bacterial species including members of the *bacteroides* spp., *eubacterium* spp., and the *bifidobacteria*.[22] The remainder of the carbohydrate entering the colon comprises nonstarch polysaccharides (about 8–18 g per day), unabsorbed sugars, for example, raffinose, stachyose, and lactose (about 2–10 g per day) and oligosaccharides such as fructooligosaccharides, xylooligosac-charides, galactooligosaccharides (about 2–8 g per day).[12] The degree to which these carbohydrates are broken down by the gut microflora varies greatly. Unabsorbed sugars entering the colon are readily fermented and persist for only a short time in the proximal colon. Some sugars such as raffi-nose may have a more selective fermentation (being mainly assimilated by *bifidobacteria* and *lactobacilli*) while others support the growth of a range of colonic bacteria. Similarly, nondigestible oligosaccharides reaching the colon display different degrees of fermentation. Certain oligosaccharides such as fructooligosaccharides, galactooligosaccharides, and lactulose may be fermented preferentially by *bifidobacteria*, which has given rise to the concept of prebiotics.[25] Nonstarch polysaccharides include pectin, arabinogalactan, inulin, guar gum, and hemicellulose, which are readily fermented by the colonic microflora, and lignin and cellulose, which are much less fermentable. Endogenous carbohydrates, chiefly from mucin and chondroitin sulfate, contribute about 2–3 g per day of fermentable substrate. The main saccharolytic species in the colonic microflora belong to the genera *bacteroides, bifidobacterium, ruminococcus, eubacterium, lactobacillus,* and *clostridium.*[41] Protein and amino acids are also available for bacterial fermentation in the colon. Approximately, 25 g of protein enters the colon daily.[38] Other sources of protein in the colon include bacterial secretions, sloughed epithelial cells, bacterial lysis products, and mucins. The main proteolytic species belong to the *bacteroides* and *clostridia* groups. Infants receiving the standard infant formula (SF) show higher prevalence of anaerobic microbes and coagulase-negative staphylococci than in the preterm infant formula (PF group).[43]

5.5.8 IMPACT OF ANTIBIOTICS

It is well known that the microbiota can be strongly affected by antibiotic treatment. Recently, it was shown that the gut microbial composition could be disrupted for a long period of time, up to 2 years in some cases depending on the antibiotics used. Scientists have showed that the *bacteroides* community never reestablished to its original composition within 2 years after a 7 day clindamycin treatment. Still, more studies are necessary to determine what effect a population shift of the microbiota will have over a long period of time on host physiology. Various studies have been reported regarding the effect of antibiotics on the human intestinal microflora, including parameters such as suppression and proliferation of human intestinal microflora by different scientists.[5,6,36,42,52]

5.5.9 HOSPITALIZATION

Hospitalization and antibiotic treatment exacerbate gut dysbiosis in older people, as shown by Barstoch et al.[8] Selected fecal bacterial populations were investigated in 35 healthy elderly subjects (63–90 years) living in the local community, 38 elderly hospitalized patients (66–103 years), and 21 elderly hospitalized patients receiving antibiotic therapy (65–100 years). The main difference found between the healthy elderly and the two patient groups was a large reduction in *bacteroides* number after hospitalization.

In hospitalized newborns, intestinal colonization by *Klebsiella, Proteus, Pseudomonas* as well as *E. coli* occurred more frequently.

5.5.10 EFFECT OF STRESS

Psychological stress has profound effects on intestinal microbiota in humans, these changes have been attributed among other factors, to a catecholamine bacterial exposure that produces a significant decrease in *lactobacilli* and *bifidobacteria* and an inverse effect on potentially pathogenic bacteria such as *Yersinia enterocolitica, E. coli, Bacteroides* spp., *and Pseudomonas aeruginosa.*

5.5.11 PREBIOTICS, PROBIOTICS, AND SYNBIOTICS

Probiotics are living bacterial food ingredients with beneficial effects on human health. *Bifidobacteria* and lactic acid bacteria (LAB), especially *lactobacilli*, are commonly used as probiotics agents. In addition, specific dietary compounds have been shown to alter the metabolism or stimulate growth of beneficial groups of bacteria, such as inulin or fructooligosaccarides that stimulate the growth of *bifidobacteria* and *lactobacilli* in the gut.[17,51,56] These compounds are collectively called prebiotics.[25] The effect of supplementation was characterized as a decrease in the numbers of *E. coli* and protection against an increase in bacteroides number during weaning.[32] Lactulose and lactitol a disaccharide sugar have been shown to affect the microflora in humans by decreasing populations of *Bacteroides* spp., *clostridia, coliforms,* and *eubacteria* and increasing numbers of *bifidobacteria, lactobacilli,* and *streptococci.* The phase II anticancer study randomized double blind and placebo controlled in 80 patients with a history of colon cancer when supplemented with a synbiotic oligofructose enriched with inulin, and *Bifidobacterium lactis* Bb12 and *Lactobacillus rhamnosus* GG for 12 weeks showed increased levels of *Bifidobacteria* and *lactobacilli* and decrease in the number of pathogens (coliforms and *Clostridium perfrigens*).[14] Inulin increased *Bifidobacteria* from 7.9 to 9.2 log10/g dry faces but decreased *enterococci* in numbers, whereas lactose increased *enterococci* and decrease *lactobacilli* and *clostridia* in elderly constipated persons.[33] The use of probiotics, prebiotics, and synbiotics and its effect on the human intestinal microflora of elderly persons have been studied, which include the parameters such as the change in microbial composition, bowel function, and immune function.[3,15,16,18,44,48,53]

5.5.12 MICROBIOTA AND GUT-RELATED DISORDERS

A balance between the host immune system and the commensal gut microbiota is crucial for maintaining health. When this balance is disturbed, the host–microbe relationship can progress toward a disease state. In addition, the gut microbiota has been implicated to play an important role in several other disorders of the GIT, such as inflammatory bowel diseases (IBD), including Crohn's disease (CD) and ulcerative colitis (UC), irritable bowel syndrome (IBS), as well as colorectal cancer (CRC) and allergies.

5.5.13 ALLERGY AND ASTHMA

There has been a constant increase of asthma and allergic diseases in western countries over the last few decades. Within any given country, asthma and allergies are more prevalent in urban than in rural areas. By comparing the microbiota of infants that were only 3 weeks old, the microbial community composition was found to differ between atopic and nonatopic infants, mainly with higher proportions of *Bifidobacteria* in the nonatopic infants. However, this difference in the microbial composition was not reflected when the children were 3 months old.[30] Furthermore, the microbial diversity was significantly lower in infants with atopic disease compared with nonatopic disease.

Differences in the gut microbiota have also been observed between Estonian (low prevalence of allergies) and Swedish infants (high prevalence of allergies), mainly with higher counts of *lactobacilli* and *eubacteria* in Estonian infants and instead higher counts of *clostridia* in Swedish infants. Interestingly, the Estonian infant microbiota reflected the microbiota of infants in Western European countries several decades earlier, when the prevalence of allergies was low in Western Europe.[50] In addition, the nonallergic infants within these respective countries had higher counts of *lactobacilli* and *Bifidobacteria*, whereas higher counts of *coliforms* and *Staphylococcus aureus* were reported among the allergic infants.[13] An increased prevalence of *lactobacilli* and LAB was shown in the infant microbiota of anthroposophic children.[4] Children raised according to the anthroposophic lifestyle, for example, Steiner school children, are a low-risk group for allergy development. Infants in the highly sensitized group (HSG) had greater number of *lactobacilli/enterococci* than those in the sensitized group (SG). Serum total immunoglobulin E (IgE) concentration correlated directly with *E. coli* counts in all infants and with bacteroides counts in the HSG, indicating that the presence of these bacteria is associated with the extent of atopic sensitization.[32] Infants with allergy colitis possessed significantly lower counts of *Bifidobacteria* and total anaerobes and significantly higher counts of *clostridia* in their faeces.

5.5.14 CHRONIC CONSTIPATION

Concentrations of *Bifidobacterium* and *lactobacillus* were significantly lower in constipated patients; potentially pathogenic bacteria, and/or fungi are increased level.

5.5.15 OBESITY

Obesity disorders affect many developed countries and result from an imbalance in energy (input vs output) with serious health consequences, including cardiovascular disease, type 2 diabetes mellitus, and colon cancer. The role of gut microbiota in obesity is a current public health problem. With robust real time polymerase chain reaction (PCR) tool that includes a plasmid internal control system tool, studies of the bacterial communities of the gut microbiota revealed a shift in the ratio of *Firmicutes* and *Bacteroidetes* in obese patients. Determining the variations of microbial communities in feces may be beneficial for the identification of specific profiles in patients with abnormal weights. The roles of the archaeon *Methanobrevibacter smithii* and *Lactobacillus* species have not been described in these studies.[5]

5.5.16 DIABETES

Type 2 diabetes is a metabolic disease which primary cause is obesity-linked insulin resistance. Recent studies based on large-scale 16S rRNA gene sequencing and more limited techniques, based on quantitative real time PCR (qPCR) and fluorescent in situ hybridization (FISH), showed a relationship between the composition of the intestinal microbiota and metabolic diseases such as obesity and diabetes. It is reported by several scientists that presence of *Bifidobacterium* in appropriate levels can improve glucose tolerance and low-grade inflammation in prebiotic-treated mice. Furthermore, the development of diabetes type 1 in rats was reported to be associated with higher amounts of *Bacteroides* spp. It is also proposed that the gut microbiota directed increased monosaccharide uptake from the gut and instructed the host to increase hepatic production of triglycerides associated with the development of insulin resistance. A study including 36 male adults with a broad range of age and body mass indices, among which 18 subjects were diagnosed with diabetes type 2. The proportions of phylum *Firmicutes* and class *Clostridia* were found significantly reduced in the diabetic group compared with the control group.[34]

5.5.17 AUTISM

Children with autistic spectrum disorders (ASDs) tend to suffer from severe gastrointestinal problems. Such symptoms may be due to a disruption of the indigenous gut flora promoting the overgrowth of potentially pathogenic microorganisms. The faecal flora of ASD patients contained a higher incidence of the *Clostridium histolyticum* group of bacteria than that of healthy children. However, the nonautistic sibling group had an intermediate level of the *C. histolyticum* group.[45]

5.5.18 INFLAMMATORY BOWEL DISEASE

IBD refers to chronic relapsing disorders of the intestinal tract with yet unknown etiologies. It is the collective term for a group of heterogeneous tissue reactions, with similar clinical behavior. Several studies have identified a higher prevalence of enterobacteria and in particular *E. coli* in mucosal tissues from IBD patients.[9] There was an increase in anaerobic bacteria (*Bacteroides vulgates, Streptococcus faecalis*) in adult IBD, whereas an increase in aerobic and facultative anaerobic (*E. coli*) in pediatric IBD.

5.5.19 ECZEMA

Eczema is a chronic form of childhood disorder that is gaining in prevalence in affluent societies. Previous studies hypothesized that the development of eczema is correlated with changes in microbial profile and composition of early life endemic microbiota, but contradictory conclusions were obtained, possibly due to the lack of minimization of apparent nonhealth-related confounders (e.g., age, antibiotic consumption, diet, and mode of delivery). Quantitative analysis of bacterial targets on a larger sample size and 12 months of age revealed that the abundances of *Bifidobacterium* and *Enterobacteriaceae* were different among caesarean-delivered infants with and without eczema.[59]

5.5.20 COLORECTAL CANCER

A Western diet, comprising a high intake of fat and animal protein but low fiber content is associated with a higher risk in developing colon cancer, compared with a Japanese diet and vegan diet which are associated with a low risk of developing colon cancer. For example, increased numbers of *Bacteroides* have been associated with an increased risk for CRC.[40] Also, increased numbers of both *Clostridium* and *Lactobacillus* have been found in individuals with CRC.[31]

5.5.21 DIARRHEA

Information regarding the fecal microflora in acute diarrheal states is every limited and nothing is known about the faceal flora of infants and young children with diarrhea of diverse aetiologies. The number of anaerobic bacteria flora were found to decrease in diarrheal patients then in 29 control groups, whereas the total aerobes increased in the diarrheal group than in the control group.

5.5.22 RHEUMATOID ARTHRITIS

An essential component of the bacterial cell wall is the peptidoglycan layer, which is particularly thick in Gram-positive bacteria. It consists of *N*-acetylmuramic acid and *N*-acetylglucosamine, layers of which are bound by peptide bridges. The ability of bacterial cell walls to induce chronic, erosive arthritis was first described in the rat by using *Streptococcus pyogenes*. Self-perpetuating arthritis, closely resembling human RA by histological criteria, develops in susceptible rat strains after a single intraperitoneal injection of the bacterial cell wall. In addition to *Streptococcus pyogenes*, several bacterial species representing *lactobacillus, Bifidobacterium, eubacterium, Collinsella*, and *Clostridium* have been observed to have a similar ability. Surprisingly, most of these are anaerobic Gram-positive rods belonging to the normal intestinal microbiota in man. Similarly, differences in gut microbiota have been observed in patients with RA; RA patients had decreased *Bifidobacteria* and *bacteroides*.[21,55]

5.6   SUMMARY

The intestinal ecosystem is characterized by dynamic and reciprocal interactions between the host and its microbiota. Although the importance of the gut microbiota for human health has been increasingly recognized, yet the early bacterial colonization in the neonatal gut is not yet completely understood. The mechanisms underlying these interactions are complex and influenced by many factors. The relative importance of these factors is difficult to organize into a hierarchy. A better knowledge of the microbiota and the impact of antibiotics will provide an essential step toward understanding the development of this important bacterial community. Recent research in the area of probiotics, prebiotic oligosaccharides, and symbiotic combinations is leading to a more targeted development of functional food ingredients. Improved molecular techniques for analysis of the gut microflora and its development, increased understanding of metabolisms and interaction between host and environment, and new manufacturing biotechnologies are facilitating the production of such food supplements. Thus, our increased understanding is fostering our ability to modulate the gastrointestinal microbiota for therapeutic outcomes.

KEYWORDS

- aging
- arthritis
- bowel disease
- chronic constipation
- gastrointestinal microflora
- host genetics
- inflammatory bowel disease
- prebiotics
- probiotics
- rheumatoid arthritis

REFERENCES

1. Adlerberth, I.; Strachan, D. P.; Matricardi, P. M.; Ahrné, S.; Orfei, L.; Åberg, N.; Perkin; M. R.; Tripodi, S.; Hesselmar, B.; Saalman, R.; Coates, A. R.; Bonanno, C. L.; Panetta, V.; Wold, A. E. Gut Microbiota and Development of Atopic Eczema in 3 European Birth Cohorts. *J. Allergy Clin. Immunol.* **2007,** *120,* 343–350.
2. Adlerberth, I.; Wold, A. E. Establishment of the Gut Microbiota in Western Infants. *Acta Paediatr.* **2009,** *98,* 229–238.
3. Ahmed, S.; Macfarlane, G. T.; Fite, A.; McBain, A. J.; Gilbert, P.; Macfarlane, S. Mucosal-associated Bacterial Diversity Associated with Human Terminal Ileum and Colonic Biopsy Samples. *Appl. Environ. Microbiol.* **2007,** *73,* 7435–7442.
4. Alm, J. S.; Swartz, J.; Björksten, B.; Engstrand, L.; Engström, J.; Kuhn, I.; Lilja, G.; Möllby, R.; Norin, E.; Pershagen, G.; Reinders, C.; Wreiber, K.; Scheynius, A. An Anthroposophic Lifestyle and Intestinal Microflora in Infancy. *Pediatr. Allergy Immunol.* **2002,** *13,* 402–411.
5. Armougom, F.; Henry, M.; Vialettes, B.; Raccah, D.; Raoult, D. Monitoring Bacterial Community of Human Gut Microbiota Reveals an Increase in *Lactobacillus* in Obese Patients and Methanogens in Anorexic Patients. *Plos One* **2009,** *4,* e7125.
6. Backhed, F.; Ley, R. E.; Sonnenburg, J. L.; Peterson, D. A.; Gordon, J. I. Host-bacterial Mutualism in the Human Intestine. *Science* **2005,** *307,* 1915–1920.
7. Balamurugan, R.; Janardhan, H. P.; George, S.; Chittaranjan, S. P.; Ramakrishna, B. S. Bacterial Succession in the Colon During Childhood and Adolescence: Molecular Studies in a Southern Indian Village. *Am. J. Clin. Nutr.* **2008,** *88,* 1643–1647.
8. Bartosch, S.; Woodmansey, E. J.; Paterson, J. C. M.; McMurdo, M. E.; Macfarlane, G. T. Microbiological Effects of Consuming a Synbiotic Containing *Bifidobacterium*

bifidum, Bifidobacterium lactis, and Oligofructose in Elderly Persons, Determined by Real-time Polymerase Chain Reaction and Cunting of Viable Bacteria. *Clin. Infect. Dis.* **2005**, *40*, 28–37.

9. Baumgart, M.; Dogan, B.; Rishniw, M.; Weitzman, G.; Bosworth, B.; Yantiss, R.; Orsi, R. H.; Wiedmann, M.; McDonough, P.; Kim, S. G.; Berg, D.; Schukken, Y.; Scherl, E.; Simpson, K. W. Culture Independent Analysis of Ileal Mucosa Reveals a Selective Increase in Invasive *Escherichia coli* of Novel Phylogeny Relative to Depletion of *Clostridiales* in Crohn's Disease Involving the Ileum. *ISME J.* **2007**, *1*, 403–418.

10. Benno, Y.; Endo, K.; Mizutani, T.; Namba, Y.; Komori, T.; Mitsuoka, T. Comparison of Fecal Microflora of Elderly Persons in Rural and Urban Areas of Japan. *Appl. Environ. Microbiol.* **1989**, *55*, 1100–1105.

11. Biasucci, G.; Benenati, B.; Morelli, L.; Bessi, E.; Boehm, G. Cesarean Delivery May Affect the Early Biodiversity of Intestinal Bacteria. *J. Nutr.* **2008**, *138*, 1796S–800S.

12. Bingham, S. A.; Pett, S.; Day, K. C. NSP Intake of a Representative Sample of British Adults. *J. Hum. Nutr. Diet* **1990**, *3*, 339–344.

13. Björksten, B.; Naaber, P.; Sepp, E.; Mikelsaar, M. The Intestinal Microflora in Allergic Estonian and Swedish 2-year-old Children. *Clin. Exp. Allergy* **1999**, *29*, 342–346.

14. Bosscher, D.; Breynaert, A.; Pieters, L.; Hermans, N. Food-based Strategies to Modulate the Composition of the Intestinal Microbiota and their Associated Health Effects. *J. Physiol. Pharmacol.* **2009**, *60*, 5–11.

15. Bouhnik, Y.; Achour, L.; Paineau, D.; Riottot, M.; Attar, A.; Bornet, F. Four-week Short Chain Fructo-oligosaccharides Ingestion Leads to Increasing Fecal *Bifidobacteria* and Cholesterol Excretion in Healthy Elderly Volunteers. *Nutr. J.* **2007**, *6*, 42.

16. Bunout, D.; Hirsch, S.; Pia de la Maza, M.; Munoz, C.; Haschke, F.; Steenhout, P.; Klassen, P.; Barrera, G.; Gattas, V.; Petermann, M. Effects of Prebiotics on the Immune Response to Vaccination in the Elderly. *J. Parente. Enter. Nutr.* **2002**, *26*, 372–376.

17. Campbell, J. M.; Fahey, G. C.; Wolf, B. W. Selected Indigestible Oligosaccharides Affect Large Bowel Mass, Caecal and Faecal Shortchain Fatty Acids, pH and Microflora in Rats. *J. Nutr.* **1997**, *127*, 130–136.

18. Del Piano, M.; Ballare, M.; Montino, F.; Orsello, M.; Garello, E.; Ferrari, P.; Masini, C.; Strozzi, G.P., Sforza, F. Clinical Experience with Probiotics in the Elderly on Total Enteral Nutrition. *J. Clin. Gastroenterol.* **2004**, *38*(6 Suppl.), S111–S114.

19. Dominguez-Bello, M. G.; Costello, E. K.; Contreras, M.; Magris, M.; Hidalgo, G.; Fierer, N.; Knight, R. Delivery Mode Shapes the Acquisition and Structure of the Initial Microbiota Across Multiple Body Habitats in Newborns. *Proc. Natl. Acad. Sci. USA* **2010**, *107*, 11971–11975.

20. Eckburg, P. B.; Bik, E. M.; Bernstein, C. N.; Purdom, E.; Dethlefsen, L.; Sargent, M.; Gill, S. R.; Nelson, K. E.; Relman, D. A. Diversity of the Human Intestinal Microbial Flora. *Science* **2005**, *308*, 1635–1638.

21. Eerola, E.; Mottonen, T.; Hannonen, P.; Luukkainen, R.; Kantola, I.; Vuori, K.; Tuominen, J.; Toivanen, P. Intestinal Flora in Early Rheumatoid Arthritis. *Br. J. Rheumatol.* **1994**, *33*, 1030–1038.

22. Englyst, H. N.; Macfarlane, G. T. Breakdown of Resistant and Readily Digestible Starch by Human Gut Bacteria. *J. Sci. Food Agric.* **1986**, *37*, 699–706.

23. Favier, C. F.; Vaughan, E. E.; De Vos, W. M.; Akkermans, A. D. Molecular Monitoring of Succession of Bacterial Communities in Human Neonates. *Appl. Environ. Microbiol.* **2002**, *68*, 219–226.

24. Gibson, G. R. Dietary Modulation of the Human Gut Microbiota Using Prebiotics. *Br. J. Nutr.* **1998**, *80*, S209–212.

25. Gibson, G. R.; Roberfroid, M. B. Dietary Modulation of the Human Colonic Microbiota: Introducing the Concept of Prebiotics. *J. Nutr.* **1995**, *125*, 1401–1412.

26. Grönlund, M. M.; Lehtonen, O. P.; Eerola, E.; Kero, P. Faecal Microflora in Healthy Infants Born by Different Methods of Delivery: Permanent Changes in Intestinal Flora After Cesarean Delivery. *J. Pediatr. Gastroenterol. Nutr.* **1999**, *28*, 19–25.

27. Guarner, F.; Malagelada, J. R. Gut Flora in Health and Disease. *Lancet* **2003**, *361*, 512–519.

28. Hayashi, H.; Sakamoto, M.; Benno, Y. Faecal Microbial Diversity in a Strict Vegetarian as Determined by Molecular Analysis and Cultivation. *Microbiol. Immunol.* **2002**, *46*, 819–831.

29. Hopkins, M. J.; Macfarlane, G. T. Changes in Predominant Bacterial Populations in Human Faeces with Age and *Clostridium difficile* Infection. *J. Med. Microbiol.* **2002**, *51*, 448–454.

30. Kalliomäki, M.; Kirjavainen, P.; Eerola, E.; Kero, P.; Salminen, S.; Isolauri, E. Distinct Patterns of Neonatal Gut Microflora in Infants in Whom Atopy Was and Was not Developing. *J. Allergy Clin. Immunol.* **2001**, *107*, 129–134.

31. Kanazawa, K.; Konishi, F.; Mitsuoka, T.; Terada, A.; Itoh, K.; Narushima, S.; Kumemura, M.; Kimura, H. Factors Influencing the Development of Sigmoid Colon Cancer. Bacteriologic and Biochemical Studies. *Cancer* **1996**, *77*(8 Suppl), 1701–1706.

32. Kirjavainen, P. V.; Arvola, T.; Salminen, S. J.; Isolauri, E. Aberrant Composition of Gut Microbiota of Allergic Infants. A Target of Bifidobacterial Therapy at Weaning? *Gut* **2002**, *51*, 51–55.

33. Kleessen, B.; Sykura, B.; Zunft, H. J.; Blaut, M. Effects of Inulin and Lactose on Fecal Microflora, Microbial Activity, and Bowel Habit in Elderly Constipated Persons. *Am. J. Clin. Nutr.* **1997**, *65*, 1397–1402.

34. Larsen, N.; Vogensen, F. K.; van den Berg, F. W.; Nielsen, D. S.; Andreasen, A. S.; Pedersen, B. K.; Al-Soud, W. A.; Sorensen, S. J.; Hansen, L. H.; Jakobsen, M. Gut Microbiota in Human Adults With Type 2 Diabetes Differs From Non-diabetic Adults. *PLoS One* **2010**, *5*, e9085.

35. Ley, R. E.; Peterson, D. A.; Gordon, J. I. Ecological and Evolutionary Forces Shaping Microbial Diversity in the Human Intestine. *Cell* **2006**, *124*, 837–848.

36. Lode, H.; Vonder Höh, N.; Ziege, S.; Borner, K.; Nord, C. E. Ecological Effects of Linezolid Versus Amoxicillin/Clavulanic Acid on the Normal Intestinal Microflora. *Scand. J. Infect. Dis.* **2001**, *33*, 899–903.

37. Macfarlane, G. T.; Macfarlane, L. E. Acquisition, Evolution and Maintenance of the Normal Gut Microbiota. *Digestive Dis.* **2009**, *27*, 90–98.

38. Macfarlane, G. T.; Macfarlane, S. Human Colonic Microbiota: Ecology, Physiology and Metabolic Potential of Intestinal Bacteria. *Scand. J. Gastroenterol.* **1997**, *222*, 3–9.

39. Marques. M. T.; Wall. R.; Ross. P. R.; Fitzgerald. F. G.; Ryan. A. C.; Stanton, C. Programming Infant Gut Microbiota: Influence of Dietary and Environmental Factors. *Curr. Opin. Biotechnol.* **2010**, *21*, 149–156.

40. Moore, W. E.; Moore, L. H. Intestinal Floras of Populations That Have a High Risk of Colon Cancer. *Appl. Environ. Microbiol.* **1995**, *61*, 3202–3207.

41. Mueller, S.; Saunier, K.; Hanisch, C.; Norin, E.; Alm, L.; Midtvedt, T.; Cresci, A.; Silvi, S.; Orpianesi, C.; Verdenelli, M. C.; Clavel, T.; Koebnick, C.; Zunft, H. J.; Dore, J.; Blaut, M. Differences in Faecal Microbiota in Different European Study Populations in Relation to Age, Gender, and Country: A Cross-sectional Study. *Appl. Environ. Microbiol.* **2006**, *72*, 1027–1033.

42. Nord, C. E.; Sillerstrom, E.; Wahlund, E. Effect of Tigecycline on Normal Oropharyngeal and Intestinal Microflora. *Antimicrob. Agents Chemother.* **2006**, *50*, 3375–3380.

43. Ormisson, A.; Sepp, E.; Siigur, U.; Varendi, H.; Mikelsaar, M. Impact of Early Life Nutrition on Growth and Intestinal Microflora Composition in Low-birth-weight Infants. *Scand. J. Nutr.* 1997, *41*, 71–74.

44. Ouwehand, A. C.; Tiihonen, K.; Saarinen, M.; Putaala, H.; Rautonen, N. Influence of a Combination of *Lactobacillus acidophilus* NCFM and Lactitol on Healthy Elderly: Intestinal and Immune Parameters. *Br. J. Nutr.* **2009**, *101*, 367–375.

45. Parracho, H., Bingham, M. O.; Gibson, G. R.; McCartney, A. L. Differences Between the Gut Microflora of Children with Autistic Spectrum Disorders and That of Healthy Children. *Med. Microbiol.* **2005**, *54*, 987–991.

46. Peltonen, R.; Ling, W. H.; Hanninen, O.; Eerola, E. An Uncooked Vegan Diet Shifts the Profile of Human Fecal Microflora: Computerized Analysis of Direct Stool Sample Gas-liquid Chromatography Profiles of Bacterial Cellular Fatty Acids. *Appl. Environ. Microbiol.* **1992**, *58*, 3660–3666.

47. Penders, J.; Thijs, C.; Vink, C.; Stelma, F. F.; Snijders, B.; Kummeling, I.; van den Brandt, P. A.; Stobberingh, E. E. Factors Influencing the Composition of the Intestinal Microbiota in Early Infancy. *Pediatrics* **2006**, *118*, 511–521.

48. Sairanen, U.; Piirainen, L.; Nevala, R.; Korpela, R. Yoghurt Containing Galactooligosaccharides, Prunes and Linseed Reduces the Severity of Mild Constipation in Elderly Subjects. *Eur. J. Clin. Nutri.* **2007**, *61*, 1423–1428.

49. Salminen, S.; Gibson, G. R.; McCartney, A. L.; Isolauri, E. Influence of Mode of Delivery on Gut Microbiota Composition in Seven Year Old Children. *Gut* **2004**, *53*, 1388–1389.

50. Sepp, E.; Julge, K.; Vasar, M.; Naaber, P.; Björksten, B.; Mikelsaar, M. Intestinal Microflora of Estonian and Swedish Infants. *Acta Paediatr.* **1997**, *86*, 956–961.

51. Sghir, A.; Chow, J. M.; Mackie, R. I. Continuous Culture Selection of *Bifidobacteria* and *Lactobacilli* from Human Faecal Samples Using Fructooligosaccharide as a Selective Substrate. *J. Appl. Microbiol.* **1998**, *85*, 769–777.

52. Sullivan, A.; Edlund, C.; Svenungsson, B. Effect of Perorally Administered Pivmecillinam on the Normal Oropharyngeal, Intestinal and Skin Microflora. *J. Chemother.* **2001**, *13*, 299–308.

53. Tiihonen, K.; Tynkkynen, S.; Ouwehand, A.; Ahlroos, T.; Rautonen, N. The Effect of Ageing with and Without Non-steroidal Anti-inflammatory Drugs on Gastrointestinal Microbiology and Immunology. *Br. J. Nutr.* **2008**, *100*, 130–137.

54. Tomkins, A. M.; Bradley, A. K.; Oswald, S.; Drasar, B. S. Diet and the Faecal Microflora of Infants, Children and Adults in Rural Nigeria and Urban UK. *J. Hyg.* **1981,** *86*, 285–293.

55. Vaahtovuo, J.; Munukka, E.; Korkeamäki, M.; Luukkainen, R.; Toivanen, P. Fecal Microbiota in Early Rheumatoid Arthritis. *J. Rheumatol.* **2008,** *35*, 1500–1505.

56. Van Loo, J. A. Prebiotics Promote Good Health: The Basis, the Potential, and the Emerging Evidence. *J. Clin. Gastroenterol.* **2004,** *38*, S70–75.

57. Wong, J. M.; de Souza, R.; Kendall, C. W.; Emam, A.; Jenkins, D. J. Colonic Health: Fermentation and Short Chain Fatty Acids. *J. Clin. Gastroenterol.* **2006,** *40*, 235–243.

58. Woodmansey, E. J. Intestinal Bacteria and Ageing. *J. Appl. Microbiol.* **2007,** *102*, 1178–1186.

59. Ying Hong. P.; Lee. W. B.; Aw. M.; Shek. C. P. L.; Yap. C. G.; Chua. Y. K.; Tsoliu. W. Comparative Analysis of Faecal Microbiota in Infants with and Without Eczema. *PLoS One* **2010,** *5*, 9964.

CHAPTER 6

NOVEL DRYING TECHNOLOGIES IN THE DAIRY INDUSTRY

RUPESH S. CHAVAN[1], SHRADDHA BHATT[2,*], and
SHUBHNEET KAUR[3]

[1]*Department of Quality Assurance, Mother Dairy Junagadh,
Junagadh 362001, Gujarat, India*

[2]*Department of Biotechnology, Junagadh Agricultural University,
Junagadh 362001, Gujarat, India*

[3]*Department of Food and Nutrition, Lady Irwin College, Sikandra
Road, New Delhi 110001, India*

Corresponding author. E-mail: rschavanb_tech@rediffmail.com

CONTENTS

ABSTRACT

Spray- and freeze-drying is used to produce a wide range of products including heat sensitive materials. The products produced by spray-drying include: pharmaceutical, such as antibiotics, analgesics, vaccines, vitamins, and catalysts; and foods such as milk and milk products, food color, food supplement, soup mixes, spice and herb extracts, coffee, tea, and sweetener. Spray-dried food products are appealing, retain nutritional qualities, and are convenient to consume. Drying method affects the properties of milk powder such as its solubility in cold water, its flavor, nutrient content, and its bulk density. Functional properties of milk powder can be manipulated by specific heat treatment and are discussed in detail in present chapter.

6.1   INTRODUCTION

The most common method used today for drying milk is spray-drying. The milk is thereby usually first evaporated in order to decrease the energy consumption in the drying step. The main reason for drying milk is to increase its shelf life but also to reduce costs for transportation due to a lower volume and weight of the powder. Milk powders can be reconstituted with water into liquid milk whereas the reconstituting ability is an important requirement on milk powders. The high nutritional value, high protein content, and low fat content make skim milk powder a favorable ingredient in dry mixes, bakery products, confections, infant formulas, sports and nutrition foods, ice cream, cheese, and meat products wherein the functional properties such as emulsifying, thickening, foaming, browning, or its solubility are utilized.

This chapter discusses novel technologies for drying process in dairy industry.

6.2   SPRAY-DRYING

Spray-drying is used to produce a wide range of products including heat-sensitive materials. The flexibility of dryer designs provides opportunities to produce the powders that consistently meet industrial specifications. Spray-dried food products are appealing, retain nutritional qualities, and are convenient to consume.[6,8,10] The products produced by spray-drying include:

- Pharmaceutical such as antibiotics, analgesics, vaccines, vitamins, and catalysts
- Chemicals such as carbides, ferrite, nitrides, tannins, fine organic/inorganic chemicals detergent, and dyestuffs
- Ceramic, including advanced ceramic formulations
- Foods such as milk and milk products, food color, food supplement, soup mixes, spice and herb extracts, coffee, tea, and sweetener

The process of spray-drying is used almost for drying all kind of food products especially dairy products. The fluid which is to be dried is pumped through atomizer/nozzle into the drying chamber where simultaneously compressed, filtered, dry, and hot air comes in contact with the atomized liquid and dries the liquid in small spherical-shaped particles. The evaporation of moisture takes place very quickly and the particles so formed are separated by cyclone separator, whereas lighter particles or dust form goes along with air/gas. Lighter particles are then collected by allowing the gas through scrubbers/bag filter and the air/gas without particles is allowed to be discharged into the atmosphere.

Spray dryers processes consist of primary three processes, namely, atomization, air–liquid contact with evaporation, and removal of particles from gas/air. The major problems faced during spray-drying are: stickiness during drying of whey, may be due to low pH and high content of lactic acid and stickiness due to the presence of lactose and protein, which causes the formation of liquid bridges between particles as an effect of temperature and pressure. In spite of these challenges, spray-drying technique is the most promising techniques used at commercial scale to meet the market requirements by producing large quantity of powders. However, major attention is given to design the drying chamber. Power handling and packing as most powders are both thermoplastic and hygroscopic in nature. The nature of such powders gives rise to problems such as adhesion to dryer walls, difficult handling, caking, etc.[9,11–13]

6.3 TYPES OF SPRAY-DRYING

Different designs are available for drying of milk and milk constitutes. The primary concern in the design of a spray-drying system is that the products should be sufficiently dry before contacting the chamber wall. Design to achieve a specific type of spray-dried product, the operation of the atomizer, air disperser, drying chamber, and powder collection systems should be considered during selecting a spray dryer.

6.3.1 SINGLE-STAGE DRYING

The simplest installation for making milk powder is a single-stage spray dryer. In this system, moisture removal from the concentrate to the required final moisture takes place in the spray-drying chamber and is usually fitted with a pneumatic conveying system for powder conveying. To understand better single-stage drying principle, a single-stage laboratory scale spray dryer can be used for manufacturing of milk powder (Fig. 6.1).[17,18] The spray dryer is fitted with a nozzle atomizer with inside diameter of 0.5 mm. The feed can be metered into the dryer by means of a peristaltic pump. Inlet drying air, after passing through an electrical heater, flows concurrently with the spray through the main chamber.

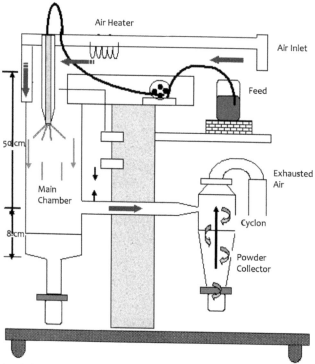

FIGURE 6.1   Schematic view of laboratory scale spray dryer (Source: SprayMate).

The main chamber is made of thick transparent glass with an inside diameter of 10.50 cm and outside diameter of 12.50 cm and a total height of 58 cm. The distance between the tip of the atomizer and the axis of the side exit tube is 31.50 cm. The bottom of the chamber is cone shaped and makes an angle of 60° with the walls. A thick transparent glass cyclone air separator/powder recovery system is used. The cyclone separator has a height of 22 cm and an inlet of 4 cm diameter tube, which is connected to the main chamber. Dried powder samples can be collected in a thick transparent glass bottle from the base of the cyclone with a height of 13 cm and outer diameter of 10 cm.

6.3.2   TWO-STAGE DRYING

In a two-stage drying system, the powder is produced in the same fashion as in single-stage drying system, but the major difference is the replacement of the pneumatic conveying system with a fluid bed dryer. The moisture content of the powder leaving the drying chamber is usually kept between 2% and 3% higher than the final moisture content, which is removed in the fluid bed dryer. Drying of milk in two stages was originally developed to obtain agglomerated powders in straight-through processing, so that the advantage of product quality improvement could be combined with the better process economy. The milk powder obtained by using either single-stage or two-stage installations is predominantly dusty and is difficult to reconstitute. Two-stage dried milk powder is however coarser due to bigger primary particles and the presence of some agglomerates, which make it slightly higher reconstituable as compared to single-stage milk powder. However, the biggest difference between these two powders is in the properties that are influenced by heat exposure during drying including solubility index, content of occluded air, and bulk density. The advantages of the two-stage drying over single-stage drying are: higher capacity/kg drying air, better economy, low powder emission, and better product quality in respect to solubility, high bulk density, lower amount of free fat, and lower amount of occluded air.

6.3.3   THREE-STAGE DRYING

Three-stage drying is an extension of the two-stage concept developed to achieve even greater savings in plant operation costs. The three-stage dryer involves transfer of the second drying stage into the base of the spray-drying chamber and having the final drying and cooling conducted in the third stage located outside the drying chamber. There are two main types of three-stage dryers: spray dryers with integrated fluid bed and spray dryers with integrated belt.

6.4   EVAPORATION

Falling film evaporator is preferred for the concentration of milk, as the liquid is evenly distributed on the inner surface of a tube, which flows downwards forming a thin film. Steam formed during evaporation condenses and flows downwards on the outer surface of the tube. The tubes are installed side by side, which is jacketed with a stainless steel jacket forming a calandria. The steam is introduced through the jacket in the space between the tubes thus forming the heating section. Concentrated milk and the vapor leave the calandria at the bottom part, from where the main proportion of the concentrated liquid is discharged. The remaining part enters the subsequent separator tangentially together with the vapor. The separated concentrate is discharged, and the vapor leaves the separator from the top. The heating steam, which condenses on the outer surface of the tubes, is collected as condensate at the bottom part of the heating section, from where it is discharged by means of a pump. Before starting an evaporator, following considerations are necessary:

- Start the vacuum pump with cooling tower and condensing unit in the circuit. Ensure seal water in vacuum pump.
- Check the direction of motor, whether vacuum break valve is closed and suction inlet is open.
- Start all the feed line pumps in descending order of calandria installed, that is, from the last calandria to the first. This avoids flooding in calandria.
- Open the steam valve followed by water inlet valve and allow formation of condensate.
- After condensate has formed, start the feed pump and stabilize up to required total solids.

6.4.1   THERMOVAPOR RECOMPRESSION

Another way of saving energy is by using a thermocompressor (TVR), which increases the temperature/pressure level of the vapor, that is, compresses the vapor from a lower pressure to a higher pressure by using steam of a higher pressure than that of the vapor. Thermocompressors operate at very high steam flow velocities and have no moving parts. The construction is simple, the dimensions are small, and the costs are low. The best efficiency in the thermocompressor, that is, the best suction rate, and thereby a good economy, is obtained when the temperature difference (pressure difference) between the boiling section and the heating section is low.

6.4.2   MECHANICAL VAPOR RECOMPRESSION

As an alternative to the thermocompressor, the mechanical vapor compressor is used extensively in dairy industry. The applied energy for the compressor is usually electricity, but also diesel motors are used. Mechanical vapor recompression (MVR) compressor is a fast-revolving high-pressure fan (3000 rpm) capable of operating under vacuum. At low boiling temperatures, the volume of the vapors is enormous. As it is essential to operate an MVR unit at a low overall temperature difference between the vapor evolved from the product and the heating medium as a result of the compression, it is a must that the boiling point elevation of the product is kept to a minimum, as this would otherwise even further minimize the temperature difference available for the evaporation. The MVR evaporator is in this context very often used as precondenser of milk products for transport purposes, where the required solids content is in the range of 30–35% and thus the boiling point elevation is limited. With the concentrate leaving the plant at low-temperature, this kind of installation is a strong competitor to hyperfiltration.

A flowchart for five-effect evaporation fitted with one thermovapor recompression (TVR) followed by spray-drying is explained in Figure 6.2. The main components and their functions are explained as below:

a. Drying chamber

The drying chamber is usually compact type with a centrifugal atomizer (17,000 rpm). The ring-formed back-mix bed is placed at the bottom of a conventional chamber cone round the exhaust duct placed in the center.

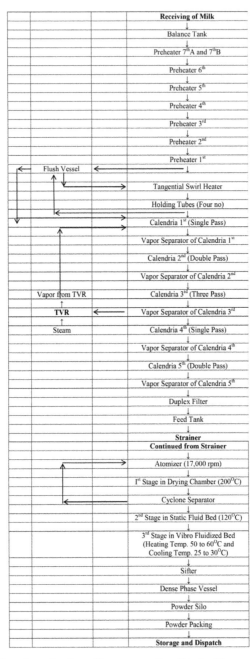

FIGURE 6.2 Flowchart for five-effect evaporation fitted with one TVR followed by spray-drying.

Thus, there are no parts in the cone obstructing the airflow, and this together with the spouts from the fluidized powder layer keeps the chamber cone free from deposits even when handling sticky powders with high moisture contents. The cylindrical part of the chamber is kept clean by a wall sweep system, where a small amount of air is introduced at high velocity tangentially into the cylindrical part of the chamber through specially designed air nozzles pointing in the same direction as the rotation of the primary drying air. The powder is discharged continuously from the static fluid bed by overflowing an adjustable powder weir, thus maintaining a certain fluidized powder level. Due to the low air outlet temperature, the drying economy is greatly improved.[37,38,40]

b. Air disperser

The main function of air dispenser is to dispense the air uniformly. Square cross section diminishes gradually to the direction of the gas flow. Baffles help for better gas distribution and minimize the pressure drop.

c. Fine returns air disperser

The main purpose is to reintroduce the fine returns from the cyclone into the main stream of product deriving from the atomizer. Bulk density can be adjusted by adjusting the deflector of the fine returns air disperser (FRAD) unit.

d. Wall sweep arrangement

It helps to prevent deposition of powder on the wall of drying chamber.

e. Electric hammer

It prevents deposition of powder at the inside wall of drying chamber.

In similar fashion, a spray-drying fitted with four-effect evaporation with an incorporated MVR and TVR is explained in Figure 6.3. The process is explained below:

a. Preheating

Skim milk coming from the silo is first heated with the outgoing condensate to 45°C followed by preheating in a thermophilic preheater by

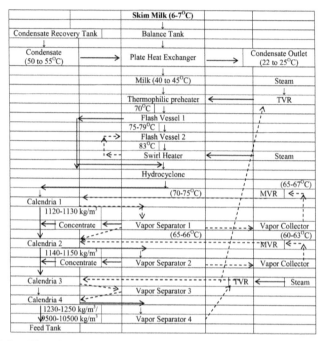

FIGURE 6.3a   Flowchart for four-effect evaporation with an incorporated MVR and TVR.

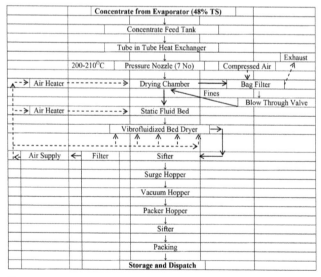

FIGURE 6.3b   Flowchart for four-effect evaporation with an incorporated MVR and TVR.

steam infusion. This inhibits the growth of thermophilic bacteria, which can proliferate in the milk when it is held for a longer duration at high temperatures. To save energy, the hot product is used to preheat the colder incoming product in a direct contact flash regenerative vessel at lower Δt.

b. Evaporation

The evaporator is a four-effect evaporator with wrap around vapor separators, which are integrated into the base of calandria which saves a lot of space (around 30%). The first effect consists of two passes, second has four, and the remaining two have one each. The first two evaporators have tubes of different lengths in different passes. The vapors generated in the first pass and the second pass have two different entrances to the vapor collector. The vapors (in case of first and second calandria) that are collected in vapor collector are compressed by the MVR and sent back to the shell side of the same effect. The third and fourth calandrias are called the *finishers* as they help to adjust the final total soil content to be subjected for drying.

c. Tubular heating

The concentrate is then preheated to 70°C before entering the spray dryer in a tubular heat exchanger. Preheating helps to lower the viscosity and density of the concentrate, keeping its total solids intact.

d. High-pressure reciprocating pump

After tubular heating, the concentrate is fed into a five-piston reciprocating pump from where it is pumped to the nozzles at a pressure of 150–200 bars which help to manufacture milk powder with low amount of occluded air, higher bulk density, and larger particle size.

e. Pressure nozzle atomizer

Usually, five to eight pressure nozzles set is used to spray the concentrate and the hole size of each nozzle is about 2.8 mm. The angle of swirl chamber and orifice plate can be changed so as to adjust the extent of agglomeration.

f. Drying chamber

In order to improve the dryer efficiency even further without an increased rate of powder deposits, a completely new spray dryer concept named as multistage dryer (MSD) is designed. The dryer operates with three drying stages, each adapted to the moisture content prevailing during the drying process. In the preliminary drying stage, cocurrent nozzles placed in the hot drying air duct atomize the concentrate. The air enters the dryer vertically through the air disperser at high velocity, ensuring optimal mixing of the atomized droplets with the drying air. The particles reach moisture content of 6–15%. The high-velocity air inlet is creating a venturi effect, the vacuum of which will suck surrounding air with entrained fines particles into the wet atomizer cloud. This will result in a "spontaneous secondary agglomeration." The circular static fluid bed is supplied with air at a sufficient velocity and temperature for the second-stage drying. The powder is then discharged into a Vibro-Fluidizer for the final drying and subsequent cooling. The drying chamber is fitted with removable insulations which is an added advantage. The older insulation material is sometimes the main reason for causing bacteriological hazards due to crack in the chamber wall, and old wet insulation material is often the reason for powder deposits in the drying chamber due to cold spots. The most radical way to avoid any problems is to remove the insulation completely; however, the drawbacks are: the evaporative capacity drops by up to 10% due to higher heat loss, the temperature in the dryer room can in some cases exceed 50°C and is thus to be considered a very unpleasant working environment.

In view of the above, Niro has developed a removable insulation panel with an air gap between the chamber wall and the panels. This makes it possible to inspect, on a regular basis, the outside of a spray-drying chamber in order to detect any leaks or cracks.[30,33]

g. Bag filter

The drying air from the preliminary drying stage and the static fluid bed leaves the chamber from the top passing through a bag filter, thus helping in reduction of powder dust. Around 240 bags are placed in the bag filter unit, which is CIP-able and has recently been introduced. The exhaust from the spray dryers enters the bag filter tangentially, thus the housing of the bag filter acts as a cyclone adding to the efficiency. The stuck powder can be removed by a stream of compressed air (6 bars) which enters each bag through a reverse jet nozzle which helps to keep lower down the pressure drop. The powder, which gets stuck to the portion between the bags, can also be removed by the compressed air entering from different parts of the bag filter unit. The powder which falls down from the bags can is collected on a perforated plate, which is constantly supplied with hot air (cone heating), which helps to maintain the powder in the fluidized state which is then discharged and pneumatically conveyed for agglomeration.

h. Carbon monoxide system for fire and explosion safety

The carbon monoxide (CO) system for fire safety is the latest technology for safety in spray-drying systems. Smoldering milk powder lumps deposited on dryer walls or falling through the hot air could potentially start a fire or even an explosion in milk powder dryers. Especially where high production throughputs are desired, it is well known that powder emits CO. The exhaust air of the dryer is continuously measured for CO concentration using a highly sensitive CO analyzer. If the difference of concentration of CO in the inlet and outlet rises above 1 ppm, fire alarms start, and water is sprayed automatically into the chamber. Safety vents are also provided for explosion safety. Thin stainless steel plates which can bear a pressure (min. 80 mbar and max. 115 mbar) are attached to the drying chamber which burst out if the pressure inside the drying chamber increases which avoids fire explosion.

6.5   ATOMIZATION

The atomization is a very complex process in spray-drying. The atomizer sprays the liquid feed into a stream of hot air or other gas. The droplet size depends on the liquid properties and the atomizer's operating parameters such as pressure drop and flow rate. The choice of atomizer type depends on the properties of liquid such as concentration and viscosity as well as the droplet size distribution desired.[16,24,25,42]

The liquid feed and drying gases can enter the chamber in the same direction (cocurrent) or in opposite directions (countercurrent) or a combination of both directions. The countercurrent option is not applicable to food drying because the hottest air will contact the driest particles and will result in unacceptable heat damage in the product.

There are three basic atomizer types: a single-fluid pressure nozzle, a two-fluid (also known as pneumatic) nozzle, and a rotary atomizer (also known as a spinning disc or centrifugal atomizer).

a. Single-fluid pressure nozzle

A single-fluid pressure nozzle is the most common atomization device, which produces a small-sized distribution of droplets. The pressure energy within the liquid feed is converted into the kinetic energy of a moving thin liquid sheet, which is then forced through an orifice. Variation of feed pressure and nozzle type can control the particle size of liquid feed. Single-fluid nozzles can be selected and installed to provide a range of spray angles and directions.

b. Two-fluid or pneumatic nozzle

The two-fluid or pneumatic nozzle is suitable for a more viscous or abrasive feeds. High-velocity compressed air or gas impacts the liquid feed. The particle size varies depending on the ratios of the atomizing gas and feed.

c. Rotary or centrifugal atomizer

The rotary atomizer is considered to be the most flexible atomizer as it is suitable for a wide range of products including fruit juices with discrete pieces of pulp. The liquid feed is accelerated to speed between 5000–25,000 rpm in the atomizer wheel and is ejected from the rim of the centrifugal atomizer at 60–200 m/s and droplets break down into smaller particles. Rotary atomizer produces large, easily dissolved dried particles. However, the liquid feed needs to be homogenized before atomizing to reduce clogging problems. The advantages of pressure nozzles can be summarized as follows:

- Powder contains low amount of occluded air.
- Powder with high bulk density can be manufactured.
- Improved flowability, especially in whole milk.
- Tendency to give less deposits in the drying chamber when sticky products like whey powder are produced.
- Helps to produce large-sized particles in powder.

- If a dual feed/nozzle system is used, the drying plant can operate continuously 24 h/day for weeks without stop; only the feed line/nozzles are wet-cleaned after 20 h.

The rotary atomizer has been used in the dairy industry for many years. The main advantages are:

- Flexibility as to throughput
- Ability to handle large quantities
- Ability to handle highly viscous concentrates
- Different wheel designs giving different powder characteristics
- Ability to handle products with crystals
- Higher solids content in the feed is possible, therefore, better economy

6.6 MIXING OF SPRAY AND DRYING MEDIUM

The feed is atomized directly into the hot air stream. As the droplets pass through the hot air flow, the moisture evaporates rapidly. The time and distance required to complete the drying of the droplet spray depend on the rate of heat and mass transfer between the droplets and the drying medium. Heat and mass transfer during drying occurs in the air and vapor films surrounding the droplet. There is the protective envelope of vapor, which keeps the particle at the saturation temperature, and as long as the particle does not become completely dry, evaporation still takes place. Thus, heat sensitive products can be spray dried at relatively high air temperatures without being harmed.[34]

6.7 SEPARATION OF DRIED PRODUCT

Separation of dried product from the air is the final phase of spray-drying. After evaporation, the large particles fall to the bottom of the chamber and are collected while the fine particles are entrained with the exhaust air and are generally collected by passing the air through a series of external cyclones, electrostatic precipitator, scrubbers, or bag filters. Fine particles are bagged or returned to an agglomeration process in the dryer. Due to variations in air temperature, air velocity, and air humidity throughout the drying chamber volume, all droplets are subjected to different local drying conditions and thus have individual drying histories; some particles being spherical or near spherical in shape, others more misshapen, collapsed, or expanded.

The SANICIP™ CIP-able bag filter is replacing the cyclones and has now reached a point where it is setting the standard for almost all dryers. The SANICIP bag filter is of reverse jet type made of stainless steel. It consists of cylindrical bag housing with scrolled air inlet, clean air plenum on top, and a conical bottom with fluidized powder discharge. During operation, the product collected on the outside of the filter material is removed by a compressed air jet stream blown into each bag via a special reverse jet air nozzle (patented) positioned above each bag. The bags are clean-blown individually or four together, resulting in even discharge of powder and using higher air-to-cloth ratios. Advantages of the SANICIP filter are:

- Low-pressure loss across the bag filter and thus the entire exhaust system which helps to reduce energy requirement and lower noise pollution.
- Air-to-cloth ratio is higher which helps maximum amount of powder to be blown down and recollected for further agglomeration.
- Increased powder recovery from spray-drying operations with a uniform quality and less amount of cyclone sweep powder is generated.
- Wear and tear of bags is avoided by lower mechanical impact due to usage of pressurized air for blowing down the deposited/separated powder from the exhausted gas.
- Variable bag length is available to suit specific building and production requirements.

6.8   FREEZE-DRYING

During freeze-drying, the product is dried via sublimation. The product is frozen before freeze-drying in order to allow ice vaporization. This is accomplished by creating a certain condition below the triple point, that is, in vacuum. Usually, a temperature of or below $-10°C$ at a pressure of or below 2 mmHg (266 Pa) is used during freeze-drying. In comparison with drum drying and spray-drying, this drying method is the best alternative for maintaining powder quality such as aroma, flavor, nutritional values, and furthermore limitation of the Maillard reaction. Freeze-dried milk powder is also the one with the best reconstitution ability due to its porosity and the rigid surface which arises after sublimation.

A freeze dryer consists of a freezing chamber, freeze-drying chamber with several heatable shelves where the product is placed, an ice condenser where the lyophilized vapor becomes ice, a vacuum pump to maintain vacuum, and a heating/cooling system. Freeze-drying is an expensive drying method due to slow drying rate and is therefore not applicable for production of skim milk powder. However, it is used for drying other foodstuffs which otherwise are difficult to dry such as soups, coffee, and onions. Freeze-drying is also used to dry pharmaceutical products.[49–51]

In recent years, spray freeze-drying of milk is done because of the advantages of freeze-drying. However, it is not widely used because of its energy demand.

Pulse combustion spray-drying has been reported as a relatively new but not yet popular development of spray-drying. It can be used to dry milk. Also, other drying techniques such as Filtermat™ drying, tall-form drying, para-flash drying, and integrated filter dryer have been used. All these towers have characteristics related to the specific properties of the product being dried (e.g., high fat content, starch, maltodextrin, egg, and hygroscopic products). The Filtermat spray dryer combines a cocurrent nozzle tower dryer with a built-in conveyor belt. The residence time while the powder is moved along with the belt is several minutes, offering sufficient time to complete powder drying and cooling while maintaining the required powder temperatures.

6.9 FUNCTIONAL PROPERTIES OF MILK POWDERS

Milk powders are frequently used for convenience in applications for transportation, handling, processing, and for product formulations. Powders possess physical and functional properties, including powder structure, particle size distribution, powder density, bulk density, particle density, occluded air, interstitial air, flowability, rehydration (wettability, sinkability, dispersibility, and solubility), hygroscopicity, heat stability, emulsifying properties, glass transition temperature, water activity, stickiness, caking, and even scorched particles.[1–5,20,28,54]

The properties of powder products are dependent on various factors, namely, equipment and processing conditions, heat stability and pretreatment of the concentrate, outlet air temperature, and contact time, the extent of precrystallization, agglomeration, etc. Physical properties of a powder notably density, flowability, and reconstitutability are determined by the interrelationship between particle properties, such as chemical composition, shape, size, porosity, and surface roughness.[32,35,36]

6.9.1 BULK DENSITY

Bulk density is a property of economical, commercial, and functional importance. Also known as apparent or packing density, bulk density is regarded as mass of powder, which occupies a fixed volume and normally expressed in g/mL or g/cc. The total volume includes particle volume, interparticle void volume, and internal pore volume. Bulk density is a very complex property of powders and depends upon particle density, amount of interstitial air content, occluded air content, and powder flowability. A high degree of whey protein denaturation has been reported to result in low occluded air content giving a high bulk density powder. Similarly, increase in feed solids also result in high bulk density, whereas increase in inlet air drying temperature and feed aeration decreases the bulk density. According to Jayaprakasha et al.,[21] packed bulk density of spray-dried whey powder varied from 0.51 to 0.58 g/cc.

Buma[7] reviewed the available literature on densities of milk constituents and found that the particle density of casein varied from 1.25 to 1.46 g/cm[3]. Munro[39] studied the densities of various types of casein using gravimetric method. The reported values of lactic, sulfuric, and rennet casein were 1.36, 1.37–1.40, and 1.44–1.47 g/cm, respectively. It is also reported that high particle density contributes to high bulk density, improved reconstitution quality of powders in water, better shelf life of fat-filled products such as whole milk powder, and reduction in stack losses of spray dryers. A two-stage spray-drying system is reported to enable operation with high feed concentration and still keeping a low particle temperature distribution which is favorable for obtaining high bulk density.[41] Buma[7] explained that the presence of air in the atomized droplets causes occluded air in dried powder particles. The interstitial air in the powder is related to the bulk density of powder.[52] The more the interstitial air in a powder, the lower is its bulk density. The greater the degree of agglomeration, the more interstitial air is contained in the powder and the lower is the bulk density.[41] The bulk density of the powdered products is also affected by its storage.

6.9.2   DISPERSIBILITY

It reflects the ability of the wetted aggregates of powder particles to become uniformly dispersed when in contact with water. This property also helps in determining the instantaneous of a powder. It is related to the ease at which lumps and agglomeration fall apart. Large particle sizes of dried products are generally associated with good dispersibility.[7] A high free-fat content is usually considered unfavorable for better dispersibility of powders.[53] Dispersibility of whey-mushroom soup powder and whey-kinnow juice powder[26] was declined during storage, the decline being more pronounced at higher storage temperatures.

6.9.3   WETTABILITY

The wettability of a powder is primarily a measure of its hydrophilic property, that is, the ability of the powder particles to be wetted by water. The tendency of powders to form lumps on adding water indicates lack of wettability. Wettability is measured as the time necessary for a given amount of powder to pass through the water surface under specified conditions. Wettability of whole milk powder is poor at temperatures below the melting point of fat as particle surface is always covered by fat. The wettability of dry powder particles, generally, depends on the surface charge, particle size, density, porosity, and the presence of amphiphilic substances.[22,31,43] For instant, the wettability of whey powders was observed to be 5–6 s.[22,23] Wettability of whey-based mushroom soup powder was reported to be 6.5 s. Wettability of powders has been found to decrease with storage. Whey-kinnow juice powder had excellent wettability; however, the wetting time increased from initial 18.3 to 36.1 s during storage at 45°C for 6 months.[26]

6.9.4   FLOWABILITY

The flow property of a powder refers to the ease with which powder will flow when a shearing force is applied. The flowability is one of the major functional properties of dried products, with considerable implications in packaging, metering into a process vessel, etc. The flowability of powder is measured in terms of the angle of repose of a powder heap with the horizontal plane. The higher the angle of repose, the lower is the flowability. Cow skim milk powder has flowability from 44° to 45° and whole milk powder 65°. The angle of internal friction for skimmed milk powder was reported to vary between 36° and 44°.[19] The flowability of cheddar cheese whey powder and whey-based mushroom soup powder were reported as 32° and 58° by Jayaprakasha.[21]

Fitzpatrick et al.[14,15] studied the effect of powder properties and storage conditions on the flowability of milk powders with different fat contents and reported that whole milk powder is more cohesive than skim milk powder. Kim et al.[26,27] investigated the effect of surface composition on the flowability of four industrial spray-dried dairy powders, namely, skim milk powder, whole milk powder, cream powder, and whey protein concentrate. The results indicated that the flowability is strongly influenced by the surface composition of powders. It was found that skim milk powder flows well compared to other powders because the surface is made of lactose and protein with a small amount of fat, whereas the high surface fat composition inhibits the flow of whole milk, cream, and whey protein concentrate powders. Long-term storage of free-flowing powders can result in flow problems due to a phenomenon known as time consolidation, where a powder consolidates (solidifies) under its own weight over time. Studies on spray-dried whey permeate powder demonstrated time-consolidation effects, where its flowability was reduced with increasing time.[47,48] Increasing relative humidity and temperature of storage also resulted in a decreasing flowability of whey permeate powder.

6.9.5 INSOLUBILITY INDEX

Insolubility index is an indication of the degree of denaturability and status of proteins as well as the amount of scorched particles present in the powders, which vary with the severity of heat treatments given during various processing steps.[44–46] The insolubility index of various whey powders varied between 0.60 and 1.10 mL.[21] According to Indian Standard Institution, the maximum insolubility index of spray-dried skim milk powder is 2.0 mL. Changes in insolubility index during storage were found to depend strongly on holding temperature, moisture content, and heat treatment of milk.[29]

6.10 SUMMARY

During the last 25 years, consumer demands for a variety of food products with more convenience have grown exponentially; together with the need for faster production rates, it has improved quality and extension in shelf life. Drying is the oldest method of food preservation. It is the amalgamation of material science and transport phenomena. Scale-up of most of the dryer types continues to be complex and empirical and is often equipment and product specific because of the highly nonlinear nature of the governing conservation equations of transport processes. Drying mechanisms are very complex to describe. There is no doubt that conventional dryers have well-established records of performance for drying most materials. But not all of these drying technologies are necessarily optimal in terms of energy consumption, quality of dried product, safety in operation, ability to control the dryer in the event of process upsets, ability to perform optimally even with large changes in throughput, ease of control, and minimal environmental impact due to emissions or combustion of fossil fuels used to provide energy for drying.

The emergence of novel drying technologies allows producing high-quality milk products with improvements in terms of heating efficiency and, consequently, in energy savings. Novel drying technologies are gaining the attention of food processors as they can provide high-quality milk products while reducing processing costs and improving the added value of the products. Drying is still a common practice of the food industries in order to guarantee the microbiological safety of their products. The traditional drying methods rely essentially on the generation of heat outside the product to be dried, by combustion of fuels or by an electric resistive heater, and its transference into the product through conduction and convection mechanisms.

KEYWORDS

- **atomization**
- **dispersibility**
- **evaporation**

- freeze-drying
- hygroscopicity
- insolubility index
- pressure nozzle atomizer
- single-stage drying
- sinkability
- solubility
- spray-drying
- wettability

REFERENCES

1. Aguilera, J. M.; del Valle, J. M.; Karel, M. Caking Phenomena in Amorphous Food Powders. *Trends Food Sci. Sci. Technol.* **1995,** *6,* 149–155.
2. Attaie, H.; Breitschuh, B.; Braun, P.; Windhab, E. J. The Functionality of Milk Powder and Its Relationship to Chocolate Mass Processing, in Particular the Effect of Milk Powder Manufacturing and Composition on the Physical Properties of Chocolate Masses. *Int. J. Food Sci. Technol.* **2003,** *38,* 325–335.
3. Augustin, M. A.; Clarke, P. T. Skim Milk Powders with Enhanced Foaming and Steam Frothing Properties. *Dairy Sci. Technol.* **2008,** *88,* 149–161.
4. Banavara, D. S.; Anupama, D.; Rankin, S. A. Studies on Physico-chemical and Functional Properties of Commercial Sweet Whey Powders. *J. Dairy Sci.* **2003,** *86,* 3866–3875.
5. Bhandari, B. R.; Howes, T. Implications of Glass Transition for the Drying and Stability of Dried Foods. *J. Food Eng.* **1999,** *40,* 71–79.
6. Boersen, A. C. Spray Drying Technology. *J. Soc. Dairy Technol.* **1990,** *43,* 5–7.
7. Buma, T. J. The Cause of Particle Porosity of Spray Dried Milk. *Milk Dairy J.* **1972,** *26,* 60–67.
8. Chegini, G. R. Spray Dryer Parameters for Fruit Juice Drying. *World J. Agric. Sci.* **2007,** *3*(2), 230–236.
9. Christensen, K. L.; Pedersen, G. P.; Kristensen, H. G. Physical Stability of Redispersable Dry Emulsions Containing Amorphous Sucrose. *Eur. J. Pharm. Biopharm.* **2002,** *53,* 147–153.
10. Dhanalakshmi, K.; Ghosal, S.; Bhattacharya, S. Agglomeration of Food Powder and Applications. *Crit. Rev. Food Sci. Nutr.* **2011,** *51,* 432–441.
11. Downton, G. F. L. Mechanism of Stickiness in Hygroscopic, Amorphous Powders. *Ind. Eng. Chem. Fundam.* **1982,** *21*(4), 447–451.
12. Fäldt, P.; Bergenståhl, B. Fat Encapsulation in Spray-dried Food Powders. *J. Am. Oil Chem. Soc.* **1995,** *72,* 171–176.

13. Fäldt, P.; Bergenståhl, B. Spray-dried Whey Protein/Lactose/Soybean Oil Emulsions. 2. Redispersibility, Wettability and Particle Structure. *Food Hydrocoll.* **1996,** *10,* 431–439.

14. Fitzpatrick, J.; Barry, K.; Delaney. C.; Keogh, K. Assessment of the Flowability of Spray-dried Milk Powders for Chocolate Manufacture. *Lait* **2005,** *85,* 269–277.

15. Fitzpatrick, J. J.; Iqbal, T.; Delaney, C.; Twomey, T.; Keogh, M. K. Effect of Powder Properties and Storage Conditions on the Flowability of Milk Powders with Different Fat Contents. *J. Food Eng.* **2004,** *64,* 435–444.

16. Gianfrancesco, A.; Turchiuli, C.; Dumoulin, E. Powder Agglomeration During the Spray Drying Process: Measurements of Air Properties. *Dairy Sci. Technol.* **2008,** *88,* 53–64.

17. Goula, A. Spray Drying Performance of a Laboratory Spray Dryer for Tomato Powder Preparation. *Dry. Technol.* **2003,** *21,* 1273–1289.

18. Huntington, D. H. The Influence of the Spray Drying Process on Product Properties. *Dry. Technol.* **2004,** *22,* 1261–1287.

19. Ilari, J. L. Flow Properties of Industrial Dairy Powders. *Lait* **2002,** *82,* 383–399.

20. Isleten, M.; Karagul-Yuceer, Y. Effects of Dried Dairy Ingredients on Physical and Sensory Properties of Nonfat Yogurt. *J. Dairy Sci.* **2006,** *89,* 2865–2872.

21. Jayaprakasha, H. M.; Brueckner, H. Whey Protein Concentrate: A Potential Functional Ingredient for Food Industry. *J. Food Sci. Technol.* **1999,** *36,* 189–204.

22. Jensen, G. K.; Nielsen, P. Factors Affecting Reconstitution of Milk Powders. *J. Dairy Res.* **1982,** *49,* 515–544.

23. Jensen, G. K.; Oxlund, C. O. Moisture Uptake by Milk Powders. *Powder Technol.* **1988,** *43*(4), 76–79.

24. Kelly, J.; Kelly, P. M.; Harrington, D. Influence of Processing Variables on the Physico-chemical Properties of Spray Dried Fat-based Milk Powders. *Lait* **2002,** *82,* 401–412.

25. Keogh, K.; Murray, C.; Kelly, J.; O'Kennedy, B. Effect of the Particle Size of Spray Dried Milk Powder on Some Properties of Chocolate. *Lait* **2004,** *84,* 375–384.

26. Khamurai, K. Development of Technology for Concentrated and Dried Whey Based Fruit Juice Mixes. Ph.D. Thesis, NDRI (Deemed University), Karnal, India, 2000; pp 5–25.

27. Kim, E. H. J.; Chen, X. D.; Pearce, D. On the Mechanisms of Surface Formation and the Surface Compositions of Industrial Milk Powders. *Dry. Technol.* **2003,** *21,* 265–278.

28. Kim, E. H. J.; Chen, X. D.; Pearce, D. Effect of Surface Composition on the Flowability of Industrial Spray Dried Powders. *Coll. Surf. B Biointerfaces* **2005,** *46*(3), 182–187.

29. Kudo, N.; Hols, G.; Mil-p, J. J. M-van. The Insolubility Index of Moist Skim Milk Powder Influence of the Temperature of the Secondary Dry Air. *Milk Dairy J.* **1990,** *44*(2), 89–98.

30. Langrish, T. C. Comparison of Maltodextrin and Skim Milk Wall Deposition Rates in a Pilot-scale Spray Dryer. *Powder Technol.* **2007,** *179,* 84–89.

31. Lascelles, D. R.; Baldwin, A. J. Reconstitution of Whole Milk Powder. *N. Z. J. Dairy Sci. Technol.* **1976,** *11,* 283–284.

32. Liang, B.; Hartel, R. W. Effects of Milk Powders in Milk Chocolate. *J. Dairy Sci.* **2004,** *87*, 20–31.

33. Listiohadi, Y. D.; Hourigan, J. A.; Sleigh, R. W.; Steele, R. J. An Exploration of the Caking of Lactose in Whey and Skim Milk Powders. *Aust. J. Dairy Technol.* **2005,** *60*, 207–213.

34. Mermelstein, N. H. Products and Technologies Processing; Spray Drying. *Food Technol.* **2001,** *55*(4), 155–167.

35. Millqvist-Fureby, A.; Smith, P. In-situ Lecithination of Dairy Powders in Spray Drying for Confectionery Applications. *Food Hydrocoll.* **2007,** *21*, 920–927.

36. Mistry, V. V.; Pulgar, J. B. Physical and Storage Properties of High Milk Protein Powder. *Int. Dairy J.* **1996,** *6*, 195–203.

37. Mistry, V. V. Manufacture and Application of High Milk Protein Powder–Review. *Lait* **2002,** *82*, 515–522.

38. Mounir, S.; Schuck, P.; Allaf, K. Structure and Attribute Modifications of Spray-dried Skim Milk Powder Treated by DIC (Instant Controlled Pressure Drop) Technology. *Dairy Sci. Technol.* **2010,** *90*, 301–320.

39. Munro, P. A. The Densities of Casein Curd Particles and Caseinate Solution. *N. Z. J. Dairy Sci. Technol.* **1980,** *15*(3), 225–238.

40. Nijdam, J. J.; Langrish, T. A. G. The Effect of Surface Composition on the Functional Properties of Milk Powders. *J. Food Eng.* **2006,** *77*, 919–925.

41. Pisecky, J. Technological Advances in Production of Spray Dried Milk. *J. Soc. Dairy Technol.* **1985,** *38*(2), 60–62.

42. Refstrup E. Evaporation and Drying Technology Developments. *Int. J. Dairy Technol.* **2000,** *53*, 163–167.

43. Sanderson, W. B. Instant Milk Powders: Manufacturing and Keeping Quality. *N. Z. J. Dairy Sci. Technol.* **1978,** *13*, 137–143.

44. Schober, C.; Fitzpatrick, J. J. Effect of Vortex Formation on Powder Sinkability for Reconstituting Milk Powders in Water to High Solids Content in a Stirred Tank. *J. Food Eng.* **2005,** *71*, 1–8.

45. Shinga, H.; Yoshii, H.; Nishiyama, T.; Furuta, T.; Forssele, P.; Poutanen, K.; Linko, P. Flavor Encapsulation and Release Characteristics of Spray-dried Powders by the Blended Encapsulant of Cyclodextrin and Gum Arabic. *Dry. Technol.* **2001,** *19*(7), 1385–1395.

46. Sikand, V.; Tong, P. S.; Walker, J. Heat Stability of Reconstituted, Protein-standardized Skim Milk Powders. *J. Dairy Sci.* **2010,** *93*, 5561–5571.

47. Teunou, E. Effect of Relative Humidity and Temperature on Food Powder Flowability. *J. Food Eng.* **1999,** *42*(2), 109–116.

48. Teunou, E. Effect of Storage Time and Consolidation on Food Powder Flowability. *J. Food Eng.* **2000,** *43*(2), 97–101.

49. Tsourouflis, S. F. Loss of Structure in Freeze Dried Carbohydrate Solutions: Effect of Temperature, Moisture Content and Composition. *J. Sci. Food Agric.* **1976,** *27*, 509–519.

50. Vega, C.; Goff, H. D.; Roos, Y. H. Spray Drying of High Sucrose Dairy Emulsions: Feasibility and Physicochemical Properties. *J. Food Sci.* **2005,** *30*, 244–251.

51. Vega, C.; Kim, E. H. J.; Chen, X. D.; Roos, Y. H. Solid State Characterization of Spray-dried Ice Cream Mixes. *Coll. Surf. B* **2005,** *45*, 66–75.

52. Verhi, J. G. P.; Lammers, W. L. A Method for Measuring the Particle Density Distribution of Spray Dried Powders. *Neth. Milk Dairy J.* **1973**, *27*, 19–29.

53. Vilder, J. D.; Moermans, R.; Martens, R. The Free Fat Contents and Other Physical Characteristics of Whole Milk Powder. *Milchwissenschaft* **1977**, *32*(6), 347–350.

54. Yetismeyen, A.; Deveci, O. Some Quality Characteristics of Spray Dried Skim Milk Powders Produced by Two Different Atomizers. *Milchwissenschaft* **2000**, *55*, 210–212.

CHAPTER 7

BIOTECHNOLOGICAL ROUTE OF DAIRY FLAVOR PRODUCTION

ANIT KUMAR[1], ASHUTOSH UPDHYAY[1], RUPESH CHAVAN[2], USMAN ALI[3], and RACHNA SEHRAWAT[4,*]

[1]*Department of Food Science and Technology, National Institute of Food Technology Entrepreneurship and Management (NIFTEM), Kundli, Haryana 131028, India*

[2]*Department of Quality Assurance, Mother Dairy Junagadh, Junagadh 362001, Gujarat, India*

[3]*Dr. S. S. Bhatnagar University Institute of Chemical Engineering & Technology Panjab University, National Agricultural Food Biotechnology Institute, Mohali, Punjab, India*

[4]*Department of Food Engineering, National Institute of Food Technology Entrepreneurship and Management (NIFTEM), Kundli, Haryana 131028, India*

*Corresponding author. E-mail: sehrawatrachna@gmail.com

CONTENTS

ABSTRACT

Consumers enjoy the delicate flavor of milk and other milk-derived dairy products and it is a key parameter for product acceptance. Fresh milk is a rather bland product: it has a pleasant, slightly sweet aroma and flavor, and a pleasant mouthfeel and aftertaste. The flavor of milk and milk products is largely influenced by genetic and environmental factors which include the physical and physiological condition of mulching animal, the type of feed and fodder, the environment around the cow and the milking area, and biological, chemical, and enzymatic changes in milk during production and distribution. More than 400 volatile compounds have been reported in milk, covering a wide range of chemical classes including lactones, acids, esters, ketones, aldehydes, alcohols, furans, carbonyls, pyrazines, sulfur compounds, and aliphatic and aromatic hydrocarbons. The chapter discusses in detail the various routes for flavor production in milk and milk products via different biotechnological routes.

7.1   INTRODUCTION

Flavor mostly contributes an important role in dairy and food industries. Flavor is a unique attribute for dairy industries, which determines its acceptability. Flavor is the sum of those characteristics of any material taken in the mouth, perceived principally by the senses of taste and smell, and also the general pain and tactile receptors in the mouth, as received and interpreted by the brain. It may be defined as a combination of taste and aroma. It attracts consumer response toward "natural" ingredient(s). Different types of flavor compounds present in different types of milk are shown in Table 7.1.

This chapter discusses the biotechnological route of flavor production.

TABLE 7.1   Important Flavor Compounds Present in Different Types of Milks.

Compounds	Odor descriptors[2,6,7]	Type of milk
1-octen-3-ol	Mushroom-like	Buffalo raw milk
		Ovine raw milk
		Pasteurized bovine milk
2-heptanone	Blue cheese, spicy	UHT bovine milk

TABLE 7.1 *(Continued)*

Compounds	Odor descriptors[2,6,7]	Type of milk
2-nonanone	Mustard-like, spicy	UHT bovine milk
2-undecanone	Vegetable, floral, rose-like	UHT bovine milk
Benzothiazole	Burning smell, rubbery	UHT bovine milk
Dimethylsulfone	Sulfurous, hot milk, burnt	Bovine raw milk
		Ovine raw milk
		Buffalo raw milk
		Pasteurized bovine milk
		UHT bovine milk
Ethyl butanoate	Fruity, sweet, banana, fragrant	Bovine raw milk
		Ovine raw milk
		Caprine raw milk
		Buffalo raw milk
Ethyl hexanoate	Fruity, pineapple, apple, unripe fruit	Bovine raw milk
		Ovine raw milk
		Caprine raw milk
Heptanal	Green, sweet, herbaceous	Ovine raw milk
		Buffalo raw milk
Hexanal	Freshly cut grass, green	Pasteurized bovine milk
Indole	Fecal, putrid, musty, floral in high dilution	Buffalo raw milk
		Caprine raw milk
		Ovine raw milk
Nonanal	Sweet, floral, green, grass-like	Buffalo raw milk
		Ovine raw milk
		Pasteurized bovine milk

7.2 PROCESS OF FLAVOR PRODUCTION THROUGH BIOTECHNOLOGY

The flow of manufacturing processes for flavor is presented in Figure 7.1. Generally, there are three distinct processes of flavor production through biotechnology such as: enzymatic, microbiological, and plant cell cultures or isolated enzymes.[5]

7.2.1 FLAVOR PRODUCTION BY MICROBIOLOGICAL METHOD

Microbiological method for flavor production has several advantages, such as: less drastic process conditions, lower environmental pollution as compared to chemical process, and a good alternative for plant extracts in terms of economy. There are two methods of microbiological process, which are as follows:

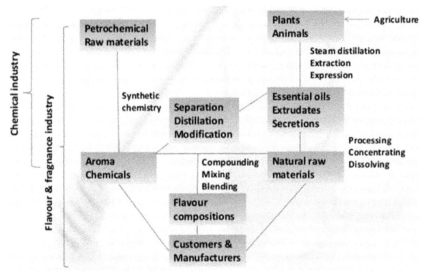

FIGURE 7.1 The flow of manufacturing processes for flavor. (Unpublished Slide 14 of 58, PPT Presentation by Supta Sarkar at Central Food Technological Research Institute, PJTSAU: http://www.slideshare.net/Supta2013/flavour-copy)

Biosynthesis (de novo synthesis): When a specific compound is synthesized with the help of metabolizing cells, it is called as biosynthesis. Generally, the batch size is small and has lower product yield and hence it is not a preferred technique by flavor industries which has made it less exploitable. The technique is more economical as compared to the other methods which can make it one of the best-suited techniques for industries engaged in flavor production in forthcoming years.

Biotransformation: Modification of chemical substrate with the help of microbial cells is known as biotransformation.

Different flavors are produced with the help of microbiological process using different microbial sources such as yeasts, filamentous fungi, and bacteria. Generally, it is cost-effective and has great application in food industries.

7.2.1.1 DIACETYL

Diacetyl is an organic compound which is a basic flavor of butter. The scientific name of diacetyl is butanedione or butane-2,3-dione. It is mostly used during the manufacturing of butter-flavored products in dairy and food industries. It is produced in different milk and milk products with the help of lactic acid-forming bacteria such as *Lactobacillus* sp., *Streptococcus thermophilus*, *Leuconostoc mesenteroide*, etc.

Ibragimova[3] reported that diacetyl may be produced by mixed culture of *Streptococcus lactis*, *Streptococcus cremoris*, and *Streptococcus diacetilactis*. Other scientists reported that subspecies *lactis* biovariety *diacetylactis* (*Lactococcus diacetylactis*) and other species of *Leuconostoc* and *Weissella* genera are availed as production of diacetyl in different dairy industries. It is also reported that *L. diacetylactis* and species of *Leuconostoc* and *Weissella paramesenteroides* not only contribute in the production of aroma but also help in the development of "eyes" by release of CO_2 in cheese.

7.2.1.2 LACTONES

Lactones occur in the form of cyclic esters of hydroxycarboxylic acids, which give buttery, creamy, sweet, or nutty flavor in different food products. In the 1960s, biotechnological route for the production of lactones with the help of different organisms was reported.[9] Free lactones are not present in the raw milk; they appear after the heating process. The buttery flavor by lactones is generally acceptable by dairy industries.

7.2.1.3 ESTERS

Esters are compound which is generally derived from a carboxylic acid and an alcohol. Esters are generally used by different dairy industries for the formation of fruity flavor in different products such as flavored butter, cream, yogurt, and flavored cheese. *Lactococcus lactis* is responsible for the synthesis of ester in favorable environmental conditions.[8] Short-chain fatty acids of ethyl or methyl esters contribute in the development of fruity flavor in the production of cheese. Synthesis of both ethyl esters and thioesters by lactic acid bacteria has been mentioned by several scientists.

7.2.1.4   TERPENES

Terpenes are organic compounds. Generally, terpenes are produced by fungi such as *ascomycetes* and *basidiomycetes* species. Terpenes are normal constituent of fresh milk which may be obtained from such cows whose fodder is based on fresh grass abundant with terpene. Terpene generally does not contribute to aroma in milk due to lower concentrations of terpene in milk.

7.2.1.5   VANILLIN

Vanillin (4-hydroxy-3-methoxybenzaldehyde or $C_8H_8O_3$) is an organic compound, which is composed of phenolic aldehyde. It is considered as the most widely applied flavoring compound in the dairy industries. Natural vanillin is extracted from the seedpods of vanilla. As harvested, green pods do not have the flavor or odor of vanilla; vanillin occurred in the form of β-D-glucoside. Vanilla flavor is developed after curing process of green pods. Nowadays, mostly vanillin is produced by chemical method but there is an increasing demand for natural vanillin. There is a limited plant supply for direct extraction of vanilla from pods which creates another way of extraction method of vanilla called biotechnological route of the flavor production. Several scientists have produced vanillin from essential oil (such as natural eugenol and isoeugenol) with the help of microbial and enzymatic biotransformations.

7.2.1.5.1   *Production of Vanillin by Natural Method*

Production of vanillin by natural method is based upon two basic processes: harvesting and curing process. First of all, pods are passed through blanching process in which, pods are blanched with hot water, then kept in the sun and completely covered with clean cloth, and packaged into airtight boxes for some weeks. After this, green pods are converted into dark brown color and vanillin as free molecules are released due to enzymatic action.

7.2.1.5.1.1 Pathway of Vanillin Biosynthesis From Tyrosine

In vanillin biosynthesis, tyrosine is converted into 4-coumaric acid, then into ferulic acid (FA), and finally into vanillin (Fig. 7.2).

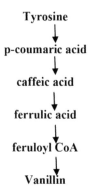

Tyrosine
↓
p-coumaric acid
↓
caffeic acid
↓
ferrulic acid
↓
feruloyl CoA
↓
Vanillin

FIGURE 7.2   Vanillin biosynthesis from tyrosine.

7.2.1.5.2   Vanillin Production Using Biotechnology

Vanillin may be manufactured by different biotechnological methods. The important substrates for vanillin production are lignin, eugenol, isoeugenol, and FA. There are some advantages of vanillin production by biotechnological method such as: condition of mild reaction, specific product which gives only one isomer product, and environment-friendly method that cause less harmful effect to the environment.

7.2.1.5.2.1   Degradation of Lignin to Vanillin

Lignin is a constituent of cell wall of plants. Degradation of lignin can be achieved by using several microorganisms such as white-rot fungi including *Pleurotus eryngii*. Depolymerization of lignin can be achieved by *Phanerochaete chrysosporium*. Six independent degradation of lignin pathways are:

- β-aryl ether cleavage pathway
- Biphenyl catabolic pathway
- Ferulate catabolic pathway
- Tetrahydrofolate-dependent-O-demethylation pathway
- Protocatechuate 4,5-cleavage pathway
- Multiple pathways involving 3-O-methylgalate catabolism

Out of six methods, β-aryl ether cleavage (with the help of *Sphingomonas paucimobilis*, *Delftia acidovorance*, and *Rhodococcus* sp.) and ferulate catabolic pathways are the most acceptable methods for a microbiologist because during this reaction, vanillin is formed as an intermediate metabolite.

7.2.1.5.2.2   Bioconversion of FA to Vanillin

Vanillin biosynthesis from FA can be also be caused by microorganism fermentation process. FA component is generally found in cell walls of woods, grasses, and corn hulls. A researcher reported a two-step process for the development of vanillin from autoclaved maize bran (which was already autoclaved) with the help of enzyme ferulic acid esterase (FAE) produced by *Aspergillus niger*.[4] First of all, FA is converted into vanillic acid and then finally into vanillin using a 72 h old starter culture of *Pycnoporus cinnabarinus* MUCL 39533. Vanillin is also produced by other microorganisms, which include *Streptomyces* sp. and *Amycolatopsis* sp., and are most widely used species for biotransformation of FA in flavor industries. FA may also be converted to vanillin and vanillic acid using *Streptomyces halstedii* GE107678. Some researchers also reported the use of *Escherichia coli* to biotransform FA into vanillin.

7.2.1.5.2.3   Catabolic Pathways of FA

Released FA can undergo dealkylation, demethylation, and decarboxylation to yield propionate, phenyl propionate, 4-vinyl phenol, and 4-ethyl phenol. Conversion of FA to vanillin occurs through several processes such as: nonoxidative decarboxylation, side-chain reduction, coenzyme-A-independent deacetylation, and coenzyme-A-dependent deacetylation.

In the last step, vanillic acid produced from vanillin either enters protocat-echuic acid pathway or guaiacyl pathway for further degradation through tricarboxylic acid (TCA) cycle.

7.2.2 FLAVOR PRODUCTION BY ENZYMATIC METHOD

7.2.2.1 HYDROLYTIC ENZYMES

7.2.2.1.1 Lipases

Serine hydrolases such as lipases are often used to catalyze the hydro-lysis of lipids to fatty acids and glycerol. Esterases on the other side show little activity in aqueous solutions, as these work at the lipid–water interface. Lipases play an essential role in organic synthesis and flavor biotechnology. Microbial lipases such as pig pancreatic extract are used for ester hydrolysis, esterification, transesterification, interesterification, and transfer of acyl groups from esters to other nucleophiles. There are numerous examples of uses of lipases in flavors, for example, lipolyzed milk fat which was one of the first flavors produced with the help of enzymes. Controlled lipase-catalyzed hydrolysis of cream forms the original fundamental basis of this process. Other common example of the use is *Mucor miehei* lipase. It is used to obtain flavor-active short-chain fatty acids. The free fatty acids (FFAs) thus produced can be extracted by steam distillation and further purified. Thus, it is possible to obtain pure short-chain fatty acids such as butanoic, hexanoic, octanoic, and decanoic acid.

Lipase-catalyzed esterification and transesterification reactions have other wide range of applications. In the production of aromatic compounds, reaction conditions have a great influence on the enzyme-catalyzed reac-tions and determine the reaction's yield and selectivity. Esters and lactones are the most common biotechnological-produced flavors. Lipase from *M. miehei* is the most widely studied fungal lipase and is used for the produc-tion of wide range of products such as hexanoic acid and alcohols extraction from methanol to hexanol. Geraniol and citronellol have been synthesized using lipases from *M. miehei*, *Aspergillus* sp., *Candida rugosa*, *Rhizopus arrhizus*, and *Trichosporum fermentans*. Fruit-like flavors can be obtained by direct esterification. Isoamyl isovalerate is produced by esterification of isoamyl alcohol and isovaleric acid in hexane with the help of *M. miehei* lipase immobilized on a weak anion exchange resin. Lipase-catalyzed intramolecular transesterification leads to enantio-selectivity to (S)-g-lactones in esters and lactones; chain length may vary from C_5–C_{11}.

7.2.2.2 OXYREDUCTASES

Oxyreductase enzymes catalyze redox processes. This includes the transfer of equivalent of two electrons by 1 two-electron step or 2 one-electron steps, for example, horse liver alcohol dehydrogenase is able to oxidize primary alcohols except methanol and reduce a large number of aldehydes. Lipoxygenase (LOX) is a nonheme iron-containing dioxygenase that catalyzes the region-selective and enantioselective dioxygenation of unsaturated fatty acids containing at least one (Z,Z)-1,4-pentadienoic system. LOX is an important factor in the large-scale use of plant enzymes for the production of natural "green note" aromatic compounds, a group of isomeric C6 aldehydes and alcohols.

7.2.2.3 PEROXIDASES

There are numerous examples of peroxidases such as soybean peroxidase which is commonly used for the production of methyl anthranilate, has a fruity odor, by enzymatic N-demethylation of methyl N-methyl anthranilate.[1]

Other example is *Lepista irina peroxida*, which cleaves β-carotene tob-cyclocitral, dihydroactinidiolide, 2-hydroxy-2,6,6-trimethylcyclohexanone, β-apo-10'-carotenal, and β-ionone.[1]

Microbial amine oxidases from *A. niger* and monoamine oxidase from *E. coli* can be used for the oxidative deamination of amines, forming corresponding aldehydes, hydrogen peroxide, and ammonia. Vanillin is commonly prepared from vanillylamine with the help of amine oxidase. Vanillyl alcohol oxidase (VAO) is a flavor enzyme from the ascomycete *Penicillium simplicissimum* that converts a broad range of 4-hydroxybenzyl alcohols and 4-hydroxybenzylamines into the corresponding aldehydes.[1]

7.2.2.4 LYASES

In the carbohydrate metabolism, four complementary aldolases can be found in nature, which show different stereoselectivity. A large variety of synthetic task has been accomplished by this broad range of enzyme. In biotechnology, Furaneol® (2,5-dimethyl-4-hydroxy-2-H-furan-3-one) is generally produced from fructose-1,6-biphosphate with the help of a three-step enzymatic process involving fructose-1,6-bisphosphate aldolase (rabbit muscle aldolase). The first step is the aldolase-catalyzed cleavage of sugar biphosphate to dihydroxyacetone phosphate and glyceraldehyde phosphate. The latter is isomerized by a coimmobilized triosephosphate isomerase to obtain dihydroxyacetone phosphate, which is the substrate for aldolase-catalyzed aldol condensation with d-lactaldehyde. The condensation's product, 6-deoxyfructose phosphate, can be easily converted to Furaneol.

7.2.2.5 GLYCOSIDASES

Raspberry ketone [4-(4'-hydroxyphenyl)-butan-2-one], the impact compound found in raspberries, can be obtained by enzymatic reactions. The first step includes β-glucosidase-catalyzed hydrolysis of naturally occurring betuloside into betuligenol.[1] The latter can be transformed into raspberry ketone by microbial alcohol dehydrogenase.

7.2.3 FLAVOR PRODUCTION BY PLANT TISSUE CULTURE

Plant tissue cultures can be used to produce the expanded range of flavors and aromas, which have characteristic of their native plant origin. Numerous chemical components are produced by the plant cell that constructs the natural flavor due to genetic makeup of the plant cell culture. The unique biochemical and genetic capacity and totipotency of the plant cell are liable for the flavor production. The production of the flavor metabolites by precursor biotransformation in the biosynthetic pathway can be enhanced by feeding intermediates.

7.3 FLAVOR PRODUCTION IN CHEESE WITH THE HELP OF MICROORGANISMS

Flavor plays the most important role in the production of cheese. However, flavor formation in raw milk cheese does not depend only on starter culture but it may also depend on presence of native microflora. The development of cheese flavor stems directly from biochemical activity during ripening. The biochemical change occurs extensively and can be categorized under three categories: (a) proteolysis and amino acid catabolism, (b) lipolysis and metabolism of fatty acids, and (c) the metabolism of lactose, lactate, and citrate.

Microbiological changes include death of the starter culture and the growth of nonstarter flora and in certain cheeses, the growth of a secondary microflora. Metabolic activity of secondary microflora dominates flavor development and texture in white-mould cheese. The freshly made cheese curd has a bland flavor and during ripening, cheese flavor develops due to the production of a wide range of compounds by the biochemical pathways.

7.3.1 METABOLIC PATHWAYS

During ripening of cheese, biochemical reactions play an important role in the formation of aroma. Generally, major milk flavor comes from three constituents such as lactose, lipids, and proteins.

7.3.1.1 LACTOSE AND CITRATE

Glucose and galactose are produced by the hydrolysis of lactose through starter culture (*lactococci*) producing galactose-6-P. Then glucose is oxidized into pyruvate through Emden–Meyerhof pathway of glycolysis and galactose is converted into glucose-6-P through galactose-positive starter bacteria and *leuconostocs* with the help of the Leloir pathway and galactose is converted into glyceraldehyde-3-P by *lactococci* by the taga-tose pathway. Pyruvate is an important compound to help in the develop-ment of short-chain flavor compounds such as diacetyl, acetoin, acetate, acetaldehyde, and ethanol. *L. lactis* subsp. *lactis* var. *diacetylactis* is mostly used by different food industries because it helps in the conver-sion of citrate into aromatic compounds. Citrate is metabolized to produce acetolactate, diacetyl, and acetoin. Lactate is an important substrate for a range of reactions which contribute to cheese ripening by metabolism to acetate and carbon dioxide by NSLAB flora through oxidative pathway. In anaerobic catabolism, lactate is converted to butyrate and hydrogen dioxide by *Clostridium tyrobutyricum* as shown in Figure 7.3.

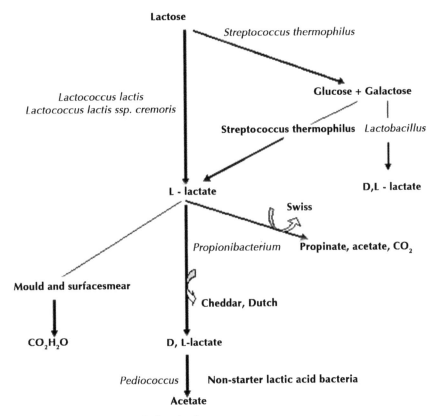

FIGURE 7.3 Lactose metabolism in cheese.

7.3.1.2 LIPOLYSIS

Lipolysis in an important event during cheese ripening and is a major contributor for flavor in cheese. Short-chain fatty acids produced during lipolysis have a considerable flavor impact but excessive fatty acids may cause defects such as rancidity. FFAs are important precursors for several volatile compounds which contribute to flavor (Fig. 7.4). FFAs act as substrates for enzymatic reactions which yield flavors. Oxidation and decarboxylation yield methyl ketones and secondary alcohols, and esterification of hydroxyl fatty acids produces lactones. Fatty acids react with alcohol groups to form esters, such as ethyl butanoate, ethyl hexanoate, ethyl acetate, ethyl octanoate, ethyl decanoate, and methyl hexanoate.

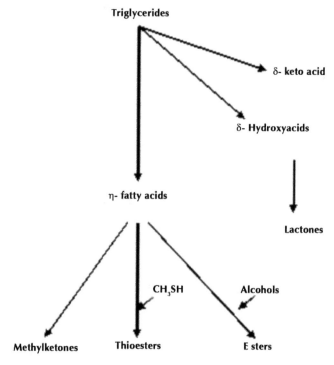

FIGURE 7.4   Summary of lipolysis in cheese.

7.3.1.3   PROTEOLYSIS AND CATABOLISM OF AMINO ACIDS

Proteolysis directly contributes to cheese flavor by releasing peptides and amino acid. Proteolysis is a two-phase phenomenon, namely primary proteolysis and secondary proteolysis. Primary proteolysis represents the extent of breakdown of the native casein. Secondary proteolysis is the further degradation that reaches to the formation of the peptides and free amino acids (FAAs). Amino acids are substrates for transamination, dehydrogenation, decarboxylation, and reduction, producing a wide variety of flavor compounds such as phenylacetic acid, phenethanol, β-cresol, methane thiol, dimethyl disulfide, 3-methyl butyrate, 3-methyl butanal, 3-methyl butanol, 3-methyl-2-butanone, 2-methyl propionate, 2-methyl-1-propanal, 2-methyl butyrate, and 2-methyl butanal. FAAs can also be catabolized by decarboxylases to form amines, some of which can cause off-flavor or adverse physiological effects or deaminases to produce ammonia (Fig. 7.5).

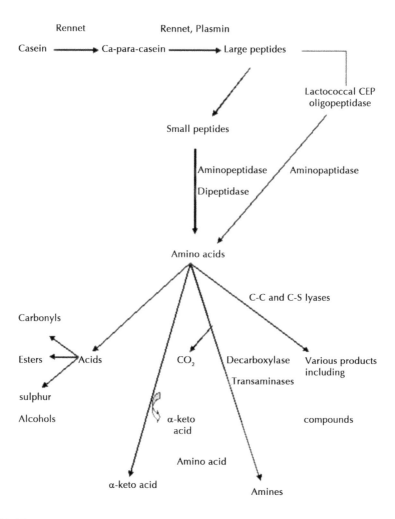

FIGURE 7.5 Summary of proteolysis and amino acid catabolism in cheese.

The catabolism of branched-chain amino acids is initiated by an aminotransferase-forming aketoisocaproate, α-keto-h-methyl valerate, and α-ketoisovalerate from leucine, isoleucine, and valine, respectively. Aminotransferase activity specific for isoleucine, valine, and leucine is reported in *Lactobacillus paracasei, Lactobacillus curvatus, Lactobacillus brevis, Lactobacillus plantarum*, and *L. lactis*.

7.4   APPLICATIONS OF BIOFLAVORS IN DAIRY INDUSTRY

There are several applications of bioflavor in dairy industries which are summarized in Table 7.2.

TABLE 7.2   Applications of Bioflavor in Dairy Industry.

Chemical component	General flavor	Applications
Furaneol	Strawberry, caramel	Dairy product, ice cream, beverages
Methylketones(2-alkanones)	Cheesy	Cheeses, blue cheeses
Nootkatone	Grapefruit	
Raspberry ketone	Raspberry	Low-fat product, dairy products, flavored milks
Vanillin	Vanilla	Dairy products, ice cream, confectionaries, beverages, tobacco, flavored milks
β-deca/dodeca-lactone	Cream, butter	Beverages, ice creams

7.5   CHALLENGES OF BIOTECHNOLOGY IN MILK FLAVOR INDUSTRY

Along with several applications, some biotechnological challenges are of paramount importance in milk flavor industry, which are mentioned below:

- Lower yields (rarely above 100 mg/L)—economically unattractive.
- A better understanding of metabolic pathways.
- Screening should be intensified (e.g., vanillin from eugenol).
- Detection, identification, and characterization of flavor compounds low threshold values (ppm to ppb).

- Needs sophisticated, sensitive, quantitative, and continuous detection systems.
- The cost of the raw materials, precursors, and the formation and elimination of unwanted side products.

7.6 SUMMARY

Biotechnological processes can be used as an alternative to natural compounds. It is essential that the use of bioflavors should enhance the safety and quality of milk products. Intensified work for downstream processing is required for increased productivity and safety.

KEYWORDS

- **bioflavor**
- **biosynthesis**
- **biotechnology**
- **biotransformation**
- **enzymes**
- **esters**
- **free fatty acids**
- **lipolysis**
- **proteolysis**
- **short-chain fatty acid**

REFERENCES

1. Dubal, S. A.; Tilkari, Y. P.; Momin, S. A.; Borkar, I. V. Biotechnological Routes in Flavor Industries. *Adv. Biotech.* **2008,** *3*, 20–31.
2. Friedrich, J.; Acree, T. Gas Chromatography Olfactometry (GC-O) of Dairy Products. *Int. Dairy J.* 1998, *8*, 235–241.
3. Ibragimova, A. Z.; Yakovlev, D. A.; Gorshkova, E. I. V. Determination of the Ratio of Diacetyl and Acetaldehyde in *Streptococcic* Cultures by Gas–Liquid Chromatography. *Molochnaya Promyshlennost* **1980,** *2*, 43–45.

4. Lesage-Meessen, L.; Lomascolo, A.; Bonnin, E.; Thibault, J. F.; Buleon, A.; Roller, M.; Asther, M.; Record, E.; Ceccaldi, B. C. A Biotechnological Process Involving Filamentous Fungi to Produce Natural Crystalline Vanillin From Maize Bran. *Appl. Biochem. Biotechnol.* **2002,** *102*(1–6), 141–153.

5. Longo, M. A.; Sanromán, M. A. Production of Food Aroma Compounds. *Food Technol. Biotechnol.* **2006,** *44*(3), 335–353.

6. Moio, L.; Etievant, P.; Langlois, D.; Dekimpe, J.; Addeo, F. Detection of Powerful Odorants in Heated Milk by Use of Extract Dilution Sniffing Analysis. *J. Dairy Res.* **1994,** *61*, 385–394.

7. Moio, L.; Langlois, D.; Etievant, P.; Addeo, F. Powerful Odorants in Bovine, Ovine, Caprine and Water Buffalo Milk Determined by Means of Gas-chromatography-olfactometry. *J. Dairy Res.* **1993,** *60*, 215–222.

8. Nardi, M.; Fiez-Vandal, C.; Tailliez, P.; Monnet, V. The Esta Esterase is Responsible for the Main Capacity of *Lactococcus lactis* to Synthesize Short Chain Fatty Acid Esters In Vitro. *J. Appl. Microbiol.* **2002,** *93*, 994–1002.

9. Okui, S.; Uchiyama, M.; Mizugaki, M. Metabolism of Hydroxy Fatty Acids, II: Intermediates of the Oxidative Breakdown of Ricinoleic Acid by Genus *Candida*. *J. Biochem.* **1963,** *54*, 536–540.

PART III

Cleaning and Novel Detection Methods in Dairy Science

CHAPTER 8

FOULING OF MILK AND CLEANING-IN-PLACE IN THE DAIRY INDUSTRY

MD. IRFAN A. ANSARI[1], RUPESH S. CHAVAN[2,*], and
SHRADDHA B. BHATT[3]

[1]Department of Agricultural Engineering, Birsa Agricultural
University, Ranchi 834006, Jharkhand, India

[2]Department of Quality Assurance, Mother Dairy Junagadh,
Junagadh 362001, Gujarat, India

[3]Department of Biotechnology, Junagadh Agricultural University,
Junagadh 362001, Gujarat, India

*Corresponding author. E-mail: rschavanb_tech@rediffmail.com

CONTENTS

ABSTRACT

Cleaning-in-place (CIP) refers to all those mechanical and chemical systems that are necessary to prepare equipment for milk processing, either after a processing run that has produced normal fouling or when switching a processing line from one recipe to another. This refers to the system of cleaning and sanitization which does not require the daily dismantling of dairy equipment. CIP is an important component in guaranteeing food safety in milk processing plants. The benefits associated with CIP are: it ensures that all equipment receives uniform treatment day after day, by eliminating the human factor; less damage to equipment; saving of total cleanup costs and man-hours; and reduces possibility of contamination through human error and improves plant utilization and appearance. Mode of action, types of CIP systems, and factors affecting the effectiveness of CIP in dairy industry are discussed in detail in current chapter.

8.1   INTRODUCTION

Cleaning-in-place (CIP) is a system for cleaning the interior surface of pipelines, vessels, filters, process equipment, and associated things without dismantling. CIP refers to all those mechanical and chemical systems that are necessary to prepare equipment for milk and milk product processing, either after a processing run that has produced normal fouling or when switching a processing line from one recipe to another. The efficiency of cleaning and sanitation of milk contact surfaces are widely influenced by many factors such as the character of contamination, micro topography of surfaces, straightness of passage ways, compatibility of surface agents, application methods, the speed of application, and related speed of penetration into biofilm structure.[10] Depending on the processing practice and load of soiling on the process equipment, the cleaning solutions may be used for a single cycle or recycled and reused for multiple uses. In multiple uses, cleaning solutions are drained after a few to several hundred cleaning cycles. CIP is an important component in guaranteeing food safety in the dairy industry. Successful cleaning between production runs avoids potential contamination and products that do not meet quality standards. Carrying out CIP correctly from design to validation helps to ensure secured barriers between milk flow and cleaning chemical flows.

Cleaning-out-of-place (COP) is a well-adapted technique in which the equipment is moved to the cleaning equipment prior to a CIP clean.

The advantages associated with CIP are increased productivity through reduction of downtime, reduction in chemical handling and usage, operations become simpler in nature, savings in cost and utilities (steam, water, and electricity), decreased load on effluent treatment plant (ETP), increased health and safety of both men and machine, reduction in labor required for cleaning and sanitization process, improved traceability, usage of stronger chemicals for tougher soils, elimination of human errors during cleaning, increased automation and improved repeatability and reliability of the process, and environment friendly.

This chapter discusses details of CIP system for the dairy industry.

8.2   FOULING IN HEAT EXCHANGERS

Heat treatment of milk and milk products is done to ensure the microbial safety and increase the shelf life. The effect of the heat treatment on the final product in terms of chemical, physical, and organoleptic properties depends on the combination of temperature and time applied. The higher the temperature, the shorter the required time to obtain a sufficiently safe product. When the temperature is between 135°C and 150°C, the heat treatment is denoted as ultrahigh temperature (UHT) process.[7] These high-temperature ranges are necessary to kill bacterial spores, such as *Bacillus stearothermophilus* and *Bacillus sporothermodurans*. Recent research has indicated that *B. stearothermophilus* attaches better to stainless steels (SS) walls than vegetative cells, stressing the need for high processing temperatures.[1]

When heat exchangers are employed to process milk for longer shelf life and to render the product safe for consumption, protein denaturation and mineral deposition (fouling) take place on the heat transfer surfaces. This deposition on heat transfer surface is called fouling. The major components in fouling are calcium phosphate ions and whey proteins, which form insoluble aggregates which can result in reducing the running time to a mere 5 h of milk pasteurizer based on plate heat exchanger. Fouling can occur similarly in tubular heat exchangers which can lead to increased pressure differential across the plant. In tubular heat exchangers, the fouling layer is having a proportionally lower effect on the total area available for product flow due to its different geometry of which makes it feasible to accommodate greater thickness of deposit without affecting the plant running time. As the gaskets are not present in the tubular heat exchangers, they can safely tolerate a high internal product pressure due to fouling but heavy deposits will hinder the exchange of heat from heating medium thereby comprehending the destruction of pathogens.

Fouling of heat exchangers by milk (raw or skimmed milk) has been studied by many researchers and they have highlighted the key role played by whey proteins, especially β-lactoglobulin. Unfortunately, milk fouling is only partially understood due to its very complex nature and it still remains the unsolved problem in the dairy industry.[19,24]

8.3   EFFECTS OF MILK FOULING

Milk fouling is recognized as a nearly universal problem in design and operation of equipment and it poses mainly three problems. First, the processing efficiency is reduced, since the deposited material disturbs both the fluid flow and the heat transfer.[19] This, in turn, increases the pressure drop and may deteriorate the product quality in terms of its microbial content, color, texture, and taste. Additionally, the necessary removal of the deposit shortens the running time between cleaning cycles and, thus, increases the operating costs. Due to fouling, cleaning at least once a day is a common practice for indirect heat exchangers. Due to the use of aggressive cleaning agents, there is also an issue of environmental concern.[12–14] It would, therefore, be rewarding to decrease the problem of fouling in the dairy industry.

A better knowledge of the fouling occurring within the indirect heat exchanger is essential for improving its design and for increasing the time of operation between cleaning. Generally, fouling is accounted for engineering design for heat exchange plants by adding extra surface area over that of the clean design. The added surface area can vary between 10% and 500% of the clean area depending on parameters such as exchanger type, application involved, etc. The equipment designers consider the effect of fouling during the desired operational lifetime and usually make provisions for sufficient extra capacity to ensure that the exchanger will meet process specifications till a scheduled cleaning is conducted. The designers also take into consideration the various mechanical arrangements to be provided for effective cleaning of the equipments.

8.4   TYPES OF SOILS

In the dairy industry, the soil that is to be removed from the plant and equipment during cleaning operations may be in one, or a combination of more than one of the forms including liquid milk film, air dried film, heat precipitated film, heat hardened film, milk stone, or miscellaneous foreign matter. Thus, the soil may range in composition from normal milk film to residues containing largely of fat and protein matter frequently mixed with hard water scale of calcium and magnesium. The science for removing of soil is based on applying the required amount of energy to the equipment to ensure that it is cleaned. The energy is primarily provided by the solution temperature (thermal energy), the use of detergent or solvent (chemical energy), and the application of suitable pipeline velocities or pressures (kinetic energy).

Fouling in milk processing is usually categorized into two types. Type A deposit which is soft, voluminous curd-like material and its formation is initiated when the temperature starts to exceed above 75°C and is maximum in a temperature range of 95–110°C which then reduces with further increase in temperature. This type of deposit is rich in protein (50–70%) and a lower in mineral (30–40%) content. At 75°C, the formation of type A deposit is started and is majority due to denaturation of protein which is predominated by β-lactoglobulin (60%) and as the temperature increases, it is dominated by casein (40%). Type A deposits are voluminous in nature which restricts the area of the flow passage in the heat exchanger and causes an increase in the operating pressure.

A second type of deposit, called type B, forms at a higher temperature and is harder and granular in structure which is higher in levels of mineral (70–80%) and the rest is protein (10–20%). Calcium phosphate makes up the majority of the deposits at a higher temperature and such deposits are reported to hinder heat transfer as compared to flow making it difficult to reach the required processing temperature without raising the temperature of the heating medium to an undesirable extent.[7,15]

Milk fat plays an insignificant role in the fouling. Protein and minerals are the main constituents during processing of cream (36% fat). The initial layer is composed prominently of minerals and calcium phosphate and its deposition is highly influenced by the charge on the wall surface, wall temperature, and fluid flow rate.[10,23] Fouling rate increases when the temperature increases above 85°C, hence it becomes a cause of concern during UHT sterilization process.

8.4.1 HYDRODYNAMICS AND KINETICS OF FOULING

Heat exchanger fouling is common in the process industries but is still poorly understood. Much of the research in milk fouling has been to study the relation between product chemistry and the processing that the material receives. Fouling at a point in a heat exchanger depends on local thermal and hydraulic conditions, together with the chemistry and process history of the fluid. In general, the stages of fouling are:

- Stage 1: The initiation of a reaction in the milk which converts one or more of its constituents into a form capable of being deposited on a surface.
- Stage 2: Transport of the depositable material to a surface by small or large scale turbulence in the liquid.
- Stage 3: Adsorption of a layer of fouling material to the surface to form an initial layer.
- Stage 4: Deposition of further fouling material on the initial layer.
- Stage 5: Buildup of the fouling layer by deposition of material, compensated by the mechanical removal of material through the shear forces caused by the flow of the product across the deposit–liquid interface.[21]

After the first layer is formed on the solid surface, other particles coming from the milk adhere on top of this layer and develop a more or less structured and compacted deposit. The structure of this deposit depends both on the structure of the first layer, which depends mainly on the surface properties, and on the particles and ions present in the solution which will contribute to the growth of the deposit (precipitation/crystallization kinetics). These factors and their interactions will also determine another important characteristic of the deposit when considering a fouling phenomenon, which is its resistance to removal, more precisely the amount of deposit kept adhered to the surface after cleaning.

8.4.2 MATHEMATICAL MODELS FOR MILK FOULING

A number of fouling models are proposed in milk as it is complex because the fouling rate at any point is a function of temperature. There are numerous models describing fouling dynamics and can be classified into four groups according to the parameters they predict (Table 8.1):

- Predicting the amount of deposit.
- Predicting the thickness of deposit.
- Changes of the overall heat transfer coefficient.
- Outlet temperature of the product.

TABLE 8.1   Summary of the Mathematical Models for Milk Fouling.

Characteristic parameters	Predictions
Deposit weight	• Rate of deposition from Reynolds number and processing temperature.[21]
	• Amount of deposit from processing temperature and holding time.[17]
	• The increase of deposit weight with processing time.[8]
Thickness of deposit or change in hydraulic diameter	• Changes of thickness of fouling deposit as a function of time and position along the length of heat exchangers.[1]
	• Distribution of the deposit in a tubular heat exchanger from wall shear stress and dynamic viscosity.[3]
Heat transfer coefficient	• Changes of the overall heat transfer coefficient from wall and bulk temperature and Reynolds number.[9]
Bulk milk outlet temperature	• Changes of outlet temperature as a function of time and position along the length of heat exchangers.[1]

8.4.3   FACTORS AFFECTING FOULING RATE

Fouling in milk and milk products is a function of many variables, both chemical and physical. Some parameters, such as temperature and flow rate, can be determined by the process designer and others, such as the chemistry of the product, cannot be changed. Some of the factors governing the rate and type of fouling are described in this section.

8.4.3.1   CALCIUM

It plays an important role in fouling of heat exchangers and equipment in the dairy industry. Deposition of calcium increases not only due to a decrease in solubility of calcium phosphate but is also expedited due to denaturation of whey proteins and the precipitation of the same. Experiments have also confirmed that increasing or decreasing the calcium concentration can lead to decreased heat stability which increases the fouling rate as compared to the normal milk. In addition to deposition of calcium, there is a shift in the protein composition of serum proteins to caseins which increases instability of the casein micelles, with denatured β-lactoglobulin at their surface, causes an increase in protein fouling.[8]

8.4.3.2 pH

Decreasing the pH of the milk (from 6.8 to 6.4) increases the fouling due to additional deposition of casein.[22] In the case of whole milk, the deposition of both protein and fat is increased, but the deposition of minerals is reduced. It appears that the increased deposit formation from pH reduced milk is mainly due to reduced stability of protein to heat.

8.4.3.3 AIR

Fouling is also believed to increase due to the presence of air bubbles as they become less soluble at increased temperatures in milk. If the local pressure is too low (below the saturation pressure), then it causes air bubbles to rise.[7] The formation of air bubbles is enhanced by mechanical forces induced by valves, expansion vessels, free falling streams, etc. It is suggested that air in milk only encourages fouling if it is separated as bubbles on the heating surface, acting as nuclei for deposit formation.[8] In addition, the composition of the deposit is influenced through drying of the membrane of air bubbles containing mainly caseins, resulting in higher casein content of the fouling layer.[12,13] Small bubbles are not entrained by the flow and stick to the heating surface resulting in overheating of the surface. On the basis of this information, degassing of dairy fluid to be heated is recommended before processing or processing under a reasonable back pressure (about 1 bar) to keep the dissolved air in the fluid to reduce fouling.

8.4.3.4   AGE OF MILK

Aged milk causes more fouling in a heat exchanger than fresh milk owing to the action of proteolytic enzymes. The increase in fouling upon storage of milk for 10 days at 5°C was the result of additional protein deposition.[8] Experiments with skim milk and whole milk showed that the action of proteolytic enzymes, produced by psychrotrophic bacteria, is responsible for the increase in deposit.[8]

8.4.3.5   COATING

Surface is one of the major participants of the fouling process. The normal material of construction used in dairy industry is SS. However, SS has a lower thermal conductivity than copper and aluminum; lower thermal conductivities give rise to higher temperature gradients, which can result in increased fouling. Covering the surface with polytetrafluoroethylene (PTFE) was found not to eliminate fouling, but cleaning was easier, as its adhesion was poorer. Several studies have reported that the nature of the surface becomes unimportant once the first layers are adsorbed during the fouling process.[8] One such study showed that different coatings on the heating surface did not affect the amount of deposit formed, but it did affect the strength of adhesion.[5] Hence, it appears that coatings may enhance the cleaning rate and not reduce the fouling rate.

8.4.3.6   TEMPERATURE

Temperature is an important operating variable to affect fouling. However, it is not totally clear whether this temperature is bulk product temperature, the surface temperature, the temperature of the heating medium (steam or hot water), or the temperature difference between the product and the heating medium. In industrial applications, attention is normally paid to the bulk product temperature. In a continuous heat exchanger, the bulk product temperature changes along the length of the heat exchanger. There is no direct relation between the bulk product temperature and the amount of deposit formed at that location. The most common method used to study the effect of product temperature on fouling involves heating the product at UHT conditions up to a predetermined temperature and studying the deposit formation at the various sections of the plant. The severity of overall fouling process increases as the processing temperature increases.

8.4.3.7   OTHER FACTORS

- Flow rate of processed fluid
- Bulk fluid temperature
- Type and nature of the surface
- Microbial contents
- Properties of the fluid in relation to temperature and pressure
- Surface conditions of the wall and/or the already present deposit
- Temperature difference over the boundary layer
- Flow conditions in the boundary layer

8.4.4   METHODS OF REDUCING FOULING

Three possible approaches to reduce or eliminate fouling are:

- Modification to the heat treatment that the fluid receives, such as changes in the temperature profiles.
- Modification to the design of the heat exchanger, either by changing its configuration or the surface finish.
- Modification to the dairy fluid, such as the addition of oxidizing agents to prevent aggregation and the addition of calcium phosphate inhibitors to prevent its destabilization and further precipitation. In many cases, this may be precluded because the product specification is constrained.

8.4.4.1   FOREWARMING OF MILK

Forewarming provides a reasonable practical solution where milk quality is poor and fouling is known to be a severe problem. Preheating leads to a substantial reduction in fouling partly because of the denaturation of whey proteins. A wide range of conditions can be used from 85°C for 10 min, up to 90°C for 15 min, and for some very poor supplies, up to more than 110°C. A reasonable compromise is about 90–100°C for 2–5 min. The main change is the denaturation of β-lactoglobulin and its association with the casein micelle, which will reduce the amount of type A deposit. A second effect is a reduction in ionic calcium, brought about by the precipitation of calcium phosphate onto the casein micelle after heat treatment. Bell and Sanders[2] preheated milk to 75°C with a holding for 10 min and reported that the amount of deposit formed in the heat exchanger was reduced by the denaturation of serum proteins and the insolubilization of some of the milk salts. Preholding is effective in reducing fouling not only in case of milk and whey but also can be used during cream processing. Hiddink et al.[10] suggested that a temperature of about 95°C with a holding period of a few minutes is suitable for cream processing. The fouling problem is less severe if raw milk is used for UHT processing without any previous heat treatment.[16]

8.4.4.2   ADDITIVES

Various additives may be used to control fouling, although in many countries they are not legally acceptable in liquid milk. Burdett[6] showed that the pyrophosphate of sodium or potassium has a marked effect on the level of fouling independent of any pH effect. The addition of iodate extends the running time considerably by interfering with the formation of type A deposit. According to Skudder et al.,[22] fouling depends on pH of milk and slight adjustment by adding NaOH or citrate can extend running time in case of indirect type UHT plants.

8.4.4.3   STORAGE CONDITIONS OF RAW MILK

Increased acidity causes an increased rate of fouling, the storage of milk before processing, without any change in pH can significantly reduce the chances of fouling. The amount of fouling is reduced to half when the milk is stored at refrigerated conditions for 12–24 h as compared to fresh milk. It is hence advisable that before subjecting the milk for UHT processing, or as part of a product to be processed, should be stored under refrigeration for about 12–24 h before it is given any heat treatment which will inactivate the native lipases.

8.4.4.4   MODIFYING HEAT TRANSFER SURFACE

As heat modified material must adhere to the heat transfer surface to form a fouling layer, it is conceivable that modification of the surface could prevent or limit this adhesion and so reduce the fouling problem. Some surface coatings applied to small areas in a heat exchanger have given local reductions in the amount of fouling. However, it seems probable that they have interfered with heat transfer and reduced the temperature at the surface milk interface. Recently, several new techniques such as direct ion implantation, magnetron sputtering, dynamic mixing, plasma enhanced vapor deposition, and autocatalytic Ni–P–PTFE coating[25] have been considered as being of interest in fouling reduction. However, very few studies have been successful in applying such novel surface treatments. Müller-Steinhagen and Zhao[18] investigated the influence of SiF_4^+ ion-implanted SS, which significantly reduces $CaSO_4$ scale formation during pool boiling. Various surface treatments such as coatings (silica, SiOX, Ni–P–PTFE, Excalibur®, and Xylan®) and ion implantations (MoS_2, SiF^+) were tested by Beuf et al.[4] for processing of milk using plate heat exchangers. After fouling, no significant difference was observed between all the modified steels and the reference, but cleaning efficiency of Ni–P–PTFE was significantly the best.

Rosmaninho et al.[20] studied the influence of the properties of SS based surfaces on the fouling behavior of different milk components (calcium phosphate and whey proteins), complex milk systems and reported that the Ni–P–PTFE surface was the most promising one, since it generally promoted less deposit build up and, in all cases, was the easiest to clean. The results suggested that the initial interaction of the deposited layer to the surfaces is more important for the cleaning process than the total amount of deposit built up.

8.4.4.5   ELECTRIC OR MAGNETIC PRETREATMENT

Some electric or magnetic devices are tested as possible ways of reducing milk fouling in heat exchangers. One device was claimed to reduce pasteurization fouling. Trials in the United Kingdom with a magnetic device attached to a UHT plant showed that it had no effect on the fouling of the heat exchanger. Mechanism of fouling can be studied by experimentation where various types of milk may be heated to produce deposit on a test section. This can verify simulation of fouling studies done purely on theoretical fundamentals.

8.5 CLEANING

Frequent cleaning of the dairy plant is needed for both microbiological reasons and to restore heat transfer characteristics in dairy equipments by removal of fouling layer and reduce the pressure drop in the plant. Complex CIP techniques have been developed empirically. Two types of CIP treatments are mainly found in milk processing:

- Two-stage cleaning: using alkali, commonly sodium hydroxide and an acid wash of nitric or phosphoric acid.
- Single-stage cleaning: using formulated detergents containing wetting and other surface agents as well as chelating compounds.

Selecting the correct cleaning strategy requires an understanding of milk fouling, which has been studied for a number of years.[4] CIP is defined as "cleaning without dismantling of equipment or systems designed to this purpose by treatment of the surfaces with chemical solutions, cleaning fluids and rinsing water and circulation within the system."[11] If the CIP principle is adhered to completely, equipment must be designed so that it is completely clean and germ-free after automatic cleaning, without additional dismantling and subsequent manual cleaning. This means that there should not be any gaps or dead zones in which the product or cleaning fluid can collect.

8.5.1 ADVANTAGES OF CIP CLEANING

Fully automatic cleaning saves time and money. It is not necessary to constantly dismantle and manually clean the components. CIP cleaning is particularly advantageous for components which are difficult to access. CIP system is reproducible, repeatable, and validatable in nature which can produce controllable results in terms of cleaning efficacy. CIP helps to increased productivity through reduction of down time. Risks associated with chemical handling in manual cleaning are reduced thereby reducing the chances of accidents. Apart from simple in operation, the system helps to save utilities including chemicals, water, and effluent, labor time, etc. Fouling, which is not easy to clean, can be cleaned by usage of stronger chemicals at higher temperatures.

8.6   TYPES OF CIP SYSTEMS

In the dairy industry, a typical CIP system is designed as a three-tank system to hold chemicals: alkaline (2% NaOH), acid (3.5% phosphoric acid), and rinse water. The dairy CIP system is defined as a utility system so that its sole purpose is to hold and transfer cleaning solutions from separate holding tanks into the process equipment to perform cleaning after termination of a production batch. The actual CIP cleaning is performed and controlled by the process equipment's vessels, pumps, and instruments that are divided into independent CIP circuits. This dairy-type CIP method relies exclusively on chemical cleaning (rinse—until clean using water and chemicals) to remove/dissolve water-soluble product residuals and are classified into: Boil out (fill/flood) system; total loss; single-use recirculation; reuse (recovery); multi channel; fixed and mobile systems; and sterilizing-in-place (SIP), wash-in-place (WIP), and COP.

8.6.1   SINGLE-USE CIP SYSTEM

In single-use CIP systems, the fixed volume of CIP solutions including acid, alkaline, and water is recalculated and then drained off after completion of the cycle (Fig. 8.1).

All rinses used during the cleaning process are pumped for a single pass and then drained out. Such type of system is best suited for critical cleaning applications and high soil conditions where detergents cannot be reused in subsequent cycles, for example, in buttermilk and condensed milk manufacturing equipments. The advantages associated with usage of single-use CIP systems are multiple cleaning regimens; minimize thermal shock; minimize potential cross contamination and smaller tank and footprint; and lower initial cost for installation of the system and detergent concentrations can be metered and monitored. Longer heat up times, higher chemical, water, and steam usage is some of the shortfalls to be considered during usage of such systems.

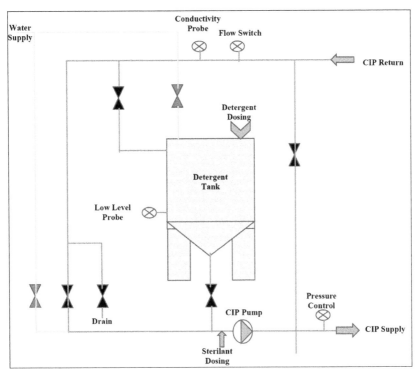

FIGURE 8.1   Single-use CIP system.

8.6.2 REUSE CIP SYSTEM

In this type of CIP systems, the wash solutions are reused in subsequent a CIP wash cycle which causes lowering of the amount required for cleaning the dairy equipment (Figs. 8.2 and 8.3).

Acid and post-rinse solutions may also be saved in separate recovery tanks and are well suited for low to moderate soil conditions where detergents can be reused in subsequent cycles, and total cost of operation must be minimized. Such types of systems are used for cleaning of silos, tanks, plate heat exchangers, milk tankers, and others in the dairy industry. Reuse CIP systems help to reduce the cost of chemical required for cleaning. They also help to reduce the amount of water and steam required for completing the cycles. Such systems are simpler to operate and having shorter heating times and cycle times which increases the running time and ultimately the productivity of the dairy plant. Risk of cross contamination

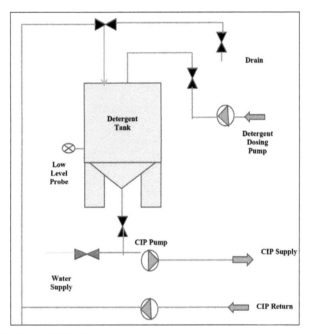

FIGURE 8.2   Reuse CIP system.

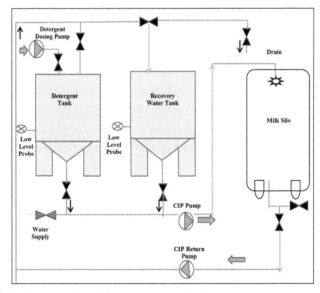

FIGURE 8.3   Reuse system with recovery water tank.

is higher in usage of systems. Apart from food safety risk, such systems have a limited cleaning regimen, higher initial cost, larger footprint, and the CIP tanks must be frequently cleaned, recharged, and checked for their concentrations.

8.6.3 SINGLE-USE RECIRCULATION CIP SYSTEM

In such systems, a fixed volume of CIP solutions is recirculated and drained after every wash cycle. The systems use a single pump for both CIP supply and vacuum return flow, eliminating return pumps (Fig. 8.4). They are best suited to tank CIP in critical cleaning applications and moderate to high soil conditions, for example, in curd manufacturing. Single-use CIP system is best suited for smaller dairy plants as they can be used for

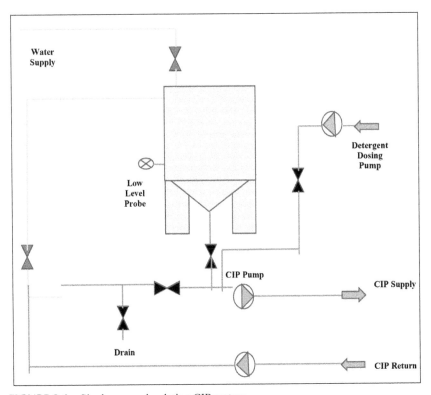

FIGURE 8.4   Single-use recirculation CIP system.

multiple cleaning regimens, minimize the thermal shock to the equipment and the pipelines. They also help to eliminate the chances of cross contamination. Detergent concentrations can be metered and lower amount of water, steam, and chemicals are required which lowers the footprint. Such systems also help to eliminate return pumps and minimize water pooling. Disadvantages of implementing such CIP systems are: usage of higher amount of chemical, water, and steam online circuits versus reuse with limited flow rates and necessity for verifying the temperature.

8.6.4 MULTICHANNEL REUSE SYSTEM

Where there is lot of different plants to be cleaned, a CIP system may be configured to clean more than one circuit at a time, a two, three, or four circuit CIP systems may be provided, serviced by the same tanks (Fig. 8.5). Multichannel reuse system consists of rinse water tank, which is used either where the flow rate of mains water in the dairy is insufficient to provide turbulent flow in the pipework, or when there is no other water break tank in the facility. The water break tank is used to store enough freshwater for at least one CIP cycle. Recovered rinse water tank is

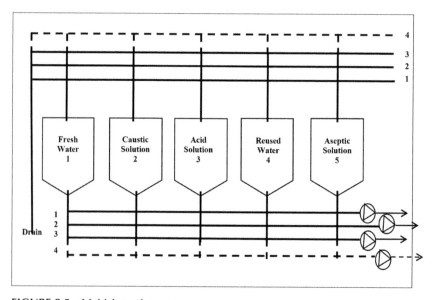

FIGURE 8.5 Multichannel reuse system.

provided to collect water after a detergent cycle when the freshwater used as a post-rinse and the terminal sanitizers are collected and reused as the pre-rinse for the next CIP cycle which saves both water usage and effluent costs. Detergent tanks are used to store a batch of dilute detergent, which is recirculated around the plant and returned back to the detergent tank again. Some larger dairies may use both caustic- and acid-based detergents in which case the CIP set may have two detergent tanks. Use of a detergent tank saves costs of chemicals when compared to single-tank total loss systems.

8.7   FACTORS AFFECTING CIP SYSTEMS

While designing cleaning procedures, the factors which are to be considered are the time, temperature, concentration of the chemical and mechanical force. Energy is required in a cleaning process in order to remove the soil and once dissolved, keep it in solution and carry it away. These four parameters are interconnected and depend on each other, that is, if any of the parameters is changed, the other three might need to be adapted so as to give the same end result as before. They are usually grouped in a diagram called Sinner's circle (Fig. 8.6). The circle diagram was originally constructed in 1959 by Dr Herbert Sinner, a chemist who worked for Henkel, a German detergent supplier.

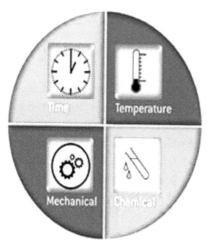

FIGURE 8.6   Sineer's circle.

8.7.1 MECHANICAL FORCE

The mechanical force in CIP is the shear forces created by the flow. Mechanical force can be a simple manual scrubbing with a brush or as complex as turbulent flow and pressure inside a pipeline. In a plant, the flow velocity of the cleaning liquids can be increased by pumping it faster. As a general CIP rule, the flow must be turbulent and its velocity should be at least 1.5 m/s to have an adequate mechanical force. For effective turbulent flow, rotameter can be placed at CIP delivery pump line. The throughput rate Q (m^3) in pipe with a radius, r (m) to meet the turbulent flow of at least 1.5 m/s for effective flushing can be calculated by the following equation:

$$Q = \pi \times r^2 \times 1.5 \times 3600 \qquad (8.1)$$

To verify the turbulent velocity (v) (m/s) in any process pipe section having a radius (r) (m) with a throughput rate Q (m^3), following equation may be used:

$$v = Q/(\pi \times r^2 \times 3600) \qquad (8.2)$$

Similarly, for effective cleaning of the tanks and silos, it is necessary to pump about 7.5–9.5 L/ft for Silo and about 0.75–1.15 L/ft in case of horizontal tanks. The flow rate can be calculated as below:

$$\text{Silos: } v = \pi D \times 9.5 \times 60 \qquad (8.3)$$

$$\text{Horizontal Tanks: } v = 2\pi r h \times 1.15 \times 60 \qquad (8.4)$$

In eqs 8.3 and 8.4: v = required flow rate in kL/h; π = 3.14; D = diameter of tank in feet; r = radius of tank in feet; and h = height of tank in feet.

The mechanical force/turbulence of the CIP solutions helps to transport the CIP liquid to the soiled surface, react with the soil and finally remove the dissolved soil and transfer it out of the equipment being cleaned. Hygienic design of the plant is a prerequisite to ensure the effectiveness of CIP. Perfect cleaning parameters and excellent CIP execution will not give good results if the equipment has design faults, such as dead ends that cannot be flushed through. Dead ends are never desirable, but if at all it is not avoidable, it is better to have flow directed into the dead end than the opposite, to avoid the risk of a stagnant zone. When the pipe is expanding, the flow velocity will reduce which creates a recirculation zone where shear forces or velocity will be lower than average.

8.7.2 TYPES OF CHEMICALS AND CONCENTRATION

The second most important factor which decides the effectiveness of cleaning is the correct choice of detergents and their concentration. Detergents to be used for cleaning in the dairy industry must be: effective on target soil; nonfoaming or include antifoam; free rinsing/nontainting; noncorrosive; controllable; and environment-friendly. Alkaline detergents are often used for cleaning the dairy equipments as they have the capability to dissolve protein, fat, and sugars (i.e., mostly organic soil). Sodium hydroxide can be used in its pure form or can be formulated to suit the special cleaning purpose. In formulated detergents along with NaOH, different aiding components are also added, which might take care of hard water, suspend the dissolved dirt, wet the surfaces more efficiently, etc.

Formulated CIP detergents usually contain active components, referred to as sequestrants, and surfactants, which are designed to keep soiling in suspension once removed from the surfaces, and to reduce foaming and to give other beneficial properties. Once the correct chemical is selected, the correct concentration must be used and sodium hydroxide is typically used at a concentration of 0.5–2%, but too high levels can cause cross-linking of milk proteins systems, making them even harder to remove. In many dairy applications, acid step is followed by the alkaline cleaning step. The role of acid is to dissolve minerals, as well as digestion of fat, sugar, and protein. Acids commonly used in dairy industry are nitric acid or phosphoric acid. Nitric acid is typically used in the concentration of 0.5–1.5% and higher concentrations need to be considered with care, since it may then attack polymer material as well as SS. Use strong concentrations of sodium hypochlorite should be avoided if possible, as they are corrosive to SS. Heat exchangers are particularly vulnerable, due to their very thin plates. When used hot, these products can contribute to the causes of stress corrosion cracking in SS tanks and must be removed from the system by using hot water or treated water.

Caustic-based detergents are the most preferred for cleaning in dairy industry due to added advantages which include: excellent detergency properties; disinfection properties, especially when used hot; effective at removal of protein soil; auto strength control through use of conductivity meter; highly effective for high levels of soil as compared to acid; and is cost-effective. Some of the limitations of using caustic detergents are: they are easily degraded by CO_2 forming carbonate; ineffective at removing inorganic scale; rinsability is poor; not suitable for treating surfaces made of aluminum; and its activity is affected by the hardness of water.

Acid-based detergents when used for cleaning are highly effective against the removal of inorganic soil and are not degraded in presence of CO_2. The effectiveness is not affected by the hardness of water and is more effective in the systems where soil level is low. Acid detergents can be easily rinsed from the system and their concentration can be easily maintained by using conductivity meter. Disadvantages associated with usage of acids as detergents are: less effective against organic soil; lower biocidal activity; limited effectiveness in high soil environments; high corrosion risk especially with nitric acid and discharges of phosphate/nitrate are not environmentally friendly.

8.7.3 TIME CONSIDERATIONS

It is important to specify the correct processing times of cleaning methods, as long-standing cleaning procedures will cause fouling on the surface of processing equipment especially in heat exchanger plate, which can result in the growth of thermophilic bacteria. The minimum time for a CIP cycle is dependent upon all the other factors: mechanical action, detergent type and concentration, and temperature. If one part of the equation is not ideal, then extending the time can sometimes compensate. Too little time and the soiling will not be removed, but there is little benefit in massively overextending cycle times. In dairy industry, depending upon the type of equipment and soil, the total time for a complete caustic acid CIP may extend up to 90–120 min.

8.7.4 TEMPERATURE

The effectiveness of the detergents increases when the solutions are used hot (70–80°C) and is of utmost important in dairy industry as the soils are fat based. Hot CIP also helps to sanitize the plant, killing off any possibility of infections.

8.8   CIP PROCEDURE

CIP procedure is a step by step protocol which is used by the dairy industry to clean the equipments and the pipelines. The cleaning procedure may be used for intermediated cleaning or final cleaning. The production cycle is always completed with CIP as it prevents the milk and milk products from getting dry which arrests the growth of microorganisms. Essentially CIP comprises of the following stages:

- Wetting of the soiled surface
- Removal of the soil from the surface by cleaning solution, emulsification, saponification, and/or mechanical action
- Dispersion of the undissolved soiling material
- Removal of the used detergent solution along with the soil
- Rinsing to remove the detergent residues
- Sanitizing the cleaned surface with the help of hot water or food grade chemical sanitizers

It is recommended that the final cleaning takes place once per day. Intermediary cleaning is used only if a short stop in production is required. CIP for dairy equipment is usually conducted in two stages, caustic cleaning followed by acid cleaning. Often one more caustic cleaning is done to remove residual acid. Thus, it becomes a five- to seven-step CIP cleaning program, comprising pre-rinse, caustic cleaning, rinse, acid cleaning, rinse, weak alkali rinse and final rinse.

- Pre-rinse: Soiled equipment surfaces are rinsed with water to remove the gross amounts of loose soils. Final rinse water from previous CIP cycle can be used as pre-rinse water to save the water consumption.
- Caustic cleaning cycle: Removal of residual soils from equipment surfaces is carried out by using alkaline chemical solutions (usually hot).
- Rinse: Rinsing of all surfaces with cold to hot water, depending upon the temperature of the cleaning cycle, to thoroughly remove all remaining chemical solution and soil residues.
- Acid cleaning cycle: A mild acid rinse of the equipment neutralizes any alkaline residues left and removes any mineral soil present.

- Weak alkali rinse: A mild alkali rinse of the equipment neutralizes any acid residues left in the line or equipments.
- Final rinse: All equipment surfaces are rinsed or flooded with treated water for conditioning the equipments.

Example of five-step (Fig. 8.7) and seven-step (Fig. 8.8) CIP program suited for cleaning of pasteurizers, clarifiers, separators, and homogenizers is explained as follows:

Pre-rinse with freshwater (45–50°C/3–5 min)
↓
Circulation with caustic solution of 1–1.5% concentration (75–80°C/30 min)
↓
Rinse with treated freshwater till complete removal of caustic (50–60°C/4–5 min)
↓
Sanitize with hot treated water (80–85°C/20–15 min)
↓
Final rinse with treated water till the equipment surface attends a temperature of <35°C

FIGURE 8.7   Five-step CIP program.

Pre-rinse with freshwater (45–50°C/3–5 min)
↓
Circulation with caustic solution of 1–1.5% concentration (75–80°C/30 min)
↓
Rinse with treated freshwater till complete removal of caustic (50–60°C/4–5 min)
↓
Circulation with nitric acid solution of 0.8–1.2% concentration (70–75°C/20 min)
↓
Rinse with treated freshwater till complete removal of acid (50–60°C/4–5 min)
↓
Rinse with caustic solution 0.5–0.8% concentration till complete removal of acid (50–60°C/4–5 min)
↓
Final Rinse with treated water till the equipment surface attends a temperature of <35°C

FIGURE 8.8   Seven-step CIP program.

Single-stage on other hand is essentially a three-step cleaning process (pre-rinse, cleaning, and rinse) using lesser time, water, and energy (Fig. 8.9). This type of cleaning process is usually used for intermediate cleaning of milk storage tank, silo, and pipelines.

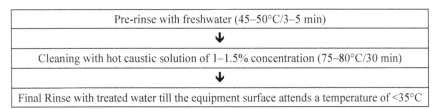

Pre-rinse with freshwater (45–50°C/3–5 min)
↓
Cleaning with hot caustic solution of 1–1.5% concentration (75–80°C/30 min)
↓
Final Rinse with treated water till the equipment surface attends a temperature of <35°C

FIGURE 8.9   Single-stage CIP program.

8.9   CONTROLS AND MONITORING OF CIP

The best-designed CIP system will fail or run inefficiently/ineffectively if not controlled and monitored properly. The CIP control system opens/ closes valves, runs pumps, performs controls functions, and has set points to ensure proper CIP cycle operation. Plant operators engaged in CIP operations must be able to select individual programs for each CIP circuit; monitor the CIP status; receive alarms in the event of malfunction; take corrective action; verify CIP performance and make changes if necessary in the recipe of the program. The control of CIP systems can vary from simple manual operation to fully integrated programmable logic controller (PLC) controls with touch screen operator interfaces. The point where control is to be placed for effective controlling of CIP is:

- Monitor flow in return pipe.
- Monitor the temperature in the return pipe during the caustic, acid and the hot water cycle.
- Monitor flow rate in the pre-rinse cycle.
- Pulsing of the pre-rinse must be monitored for effectiveness of soil removal.
- Dosing of the caustic and acid to be monitored in respect to the flow of freshwater.
- Flow rate of the freshwater in caustic, acid and hot water tanks need to be monitored.
- Time required for individual cycle is to be monitored.

8.10   ERIFICATION OF CIP PROGRAM

In dairy plants, equipment should be subjected to routine microbiological monitoring to verify the effectiveness of the CIP program. Apart from the microbiological index, many procedures are available for ensuring the effectiveness of CIP and those depending on the measurement of physical parameters include: CIP cycle times, solution temperatures, flow rates, and pressures. Chemical analyses which can be monitored for checking the effectiveness include: checking the concentration using conductivity meter, alkalinity, pH, color characteristic of return alkali and acid solution, level of suspended solids, and presence of foam. Physical verification can be done after CIP of the milk pipelines, filling bowl, balance tank, and equipment surfaces for any traces of milk solids/residues.

8.11   SUMMARY

CIP relies on the principle of applying a suitable detergent or solvent at a suitable flow, pressure, temperature, and concentration for the correct length of time. For effective removal of soil, components and fittings used for process installations must be of smooth surface and preferably of 304/316L SS or similar corrosion resistance and must meet 3A standards. The 3A or similar design and fabrication standards must be used for surface finish of tanks and pipelines during installation of the dairy plant and CIP circuit. Welding techniques must be used as mentioned in AWS D18 to avoid the buildup of soil/microorganisms. Dead legs must be avoided in the circuits and if not avoidable, more than two dead legs must not be encouraged. CIP circuit must be fitted with safety breaks and the circuit balancing should be done for effective cleaning. Newer techniques such as foam/gel cleaning and dry ice blasting can also be used for cleaning and sanitizing the dairy equipments. For effective CIP, the time, temperature, mechanical action, and detergent must be monitored and controlled effectively. The chapter also discusses mechanisms of fouling of milk.

KEYWORDS

- boil out (fill/flood) system
- calcium
- caustic soda
- CIP circuit
- CIP program
- fouling
- manual cleaning
- multichannel CIP system
- reuse (recovery)
- CIP system
- single-use recirculation CIP system
- Sinner's circle
- soil
- sterilizing-in-place

REFERENCES

1. Ansari, Md. I. A.; Sahoo, P. K.; Datta, A. K. Milk Fouling Simulation in a Triple Tube Heat Exchanger. *J. Agric. Eng.* **2012,** *49*(3), 1–11.
2. Bell, K. J.; Sanders, C. F. Prevention of Milkstones Formation in a High-temperature-short-time Heater by Preheating Milk, Skim Milk and Whey. *J. Dairy Sci.* **1944,** *27*, 499–504.
3. Belmar-Beiny, M. T.; Fryer, P. J. Preliminary Stages of Fouling from Whey Protein Solutions. *J. Dairy Res.* **1993,** *60*, 467–483.
4. Beuf, M.; Rizzo, G.; Leuliet, J. C.; Müller-Steinhagen, H.; Yiantsios, S.; Karabelas, A.; Benezech, T. Fouling and Cleaning of Modified Stainless Steel Plate Heat Exchangers Processing Milk Products. *Fouling Clean. Plate Heat Exch.* **2006,** *14*, 96–106.
5. Britten, M.; Green, M. L.; Boulet, M.; Paquin, P. Deposit Formation on Heated Surfaces: Effect of Interface Energetics. *J. Dairy Res.* **1988,** *55*, 551–562.
6. Burdett, M. The Effect of Phosphates in Lowering the Amount of Deposit Formation During the Heat Treatment of Milk. *J. Dairy Res.* **1974,** *41*, 123–129.
7. Burton, H. Deposit of Whole Milk in Treatment Plants: A Review and Discussion. *J. Dairy Res.* **1968,** *35*, 317–330.

8. De Jong, P.; Bowman, S.; Van Der Linder, H. J. L. J. Fouling of Heat Treatment Equipment in Relation to Denaturation of β-lactoglobulin. *J. Soc. Dairy Technol.* **1993**, *45*, 3–8.

9. Fryer, P. J.; Pitchard, A. M.; Slater, N. K. H.; Laws, J. F. In *A New Device for Studying the Effects of Surface Shear Stress on Fouling in the Presence of Heat Transfer Fouling in Heat Exchange Equipment*, Symposium of the American Society of Mechanical Engineers, Niagara Falls, 1984; pp 27–32.

10. Hiddink, J.; Lalande, M.; Maas, A. J. R.; Streuper, A. Heat Treatment of Whipping Cream. I. Fouling of the Pasteurization Equipment. *Milchwissenschaft* **1986**, *41*, 542–546.

11. ISO 14159. *Safety of Machinery—Hygiene Requirements for the Design of Machinery*, International Organization for Standardization, 2002, 1–30.

12. Jeurnink, T. J. M. Effect of Proteolysis in Milk on Fouling in Heat Exchangers. *Neth. Milk Dairy J.* **1991**, *45*, 23–32.

13. Jeurnink, T. J. M. Fouling of Heat Exchangers by Fresh and Reconstituted Milk and the Influence of Air Bubbles. *Milchwissenschaft* **1995**, *50*, 189–193.

14. Jeurnink, T. J. M.; Brinkman, D. W. The Cleaning of Heat Exchangers and Evaporators After Processing Milk or Whey. *Int. Dairy J.* **1994**, *4*, 347–368.

15. Koen, G.; Leen, M.; Jan De, B.; Van, R. R. Applications of Modelling to Optimize Ultra-high Temperature Milk Heat Exchangers with Respect to Fouling. *Food Control* **2004**, *15*(2), 117–130.

16. Lalande, M.; Tissier, J. P.; Corrieu, G. Fouling of a Plate Heat Exchanger Used in Ultra-high-temperature Sterilization of Milk. *J. Dairy Res.* **1984**, *51*, 557–568.

17. Mottar, J.; Moermans, R. Optimization of the Fore-warming Process with Respect to Deposit Formation in Indirect Ultrahigh Temperature Plants and the Quality of Milk. *J. Dairy Res.* **1988**, *55*, 563–568.

18. Müller-Steinhagen, H.; Zhao, Q. Investigation of Low Fouling Surface Alloys Made by Ion Implantation Technology. *Chem. Eng. Sci.* **1997**, *52*(19), 3321–3332.

19. Petermeier, H.; Benning, R.; Delgado, A.; Kulozik, U.; Hinrichs, J.; Becker, T. Hybrid Model of the Fouling Process in Tubular Heat Exchangers for the Dairy Industry. *J. Food Eng.* **2002**, *55*(1), 9–17.

20. Rosmaninho, R.; Santos, O.; Nylander, T.; Paulsson, M.; Beuf, M.; Benezech, T.; Yiantsios, S.; Andritsos, N.; Karabelas, A.; Rizzo, G.; Müller-Steinhagen, H.; Melo, L. F. Modified Stainless Steel Surfaces Targeted to Reduce Fouling—Evaluation of Fouling by Milk Components. *Food Eng.* **2007**, E-article. DOI: 10.1016/j.jfoodeng.2006.09.008.

21. Sandu, C.; Lund, D. Minimizing Fouling in Heat Exchanger Design. *Biotechnol. Prog.* **1985**, *1*, 10–17.

22. Skudder, P. J.; Brooker, B. E.; Bonsey, A. D.; Alvarez-Guerreo, N. R. Effect of pH on the Formation of Deposit from Milk on Heated Surfaces During Ultra-high-temperature Processing. *J. Dairy Res.* **1986**, *53*, 75–87.

23. Tissier, J. P.; Lalande, M. Experimental Design and Methods for Studying Milk Deposits Formation on Heat Exchange Surfaces. *Biotechnol. Prog.* **1986**, *2*, 218–229.

24. Visser, J.; Jeurnink, T. J. M. Fouling of Heat Exchangers in the Dairy Industry. *Exp. Therm. Fluid Sci.* **1997**, *14*, 407–424.

25. Zhao, Q.; Liu, Y.; Müller-Steinhagen, H. Graded Ni-P-PFE Coatings and Their Potential Applications. *Surf. Coat. Technol.* **2002**, *155*, 279–284.

NOVEL METHODS TO DETECT ADULTERATION OF GHEE OR MILK FAT

A. S. HARIYANI[1,*], A. J. THESIYA[2], TANMAY HAZRA[1], J. B. GOL[3], and RUPESH S. CHAVAN[4]

[1]Dairy Chemistry Department, College of Dairy Science, Kamdhenu University, Amreli 365601, Gujarat, India

[2]Dairy Technology Department, College of Dairy Science, Kamdhenu University, Amreli 365601, Gujarat, India

[3]Department of Quality Assurance, Gujarat Co-operative Milk Marketing Federation (GCMMF), Anand, Gujarat, India

[4]Department of Quality Assurance, Mother Dairy Junagadh, Junagadh 362001, Gujarat, India

*Corresponding author. E-mail: ashariyani@ku-guj.com

CONTENTS

This chapter is partially printed from Ashvinkumar S. Hariyani, A study on physicochemical qualities of market ghee. Unpublished M. Tech. in Dairying (Dairy Chemistry) Thesis for Division of Dairy Chemistry at National Dairy Research Institute (NDRI), Karnal 132001, India, 2010; Pages 14

ABSTRACT

Food adulteration is a global concern, and developing countries are at higher risk associated with it due to lack of monitoring, policies, and lack of infrastructure for analyzing the same. Ghee is a widely consumed dairy product in India and that prepared from cow milk is mentioned in ayurvedic texts as an ingredient of many formulations/additive as well. Ghee is almost three to four times costlier than the other edible fats and oils, and its limited supply attracts the unscrupulous manufacturers to adulterate it with cheaper alternatives in Indian subcontinent. Detection of ghee adulterated with vegetable oils/fats and animal body fats is a key concern at grassroot level during processing and marketing. This chapter presents a detailed information of common ghee adulterants as well as different methods to detect the adulterants both qualitatively and quantitatively. Among reported techniques, Baudouin test, Bieber's test, Bömer value, differential scanning calorimetry, differential thermal analysis, gas–liquid chromatographic, high-pressure liquid chromatography, Holde's test, hydroxamic acid test, iodine value, refractive index, Reichert–Meissl value, saponification value, solidification point, spectroscopic method, thin-layer chromatography, ultraviolet spectroscopy, visible spectroscopy, and other methods are discussed in current chapter.

9.1   INTRODUCTION

Lipids are one of the most important constituents of milk and milk products. Major part of milk lipids consists of triglycerides (generally called *fats*). Minor components of milk lipids include partial glycerides (mono and diglycerides), phospholipids, fat-soluble vitamins, cholesterol, squalene, waxes, etc. In India, milk fat is mostly consumed in the form of ghee (clarified butterfat). Ghee is being manufactured by mainly four methods, namely, desi (traditional) method, creamery butter method, direct cream method, and prestratification method.[39]

However, due to short supply of ghee, particularly, in the lean (summer) season and comparatively more demand, expensiveness (costing 3–4 times as much as edible vegetable oils), and variable chemical composition, ghee is prone to adulteration by the unscrupulous traders in the market. The commonly used adulterants are vegetable oils and fats, animal body fats, mineral oils, starchy material, etc. Extensive survey of the literature reveals that several methods have been developed to detect the adultera-tion in ghee. These methods were mostly based on chemical parameters such as fatty acid composition and the physicochemical constants.[59,102]

This chapter discusses different physicochemical attributes of milk fat and how it is influenced by different adulterants. Based on deviation from the normal value of physicochemical characteristics, authenticity of milk fat can be predicted. Therefore, this chapter delineates the novel methods to detect adulteration of ghee/milk fat.

9.2   METHODS BASED ON PHYSICAL PROPERTIES FOR DETECTION OF ADULTERATION IN GHEE

9.2.1   MELTING POINT

Melting point (slip point) of various oils and fats covers a wide range and this property has been employed for checking the adulteration of milk fat. Body fats (36–51.3°C) and vanaspati (37.8–38°C) have slightly higher melting point,[116] while vegetable oils (20–30°C) have slightly lower melting point than milk fat (28–41°C).[7] Although the average melting point of vanaspati ghee is reported to fall between 31°C and 37°C,[94,95] yet the melting point of vanaspati (hydrogenated oils) oils varies between 30°C and 65°C depending on the extent of hydrogenation, more the hydro-genation, higher is the melting point.[20] Singhal[104] reported that buffalo milk fat (33.4–34.2°C) has slightly higher melting point than cow milk fat (30.6–31.2°C) and ghee from cotton tract area showed considerably higher melting point (43.0–44.0°C), which resembles with that of animal body fats (43.9–51.0°C).[104] Among the animal body fats, buffalo body fat showed the highest melting point (50.5–51.0°C), while pig body fat showed the lowest (43.9–44.6°C).

The addition of animal body fats (buffalo, goat, pig, and sheep) at 5–20% level increased the melting point of pure buffalo or cow ghee. Buffalo body fat caused the largest increase in melting point of ghee (5.7°C increase at 20% level), while pig body fat caused the least increase (2.2°C at 20% level). The study, however, concluded that adulteration up to 20% level does not make significant change in the melting point of ghee and, therefore, the method is not found to be useful for the detection of adulteration. Some workers also confirmed these observations using body fats (buffalo, goat and pig) and vanaspati at 5–20% level of adulteration (irrespective of their mode of addition, either directly to ghee or through milk) and noted that the increase in melting point was more with body fats than vanaspati.[32,95]

9.2.2   APPARENT SOLIDIFICATION TIME TEST

The apparent solidification time (AST) test was developed to detect adulteration of milk fat with vegetable oils and animal body fats.[9] The AST is determined by taking 3 g of melted fat sample in a test tube (10.0 × 1.0 internal diameter) to apparently become solidified at 18°C.[9] Studies conducted on the solidification behavior of various oils and fats including milk fat in terms of AST at the selected temperature (18°C) have revealed that the average AST values for buffalo and cow milk fat were 2 min 40 s and 3 min 10 s, respectively.[9] The average AST values of pig body fat, goat body fat, and vanaspati were 1 min 30 s, 40 s, and 1 min 50 s, respectively. On the other hand, all the vegetable oils under study remained liquid for an indefinite period. Addition of vegetable oils caused an increase in the AST values of buffalo and cow pure milk fat, whereas the addition of body fats and Vanaspati (hydrogenated vegetable oils) resulted in the decrease in the AST values of buffalo and cow pure milk fat, depending on the amount of adulterant oils and fats added. Taking into account the overall range of AST values at 18°C pertaining to fresh, stored, and seasonal samples of both buffalo and cow pure milk fat as the criteria for the detection of adulteration, it was found that the technique could detect the addition of individual vegetable oils at all the levels in case of cow milk fat but not in buffalo milk fat. Adulteration of buffalo milk fat with vanaspati (hydrogenated vegetable fat) was detectable at levels ≥10%, but not in cow milk fat. Addition of goat body fat to both cow and buffalo milk fat was detectable at levels ≥10%, whereas pig body fat was detectable only in buffalo milk fat at levels ≥10%.

9.2.3   COMPLETE LIQUEFACTION TIME TEST

The complete liquefaction time (CLT) test was developed to detect adulteration of milk fat with foreign fats and oils.[3] The CLT of the fat samples was recorded by observing the time taken by the solidified fat samples to get melted completely at 44°C. At 44°C, the CLT values of pure cow ghee samples ranged from 2 min 12 s to 3 min 15 s with the mean of 2 min 52 s, while that of pure buffalo ghee ranged from 2 min 35 s to 3 min 15 s with the mean of 2 min 57 s. There were large variations between CLT values of ghee samples collected over a period of whole year both for cows and buffaloes. Further, it was observed that both (cow and buffalo) type of ghee samples showed higher CLT values in the summer months (May–September) and lower CLT values in the winter months (November–March).

Addition of vegetable oils (palm, rice bran, and soybean) caused a decrease, whereas addition of body fats (buffalo, goat, and pig) resulted in an increase in the CLT values of cow and buffalo pure ghee at 44°C. This decrease or increase observed in CLT values caused by the addition of adulterant oils/fats to ghee depended upon the amount of adulterants added. Higher the quantity of adulterant added, greater was the effect. However, in case of ghee samples containing body fats, samples with pig body fat showed slightly lower increase in CLT values than the samples containing buffalo and goat body fats. Taking into account the overall range of CLT values at 44°C, that is, from 2 min 12 s to 3 min 15 s for pure ghee (cow and buffalo) as the criteria for the detection of adulteration, a perusal of the results on CLT values of adulterated ghee samples revealed that addition of vegetable oils individually at the levels of 5%, 10%, and 15% to either cow ghee or buffalo ghee was not detectable. However, among animal body fats, buffalo and goat body fats in either of the ghee were detectable at 10% level while pig body fat was detected only at higher (15%) level of adulteration.

9.2.4 CRYSTALLIZATION TIME TEST

The detection of added animal body fats in milk fats has always been a challenging task for dairy professionals. In the past, some physical methods, which were based on the partial solidification behavior of milk fat such as opacity test,[61] crystallization test,[74] and AST test[10] were developed and applied to detect the added animal body fats in milk fats (ghee). However, each of these methods has its own limitations. Among these tests, crystallization test was considered to be an excellent method to detect animal body fats.

The crystallization time test is defined as time taken for the onset of crystallization of milk fat at 17°C when dissolved in a solvent mixture consisting of acetone and benzene (3.5:1.0). This test was used on samples of cow and buffalo pure ghee pertaining to whole of the year collected on bimonthly basis, as well as the ghee samples added with adulterants (animal body fats and vegetable oils), individually, at 5%, 10%, and 15% levels.[3] Crystallization time for the pure cow ghee and pure buffalo ghee samples ranged from 6 min 50 s to 16 min 20 s and from 6 min 30 s to 12 min 30 s with the mean of 11 min 4 s and 8 min 42 s, respectively. The crystallization time of ghee samples increased when the samples were adulterated with vegetable oils (palm, rice bran, and soybean), while it was found to decrease for ghee samples adulterated with animal (buffalo and goat) body fats. The extent of increase and decrease was dependent upon the level of adulteration with vegetable oils and animal body fats, respectively. Higher the level of adulterant oils/fats greater was the effect. However, ghee samples adulterated with pig body fat, initially, showed a decrease in crystallization time at 5% level of adulteration, which subsequently showed an increase as the level of adulteration increased. At 15% level of pig body fat adulteration, the values of crystallization time became closer to that of corresponding average values for pure ghee samples. The same trend was observed for both type of ghee (cow and buffalo). The crystallization time was highest in the month of May while it was lowest in July.

Generally, crystallization time is expected to be higher if the fat has more unsaturated fatty acid, that is, crystallization time will be increased with the increase in butyro refractometer (BR) reading and iodine value of a fat. It was concluded that the crystallization time test is a useful tool to detect addition of animal body fats particularly, buffalo and goat body fat to milk fat at 5% level.[3] However, pig body fat could not be detected by this test when added to milk fat at any of the levels studied.

9.2.5 SOLIDIFICATION POINT

Solidifying point is defined as the temperature at which fat shows first sign of appearance of solid phase on cooling. Solidification temperature of milk fat depends very much on the procedure employed for cooling.[114] Solidification point reported 19.7°C and 23.6°C for samples of the same milk fat cooled by immersion at 14°C and 20°C, respectively.[85] The solidifying point of buffalo ghee (16.0–28.0°C) is reported to be slightly higher than cow ghee (15.0–23.5°C). The solidifying points for beef tallow and lard vary between 32–37°C and 25–30°C, respectively,[89] whereas most of the vegetable oils, such as sunflower oil, maize oil, soya oil, groundnut oil, and cottonseed oil have very low solidification point in the range of −16–5°C, with the exception of palm oil (22–40°C), palm kernel oil (24–26.5°C), and coconut oil (22–23.5°C) whose values are close to that of animal fats reported above.

9.2.6   TITRE VALUE

The titre value of ghee represents the highest temperature reached when the liberated water insoluble fatty acids are crystallized under arbitrarily controlled conditions. The titre is generally taken to represent the solidification point of the fatty acids, although they actually solidify over a range of temperature. For its determination, ghee (oils and fats) is saponified with glycerol–potassium hydroxide solution. The resulting soap is decomposed with sulfuric acid and the liberated water-insoluble fatty acids are separated, washed free from mineral acid, and dried. Titre is then determined on these fatty acids.[18]

Titre value of butterfat lies between 33°C and 38°C.[42] Animal body fats such as beef tallow, lard, and mutton tallow generally have a higher range (32–48°C). On the other hand, vegetable oils such as sunflower oil, safflower oil, soybean oil, groundnut oil, palm kernel oil, etc., generally have lower titre values (15–32°C), except cottonseed oil (30–37°C) and palm oil (40–47°C). The titre value of ghee has been reported as 33.5°C,[42] whereas a survey conducted by Directorate of Marketing and Inspection, Govt. of India reported the titre value of ghee samples from cotton tract areas varying from 40.4°C to 44.6°C.[104] At one time, the titre value was used for judging the quality of ghee produced in cotton tract areas, but later dispensed with as it was considered to be of not much use. Moreover, its use is limited due to the introduction of more modern instrumental methods of analysis.

9.2.7   DENSITY AND SPECIFIC GRAVITY

Density or specific gravity of the fatty oils has long been used in connection with the analysis and identification of oils. The density of fats in their liquid form is commonly expressed as specific gravity rather than in terms of absolute density. The specific gravity of oils and fats depends upon their chemical composition and the temperature at which density or specific gravity is measured.[53]

Specific gravity values for cow ghee and buffalo ghee at 30°C are reported to be 0.9358–0.9443 and 0.9340–0.9444, respectively.[89] The liquid milk fat at 20°C in general, has a density of about 0.915 g/mL. Most of the edible vegetable oils and fats have a density in the range of 0.910–0.927 at 20°C.[113] Animal body fats have slightly lower density values ranging from 0.894 to 0.906 at 20°C.[68] With a view to detect adulteration, density, and specific gravity of ghee (including cotton tract) and animal body fats (buffalo, goat, pig, and sheep) were studied at 40°C. Based on the narrow differences in the values for different types of oils and fats, it was concluded that detection of adulteration may not be possible using this property.[104]

9.2.8 BÖMER VALUE

Bömer value (BV) is defined as the sum of the melting point of saturated triglycerides (isolated by diethyl ether method) and twice the difference between this melting point and that of the fatty acids obtained after the saponification of these triglycerides. This test, originally developed by Bomer in 1913 for the detection of lard in tallow, depends upon the triglyceride structure of the fat.[91]

The BV of both cow and buffalo ghee ranges from 63 to 64, whereas those of animal body fats, for example, goat, sheep, and buffalo ranges from 68 to 69 and that of pig body fat from 75 to 76. The BV of ghee increased on adulteration with body fats even in the presence of vegetable oils, but not when vegetable oils alone were added.[105,106] The method could be used as a confirmatory test for the detection of pig body fat in ghee. However, genuine cotton tract ghee, which behaved similar to adulterated ghee samples, could not be sorted out by this test and hence may be mistaken as adulterated ghee. This test has also been applied successfully for the detection of body fats (buffalo, goat, and pig) and vanaspati added to buffalo ghee at 20% level, irrespective of mode of adulteration either directly to ghee or through milk.[96]

9.2.9 REFRACTIVE INDEX AND BR READING

This property which concerns the degree of bending of light waves passing through a liquid or transparent solid is a characteristic for the particular liquid or solid. For oils and fats, it increases with the unsaturation and decreases with rise in temperature. In the case of milk fat, the constant may be determined readily with an Abbe refractometer at 40°C. The instrument is calibrated in BR readings instead of absolute refractive indices.

The refractive index of milk fat generally ranges between 1.4538 and 1.4578.[48,113] For animal fats at 40°C, it lies in the range of 1.4570–1.4630. For most of the vegetable oils, it ranges from 1.4600 to 1.4750 but for some vegetable oils such as lauric fats (coconut and palm kernel oils) and palm oil, it lies in the range of 1.4480–1.4580.[42,53] Refractive index and BR readings are interconvertible.[18,20,89,116] The values for BR readings of milk fat (40–45) and vegetable oils and fats (above 50) are so wide apart that this property could be safely employed as an index for milk fat adulteration with vegetable oils and fats, except coconut oil (38–39) and palm oil (39–40).[40,61] Feeding of cottonseed oil raises the BR reading by 5 units in case of ghee.[89] Normally, BR reading or refractive index of oils and fats increases with the increase in unsaturation and also chain length of fatty acids. The BR readings of animal body fats are in the range of 44–51.[105]

Adulteration of milk fat with animal body fats and vanaspati at a level of 5–20% increased its BR readings.[95,104] Recently, some workers have developed a simple platform test for the detection of vegetable oil (refined mustard oil) added to milk at a level higher than 10% of the original fat on the basis of increase in BR reading of the fat.[6,58] Using general limit of BR reading as 40–43, adulteration of vegetable oil up to 5% in cow ghee and 15% in buffalo ghee can be detected.[9] It was concluded from study that adulteration of milk fat with vegetable oils (except palm oil) added individually at 10% level in cow ghee and 15% level in buffalo ghee, using the general limit of BR reading as 40–43 for Haryana state, could be detected.[3,78] However, adulteration of cow and buffalo pure ghee with animal body fats could not be detected at any of the levels studied.

9.2.10   OPACITY TEST

This test was developed to detect the adulteration of ghee with animal body fats, based on the time taken by the melted fat sample (5.0 g) in a test tube (8 cm × 1.5 cm) to become opaque [outside diameter (OD) >0.5] at 23°C using 590 nm (yellow) filter. It was observed that the normal ghee took more than 35 min, whereas animal body fats (buffalo, goat, and sheep) took only 10–20 s to become opaque.[105] In pure state, both cow and buffalo ghee took almost similar time to become opaque, but adulterated cow ghee took more time than the adulterated buffalo ghee to show a noticeable opacity. The difference in opacity time between pure and adulterated ghee at 5%, 10%, and 20% level, respectively, was 7–8, 9–10, and 14–16 min in case of adulteration with pig body fat; 18–20, 26–29, and 33–34 min in case of adulteration with goat body fat; and 22–26, 31–32, and 35–36 min in case of adulteration with buffalo body fat. It was concluded that the adulteration with buffalo, goat, and sheep body fats at 5% level and above could be safely detected by opacity test. If the sample exhibits opacity within 20 min, it is suspected to be adulterated with animal body fats particularly buffalo, goat, or sheep body fats. However, the limitations of this test are that the detection of pig body fat up to 10% level is difficult and ghee from cotton tract area also cannot be distinguished. Test also fails to detect the body fats in ghee in the presence of vegetable oils.[106]

With the modified procedure, opacity profile of pure ghee and adulterants and recorded the opacity time as the time required by a fat sample at 23°C to acquire the OD in the range of 0.14–0.16 and consequent transmittance of 68–72.[74] It was reported that the opacity time of pure ghee (14–15 min) was much higher than that of ghee adulterated with animal body fats (2–9 min at 10% level and 3–11 min at 5% level of adulteration) and much lower than that of ghee adulterated with vegetable oils (21–25 min at 10% level and 19–21 min at 5% level of adulteration). The opacity test successfully employed to detect the presence of vegetable oils and animal body fats when added directly to milk. These workers also observed the limitations of their test that the cotton tract area ghee (opacity time; 11–12 min) resembled with ghee adulterated with animal body fat such as pig body fat at 5% level (10–11 min).[104]

9.2.11 CRITICAL TEMPERATURE OF DISSOLUTION

Critical temperature of dissolution (CTD) (temperature at which turbidity appears on gradual cooling of the fat dissolved in a warm solvent or solvent mixture) is a characteristic of a particular fat, which depends upon the nature of the solvent, and nature and amount of most insoluble glycerides (usually trisaturated glycerides) present in a fat as well as the mutual solubilizing power exerted on these glycerides by the soluble glycerides.[19,89]

The CTD values for ghee and vanaspati were in the range of 39–45°C and 62–72°C, respectively, employing a 2:1 (v/v) mixture of 95% ethanol and isoamyl alcohol, and reported that gross adulteration of ghee with vanaspati could easily be detected.[17] The CTD values for butter are 50.5–57.5°C (average 54°C), for coconut oil are 28–41°C (average 33°C), and for other oils and fats are 67–82.5°C (average 71–78°C).[29] Similarly, the presence of body fats in ghee was detected by employing either a single solvent such as absolute alcohol or a solvent mixture of 95% ethyl alcohol and isoamyl alcohol in the ratio of 2:1.[17,29] Likewise, CTD was used for the detection of adulteration of ghee with mineral oils using aniline as solvent.[51] The CTD test seems to be simple, but its efficacy is greatly affected by the free acidity (FFA) and rancidity (peroxides).[51] Using the solvent mixture (ethyl alcohol and isoamyl alcohol, 2:1), it is reported that adulteration of ghee up to 15% level with vegetable oils, vanaspati, and body fats could not be detected by CTD value.[10]

9.2.12 FRACTIONATION OF MILK FAT

Fractionation is a thermally controlled process (with or without solvent) in which the milk fat is subjected to a specific temperature/time profile to allow a portion of milk fat to crystallize. Fractionation of fat with or without the use of solvent under suitable conditions of time and temperature combinations, followed by examination of fractions thus obtained has been exploited by some workers as a tool to detect foreign fats in milk fat. Different solvents that have been used for fractionation purpose include ethyl alcohol, acetone, hexane, isopropylalcohol, and 2-nitropropane. Separation of fats has been done into solid (30%) and liquid (70%) fractions after dissolving it in the hot absolute alcohol and maintaining the same at 20°C for 2 h.[15,16] The acetone soluble fraction was iodinated and subsequently subjected to refractive index measurement.

Using this method, the presence of foreign fats at 10% level could be detected. Dry fractionation approach and AST test were used to detect adulteration of ghee with vegetable oil and animal body fats. The average AST value of solid fraction (S_{20}), solid fraction (S_{18}), and liquid fraction (L_{18}) of pure buffalo ghee were 1 min 58 s, 2 min 47 s, and 3 min 10 s, and those for cow ghee were 2 min 30 s, 3 min 21 s, and 3 min 31 s, respectively.[10] Taking overall range of AST values of the S_{20}, S_{18}, and L_{18} fraction of both pure buffalo ghee and cow ghee at 18°C as the basis for the detection adulteration in milk fat, the study revealed that fractionation technique could further help in detecting adulteration of only cow ghee especially with the mixture of goat body fat with individual vegetable oils or vanaspati even at 5% level, which otherwise not possible in the unfractionated ghee samples, thereby helping in increasing the sensitivity of detection of adulteration in case of cow ghee only.

Solvent fractionation using acetone as a solvent to get different fractions is also used for differentiating adulterated and nonadulterated fats. Three temperatures (16°C, 8°C, and 4°C) were selected for successive fractionation.[3] The solid fractions obtained at 16°C and 8°C were named as S_{16} and S_8, respectively, while solid and liquid fractions obtained at 4°C were named as S_4 and L_4, respectively. On fractionation, animal body fats got concentrated in the first fraction (S_{16}), whereas vegetable oils got concentrated in the last fraction (L_4). Hence, first fraction was analyzed for CLT test performed at 44°C and 46°C, and last fraction was analyzed for BR reading and iodine value. Results of CLT test at 44°C done for first fraction (S_{16}) of cow and buffalo pure ghee showed that the temperature of 44°C is not suitable for application of CLT test on first fractions because it has failed in case of pure cow ghee as fraction of pure cow ghee obtained in the months of summer season (May–September) did not melt completely and some opacity remained throughout in the body of the fat fraction. CLT of solid fractions (S_{16}) of pure cow ghee and pure buffalo ghee at 46°C ranged from 4 min 5 s to 9 min and from 5 min 10 s to 7 min 15 s, respectively.

Taking into account the overall range of CLT at 46°C, that is, from 4 min 5 s to 9 min for both cow and buffalo pure ghee and also considering the melting behavior of samples at this temperature, it was observed that buffalo body fat adulteration along with either of the vegetable oils could be detected even at 5% level in case of cow ghee, while it was detected only at 15% level in buffalo ghee. Goat body fat along with either of the vegetable oils was detected at 10% level of adulteration in cow ghee, while it was detected only at 15% level of adulteration in buffalo ghee. However, pig body fat was detected at 15% level of adulteration with either of the vegetable oils in both cow and buffalo ghee.

9.2.13   SPECTROSCOPIC METHODS

Spectroscopic methods using visible (400–800 mμ), ultraviolet (UV) (200–400 mμ), and infrared (IR) (2–15 μ) regions have been used by many workers for characterization of fats and oils.[119–121]

9.2.13.1   TESTS BASED ON VISIBLE SPECTROSCOPY

This technique was applied for the detection of cheuri (*Madhuca butyracea*) fat in ghee, a common adulterant in Nepal. Pure ghee showed no absorption band in visible range (600–700 nm), whereas cheuri fat showed an absorption band with maxima between 640 and 680 nm. Even 5% cheuri fat added to ghee could be detected in this range.[49]

9.2.13.2   TESTS BASED ON UV SPECTROSCOPY

UV spectroscopy has been applied for characterizing the polyunsaturated fatty acids (PUFA) content of various oils and fats including milk fat and also for the detection of butterfat adulteration with foreign fats. UV spectrum of unsaponifiable matter (USM) extracted from ghee and animal body fats examined between 200 and 320 nm and observed first absorption maxima between 215 and 220 nm for both fats,[97] whereas ghee sample showed a second maxima at 270 nm, which was shifted to 280 nm in case of animal body fats. However, on this basis, adulterated ghee could not be differentiated from pure ghee.[10,97]

9.2.13.3   TESTS BASED ON IR SPECTROSCOPY

The IR absorption has been extensively used in the analysis of lipids especially for cis- and trans-isomers. Unsaturated fatty acids of natural vegetable oils and fats are in cis-configuration and are isolated (nonconjugated). Partial hydrogenation or oxidation may result into formation of trans-isomers. Animal and marine fats may also contain small amounts of natural trans-isomers.[1,55] Bovine milk fat contains a low level (5%) of trans fatty acids in comparison with hydrogenated vegetable oils, in which the value may be as high as 50% due to nonstereospecific hydrogenation.[36]

For demonstrating the presence of hydrogenated fats in milk fat, researchers applied IR spectrophotometry and observed that the absorption maxima at 10.36 µ gets increased by the addition of hydrogenated fats containing iso-oleic acids (trans-octadecenoic acids).[13,35] The cis–trans configurations of individual fats (milk fat, animal body fat, vegetable fat, and hydrogenated fat) and their mixtures using IR spectroscopy reported that the additions up to 10% of animal, vegetable, and hydrogenated fats to milk fat could be detected.[56] IR spectroscopic method can detect as little as 3% foreign fat in milk fat.[93]

The IR spectra of cow ghee, buffalo ghee, animal body fats (buffalo, goat, sheep, and pig) and ghee adulterated with body fats in the 4000–600 cm^{-1} region and observed distinct differences between body fats and ghee in the region of 1300–1180 cm^{-1} and 1120–1100 cm^{-1}, respectively.[97] Body fats showed the presence of five to six bands, while ghee showed only two bands. Ghee samples adulterated with body fats also showed three to six extra bands. USM extracted from the ghee, body fats, and adulterated ghee samples also exhibited the similar pattern of bands as reported above for the whole fat. The above author concluded that the differences in IR spectrum of ghee and body fats could be used to detect ghee adulteration with body fats at 10% level. It has been reported that on the basis of increased level of trans-isomers in ghee, as low as 5% of vanaspati added to ghee could be detected.[10]

9.2.14 DILATATION BEHAVIOR

This property is based on the thermal expansion behavior of milk fat. Using this property, different solid and liquid fractions of milk fat from different species in the temperature range of 10–80°C observed that solid and liquid fractions in equal proportions are obtained at 33°C, 30°C, and 24.5°C in case of buffalo, cow, and goat milk fats, respectively.[50] This property, employed for detection of adulteration in ghee, studied the proportion of solid and liquid fractions of pure and adulterated ghee at 5%, 10%, and 15% level. It was observed that the ratio of solid to liquid fraction for pure cow and buffalo ghee was 2.37 and 3.10, respectively.[10] On the basis of solid/liquid ratio, it was found that vegetable oils could be detected in cow ghee while body fats and vanaspati could be detected in buffalo ghee even at 5% level.

9.2.15 MICROSCOPIC EXAMINATION OF FAT

Microscopic examination of the sterol crystals has also been employed in the detection of adulteration of milk fat with the foreign fats especially vegetable fats.[18,31,45] If the sterol crystals only show the form of a parallelogram with an obtuse angle of 100°, which is characteristic for cholesterol, the fat sample is considered to be free from vegetable fat. However, if the sterol crystals show the elongated hexagonal form with an apical angle of 108°, which is characteristic for phytosterols, or if some of the sterol crystals have a reentry angle (Swallow's tail), which is characteristic for mixtures of cholesterol and phytosterols, the fat sample is considered to contain vegetable fat. Using this parameter, it was reported that adulteration of ghee samples with 15% groundnut oil could be confirmed.[10]

9.2.16 DIFFERENTIAL THERMAL ANALYSIS AND DIFFERENTIAL SCANNING CALORIMETRY

Differential thermal analysis (DTA) and differential scanning calorimetry (DSC) are both closely related thermoanalytical techniques which measure the physical properties such as phase transition and specific heats of foods as a function of temperature. DTA measures the difference in temperature (Δt) between a sample and an inert reference material as a function of temperature. In DSC thermograms, the area delineated by the output curve is directly proportional to the total amount of energy transferred in or out of the sample.

DTA and DSC have been used by different investigators for the detection of foreign fats in milk fat. Adulterant such as coconut fat, cocoa fat, and hardened vegetable fat in butterfat detected at 5% by DSC based on the differences in the shape of melting curves.[4] However, tallow, lard, and vegetable oil added at 5% level in butterfat could not be detected by this method. DTA showed that addition of 5–10% of beef tallow in butterfat changed the solidification curve as solidification started earlier and showed two distinct minima in the curve.[92] Goat body fat (more than 10%) was detected in ghee by DTA technique on the basis of differences in melting diagram and crystallization patterns of goat body fat and ghee.[61] Using DSC, detection of foreign fats such as pig and buffalo body fats, beef suet, and chicken fat in milk fat was reported. The method, however, failed to detect coconut oil, cotton tract ghee, and other animal body fats.[2,27,60]

9.3   METHODS BASED ON CHEMICAL PROPERTIES

Certain well-known physical and chemical constants have been derived for the purpose of characterization of oils and fats. Among those constants three determinations, the Reichert–Meissl (RM) value, the Polenske value (P-value), and the iodine value measure certain specific constituents of milk fats while other two, the saponification value and BR reading, give an overall average nature of the constituent fatty acids present.[89] These physicochemical constants are described briefly in the following sections.

9.3.1   RM VALUE

RM value is the number of milliliter of 0.1 N alkali solution required to neutralize the steam volatile and water soluble fatty acids distilled from 5 g of fat under specified conditions. This constant for milk fat is quite significant since it is primarily a measure of butyric ($C_{4:0}$) and caproic ($C_{6:0}$) acid. RM value for milk fat ranges from 17 to 35 that is well above the value (generally 1) for all other fats and oils except coconut oil and palm kernel oil (4–8).[104,105] Feeding of cottonseed to milch animals lowers the RM value of ghee by 5–6 units.[89] The study revealed that adulteration of pure cow ghee with animal body fats (buffalo body fat, goat body fat, and goat body fat) and all vegetable oils (palm oil, rice bran oil, and soybean oil) individually at 10% level, whereas in case of pure buffalo ghee at 15% level of adulteration could be detected using RM value as a base.[3,69,103]

9.3.2   POLENSKE VALUE

P-value denotes the number of milliliter of 0.1 N alkali solution required to neutralize the steam volatile and water insoluble fatty acids distilled from 5 g of fat under specified conditions. This value is substantially a measure of caprylic ($C_{8:0}$) and capric ($C_{10:0}$) acid. The P-value for milk fat ranges from 1.2 to 2.4. This value for other oils and fats is also low (less than 1) except the coconut oil (15–20) and palm kernel oil (6–12).[105,116] Feeding of cottonseeds to milch animals reduces the P-value of ghee by 0.3–0.7 units.[89] Adulteration of cow and buffalo pure ghee with any of the adulterants either animal body fats or vegetable oils could not be detected at any of the levels studied, except adulteration of buffalo ghee with the palm and soybean oils which could be detected at 15% level.[3]

9.3.3   IODINE VALUE

Number of grams of iodine absorbed by 100 g of fat under specified conditions represents the iodine value. This constant is a measure of unsaturated linkages present in a fat. The iodine value for milk fat ranges from 26 to 35, which is low in comparison to most of the other fats and oils.[104] Animal body fats show slightly higher iodine value ranging from 36 to 49, whereas for vegetable oils, the value is very high (74–145) except coconut oil (6–10) and palm kernel oil (10–18). For hydrogenated fats, it lies in the range of 70–79. Feeding of cottonseed raises the iodine value of ghee up to 10 units.[89] Iodine value can be a useful parameter in detecting adulteration of cow ghee as well as buffalo ghee with various vegetable oils (except palm oil) at the level of more than 10%.[3]

9.3.4   SAPONIFICATION VALUE

Saponification value which denotes the number of milligrams of KOH required to saponify 1 g of fat gives an indication of average molecular weight of fatty acids present. For milk fat, animal body fats, vegetable oils, and hydrogenated fats, the value ranges from 210 to 233, 192 to 203, 170 to 197, and 197 to 199, respectively. Coconut oil and palm kernel oil show higher saponification value ranging from 243 to 262.[48,105] Feeding of cottonseeds to milch animals lowers this value by 7 units.[89] Adulteration of pure cow ghee with either animal body fats or vegetable oils at 10% levels of adulteration could be detected except buffalo body fat which could be detected even at 5% level of adulteration, whereas in case of buffalo ghee, only 15% level of adulteration either with animal body fats (except pig body fat) or with vegetable oils (except palm oil) could be detected.[3] However, when the data on cow and buffalo ghee was pooled, the study revealed that adulterants (body fats and vegetable oils) studied could be detected at 10% and 15% level of adulteration, except palm oil and pig body fats which could be detected only at 15% level of adulteration.

9.3.5 TESTS BASED ON GAS–LIQUID CHROMATOGRAPHY OF FATTY ACIDS

The technique of gas–liquid chromatographic (GLC), which utilizes retention time, a characteristic of a particular component under specified conditions, gives fatty acid profile of oils and fats obtained following the conversion of the triglycerides into more volatile methyl esters of their component fatty acids. Milk fat derived from ruminant animals, contains an exceptional number and variety of fatty acids from 4:0 to 26:0 (saturated) and from 10:1 to 22:5 (unsaturated). Body fats such as tallow and lard contain mostly palmitic (16:0), stearic (18:0), and oleic acid (18:1), while vegetable oils consist mainly of palmitic, stearic, oleic, and linoleic (18:2) acids. Coconut oil is the best known exception, containing lauric (14:0) and myristic (16:0) acids in very large amount.[89]

Some workers employed GLC technique and reported that the milk fat sample with a ratio of $C_{12:0}/C_{10:0}$ fatty acids >1.6 or $C_{4:0}/C_{6:0} + C_{8:0}$ fatty acids >1.8 was considered to be adulterated with margarine, coconut oil or tallow or pig body fat trans-esterified with butyric acid,[21,38,117] while other workers suggested ratios of other fatty acids in the order of $C_{4:0}/C_{12:0}$ >3, $C_{18:1}/C_{18:0}$ >2, and $C_{12:0}/C_{10:0}$ >1 for the detection of vegetable oils in butterfat.[83,90] Similarly, many other workers used the different fatty acids ratios for checking the adulteration of milk fat with vegetable oils, margarine, beef tallow, lard, goat body fat, substituted fats, synthetic fats, etc.[10,73,96,111,112] The fatty acid profile of 3 fractions separated by fractional crystallization from cow and buffalo ghee adulterated with lard and margarine at various levels reported that the amounts of 16:0, 18:0, and 18:1 acids were significantly changed with different adulteration levels and can be used as a marker to detect the admixture.[34]

9.3.6 TESTS BASED ON THE NATURE AND CONTENT OF UNSAPONIFIABLE CONSTITUENTS

The USM which can be obtained from oils and fats after saponification with alkali and subsequent extraction by a suitable organic solvent constitutes less than 2% by weight of fat. It is a repository of so many valuable constituents, such as sterols (cholesterol and phytosterols), fat-soluble vitamin (A, D, E, and K), hydrocarbons such as squalene, pigments, etc.

Mineral oil, if added to oils and fats, will appear in USM.[55] Milk fat contains USM in the range of 0.30–0.45% by weight chiefly consisting of cholesterol (0.25–0.40% by weight of fat).[48] Vegetable oils and animal body fats such as lard and tallow have USM in the range of 0–2 and 0–1.0%, respectively.[42] Sterols and tocopherols are the two most important constituents of USM, which have been used to detect the vegetable fats in milk fat by using various techniques such as GLC, thin-layer chromatography (TLC), paper chromatography, etc.

9.3.6.1   TESTS BASED ON STEROLS

Sterols which represent maximum share of the USM range from 0.24% to 0.5% in butterfat, 0.03% to 0.14% in body fats, and 0.03% to 0.5% in vegetable oils.[7] Cholesterol is the characteristic sterol of animal fats, while sterols from vegetable sources consist of a mixture collectively called as phytosterols and include β-sitosterol, stigmasterol, campesterol, brassicasterol, etc. Low concentration of cholesterol is also reported in the sterol fractions of vegetable oils and fats.[55] In addition to cholesterol, milk fat contains traces of lanosterol, dihydrolanosterol, and β-sitosterol.[114] Vegetable fats contain the sterols mainly in the ester form, while animal body fats contain mostly the free form.[89] Milk fat contains cholesterol in free and ester form in the ratio of 10:1.[37] The sterols can help to distinguish between fats of animal and vegetable origin, since the melting point of cholesterol acetate (112.76–116.40°C) is substantially lower than that of the acetates of any of the phytosterols (126–137°C). Adulteration of milk fat with vegetable oils is confirmed when melting point of sterol acetate fraction is more than 117°C.[18,45,89] A circular paper chromatographic method based on the difference in the behavior of USM isolated from fats in ghee using a solvent mixture of methyl alcohol:petroleum ether:water (80:10:10; v/v) was developed and reported that the spot of USM of ghee moved as a whole along with the solvent front, while that of ghee adulterated with animal body fats at 5% level or vanaspati at 10% level did not move at all when observed under UV light or when exposed to iodine vapors.[86]

IDF (1966) recommended a TLC method for the detection of vegetable fats in milk fat based on the appearance of a small band of β-sitosterol acetate in addition to the major band of cholesterol acetate using reversed phase system consisting of undecane/acetic acid–acetonitrile saturated with undecane.[46] Thin layers of $CaCO_3$ and soluble starch (10 + 4 g) impregnated with liquid paraffin and a solvent system consisting of methanol:acetic acid:water (20:5:1; v/v) as a developer reported that the presence of cottonseed oil, groundnut oil, sesame oil, and hydrogenated fats at 10–13% level and coconut oil at 25% level in ghee could be detected on the basis of resolving factor (Rf) values of 0.53 and 0.44 for cholesterol and phytosterols, respectively.[88] Sharma (1989) carried out TLC of USM of ghee and animal body fats using hexane:ether:glacial acetic acid:ethyl alcohol (25:20:5:1, v/v) as the solvent system and reported that ghee samples adulterated with 10% body fats resulted in the appearance of an extra spot due to dihydrocholesterol present in body fats.[97] Using GLC technique, β-sitosterol has been shown to be an index of vegetable fat addition; however, by this method, addition of body fats cannot be detected as body fats also have cholesterol.[26,44]

9.3.6.2   TESTS BASED ON TOCOPHEROL

Tocopherol, the important constituents of USM of natural oils and fats which range from 0.002% to 0.005% in butterfat, from 0.05% to 0.168% in vegetable oils except coconut oil which contains only 0.0083%, and from 0.0005% to 0.0029% in body fats.[7,11] Thus, tocopherol content of butterfat is low as compared to most vegetable oils and fats. Therefore, addition of vegetable fats to butter will result in a significant increase in tocopherol content of adulterated butterfat. Accordingly, some workers have reported that vegetable fats and oils added to ghee could be detected on the basis of tocopherol content. However, body fats and coconut oil added to milk fat could not be detected.[54,64,67,70] High-pressure liquid chromatography (HPLC) analysis of tocopherol isomers showed that all the samples of cow and buffalo pure ghee contain all the three tocopherols studied (α, γ, and δ) in appreciable proportion, and hence, analysis of tocopherol isomers did not help in the detection of vegetable oils adulteration in milk fat.[3]

9.3.6.3   TESTS BASED ON GLC OF TRIGLYCERIDES

Butterfat is composed predominately of triglycerides with 26–52 carbon number, while animal depot fats and common vegetable oils have 50–54 carbon number. Coconut and palm kernel oils contain short- and medium-chain length triglycerides with 30–52 carbon number, a range almost similar to butterfat,[76,89] using GLC, detected the presence of vegetable fat and lard in butterfat at 5–10% level based on the increase in the content of high-molecular weight triglycerides, $C_{52:0}$ and $C_{54:0}$ peaks, respectively.[57] GLC of commercial butter found that triglycerides content were in the order of $C_{36:0} > C_{38:0} > C_{40:0} > C_{42:0} > C_{44:0} > C_{50:0} > C_{52:0}$.[41] Beef tallow and lard triglycerides ranged mainly from $C_{44:0}$ to $C_{54:0}$, with $C_{52:0}$ as main triglyceride. The ratio of $C_{52:0}/C_{50:0}$ was less than 1 in pure butter and between 2 and 3–4, in case of beef tallow and lard, respectively. It was concluded that the $C_{52:0}/C_{50:0}$ ratio together with $C_{52:0}/C_{38:0}$ ratio gave a valuable indication of the possible addition of tallow or lard to butter.[41]

Adulteration of the milk fat with margarine at 5–10% level could be detected on the basis that margarine had more triglycerides with 48–54 acyl carbon atoms than milk fat.[66] Some workers compared the triacylglycerol composition of different fats as analyzed by GLC and designed a multiple linear regression equation by which foreign fats could be detected with substantially improved sensitivity.[62,63,80,81] However, the method was suitable only when a single foreign fat was added to milk fat. Currently, European Union (EU) applies the method for triglyceride analysis as an official method for evaluating the milk fat purity.[81] The official method of EU coupled with the determination of 3,5-cholestadiene content reported that the detection of beef tallow up to 0.5–1.0% using 3,5-cholestadiene analysis and up to 2% using multivariate statistical techniques could be done.[65,79]

9.4   METHODS BASED ON TRACER COMPONENTS OF FATS AND OILS

Tracer components can be defined as those compounds which are present in adulterant oils and fats, either naturally or by addition, but absent in pure ghee. Addition of some tracer component in the likely adulterant of ghee has been suggested as a rapid and reliable tool to identify them in milk fat. A tracer can be a latent color which is not detectable visually, but get identified by its color reaction with certain chemicals or a coloring matter (natural or synthetic) which may impart direct coloration distinct from that of the natural color of butterfat. Among tracers, in India, sesame oil is added (5% by weight) to vanaspati according to food laws for its detection in ghee by Baudouin test.[78] The method is based on the development of a permanent crimson color due to the reaction between furfural and sesamol formed by the hydrolysis of sesamolin (present in sesame oil) in the presence of concentrated HCl. Use of hydrofuramide or *P*-hydroxy benzaldehyde has also been suggested in place of furfural for this test. Another tracer is tannins which are assumed to be naturally present as impurities in palm oil.[52,97] Ghee samples adulterated with palm oil give prussian blue color with potassium ferricyanide and ferric chloride reagent and based on this reaction, palm oil added to ghee can be detected at the level of 5%.[14] However, limitation of this method is that the ghee samples having butalyated hydroxyanisole (BHA) as antioxidant also give positive test. Gamma oryzanol, a natural tracer, having antioxidant and cholesterol-lowering properties, is found to be present in the rice bran oil. It was revealed to be a mixture of phytosteryl ferulates comprising cycloartenyl ferulate, 24-methylenecycloartenyl ferulate, and campesteryl ferulate as major components.[23,47,118]

Crude rice bran oil contains ≤2% (v/v) oryzanol.[72] This compound has been indicated as a marker of rice bran oil in other edible oils.[108] Apart from rice bran oil, gamma oryzanol is also present as natural component of corn and barley oils. On the basis of presence of gamma oryzanol, some methods have been developed recently to detect it in other edible oils. Thin layer chromatographic (TLC) method and colorimetric method for detection of rice bran oil in other vegetable oils can be used, rice bran oil even at 2% and 5% level in ghee detected, respectively.[3]

9.5 MISCELLANEOUS METHODS

9.5.1 TEST FOR MINERAL OIL

Adulteration of common edible oils and fats including milk fat with cheaper mineral oils, such as paraffin oil, heavy and light fuel oil, petroleum jelly, etc., has become a widespread phenomenon because of the price difference. Unlike oils and fats, mineral oils are not saponifiable by alkali. This characteristic behavior of mineral oils has been used as the basis for their detection in edible oils and fats. Using Holde's test, the presence of mineral oil as low as 0.3% can be detected by saponifying 10 drops of test sample (1 mL) with 5 mL of 0.5 N ethanolic potassium hydroxide solution and adding 5 mL of water to the hot soap solution and noting the appearance of turbidity.[116] Liquid paraffin added to ghee at the rate of 0.5% was detected using Holde's test.[8,116]

9.5.2 TESTS BASED ON THE CONTENT OF SPECIFIC FATTY ACIDS

Certain specific fatty acids such as butyric acid, erucic acid, iso-valeric acid, iso-oleic acid, cyclopropenoic acids, etc. which are either characteristic or absent in milk fat or present in less quantity as compared to adulterant fats have been used as an index for the detection of foreign fats in milk fat. Unusual fatty acids such as iso-oleic acid, iso-valeric acid, erucic acid, cyclopropenoic acid, etc. are altogether absent in milk fat, but are found in vegetable fats. Butyric acid is found only in milk fat, but not in adulterant fats. Some workers reported that any decrease in butyric acid content of milk fat below 9.6 mole% would indicate its adulteration with foreign fat.[33,43] The presence of iso-valeric acid in dolphin oil has been used as the basis for its detection in milk fat by several workers using ascending paper chromatography.[5,22,28,77,82,110]

9.5.3 TESTS FOR COTTONSEED OIL

Fatty acids containing cyclopropene ring, namely, malvalic ($C_{18:1}$) and sterculic ($C_{19:1}$) acids, which are altogether absent in milk fat, but are characteristic of cottonseed oil have been used as a tool by few workers for the detection of cottonseed oil in milk fat and also to distinguish cotton tract ghee from normal ghee using Halphen test or methylene blue reduction test.[12,24,75,99]

9.5.3.1 HALPHEN TEST

This test is based on the development of a crimson color due to the reaction between cyclopropenoic acids (constituents of cottonseed oil) and Halphen reagents (1% sulfur solution in CS_2 and equal volume of isoamyl alcohol) after incubation for an hour in a boiling bath of saturated sodium chloride solution. This test finds its application for differentiating the cotton tract ghee from normal ghee as well as for the detection of cotton-seed oil in milk fat.[105]

9.5.3.2 METHYLENE BLUE REDUCTION TEST

This test is used for the identification of cotton tract ghee. The color of methylene blue dissolved in chloroform:methanol (1:1) was decolorized by cotton tract ghee or ghee added with cottonseed oil due to the presence of cyclopropenoic acids, while normal pure ghee did not reduce the color of methylene blue dye.[105]

9.5.3.3 HYDROXAMIC ACID TEST

It is a colorimetric test[98] which is used to distinguish between butterfat and other vegetable and animal fats based on the fact that fats derived from milk (cow, goat, sheep, etc.) will form water-soluble hydroxamic acid–iron complexes.[71] These complexes appear pink to purple color in water layer. The hydroxamic acid–iron complexes formed from fatty acid esters in vegetable fats except coconut oil and of animal fats are insoluble in water and do not contribute a distinctive pink to purple color to the water layer. The above test modified by separating the water soluble and water insoluble hydroxamates. The soluble fraction containing the complexes of butyric, caproic, caprylic, and capric acids is extracted with a butanol–ethanol mixture which removes the caprylic and capric acids and leaves the other two in the water.[100,101] The relative proportion of these two pairs of acids estimated by comparing the color intensity in the aqueous phase before and after extraction could be used to distinguish butterfat from other fats.

9.5.4 RAPID COLOR-BASED VEGETABLE OIL DETECTION TEST

The Bieber's test, hitherto employed for the detection of almond oil adulteration with kernel oil, was suitably modified to detect the adulteration of ghee with vegetable oils. Their results showed the presence of orange brown color in case of refined vegetable oils and fats, whereas in case of pure ghee samples no color was observed.[8] By this method, adulteration of ghee with different vegetable oils to the tune of 5–7% could be detected.

9.5.5 TESTS BASED ON ENZYMATIC HYDROLYSIS

Lipases (triacylglycerol acyl hydrolases) are the enzymes that catalyze the reversible hydrolysis of triacyl glycerols under natural conditions. Pancreatic lipase which specifically hydrolyzes the primary hydroxyl positions of glycerides can catalyze the complete breakdown of triacylglycerols to free fatty acids and glycerol involving at least one isomerization step, in addition to the three hydrolytic steps, as follows:[1,25]

$$\text{Triglyceride} \rightarrow 1,2 \text{ Diglyceride} \rightarrow 2 \text{ Monoglyceride} \rightarrow$$
$$1 \text{ Monoglyceride} \rightarrow \text{Glycerol}$$

Pancreatic lipase digests some classes of milk triglycerides more rapidly than others due to difference in their glyceride structure especially with regard to fatty acids distribution.[30,84,109,115] Further, triglycerides containing unsaturated fatty acids in 1 and 3 positions are hydrolyzed faster as compared to those having saturated fatty acids in these positions.[25] Fats such as buffalo milk fat with higher melting triglycerides and long-chain saturated fatty acids ($C_{16:0}$ and $C_{18:0}$) were hydrolyzed slowly by pancreatic lipase as compared to cow milk fat.[87] Also, triglycerides with low melting points greatly contribute to the faster rate of hydrolysis.[58]

It is expected that other fats such as body fats and vegetable fats which have the preponderance of long-chain fatty acids will also have different rates of hydrolysis vis-à-vis milk fat. Information on this aspect is obscure. Moreover, it is also not known whether the rate of hydrolysis of fat by lipase can be used for detecting milk fat adulteration.

9.6 SUMMARY

Milk fat is an important component in the dairy industry, and plays significant role in economic, nutrition, physical, and chemical properties of milk and milk products. Milk fat is one of the valuable fats that continue to be a target of unscrupulous traders for the maximization of profits. Methods presently adopted by food law enforcing agencies to ensure the quality of milk fat are mainly based on the physicochemical constants such as BR reading, RM value and P-value, phytosterol acetate test, and Baudouin test. The said methods are being used by enforcing agencies to ascertain the quality of milk fat but at the same time unscrupulous traders are developing such concoctions of adulterant fats so they pass the said tests. Therefore, in addition to these methods additional methods have been developed by the research laboratories to tackle the menace of rampant adulteration in ghee. The innovative approaches include change in the ratios of different fatty acids, sterol analysis, carbon number profiling of triglycerides, TLC, and fractionation of milk fat coupled with other parameters. In this chapter, authors attempted to compile the information on selected recent approaches to detect the adulteration of milk fat. It will be more prudent to focus on the tracer component-based techniques to counter this menace.

KEYWORDS

- animal body fat
- Baudouin test
- Bieber's test
- Bumer value
- butyro refractometer reading
- differential scanning calorimetry
- gas liquid chromatographic
- high-pressure liquid chromatography
- thin-layer chromatography
- ultraviolet spectroscopy
- visible spectroscopy.

REFERENCES

1. Akoh, C. C.; Min, D. B. *Food Lipids. Chemistry, Nutrition and Biotechnology*; Marcel Dekker, Inc.: New York, 1998; pp 110–116.

2. Amelotti, G.; Brianza, M.; Lodigiano, P. Application and Limits of Differential Scanning Calorimetry to the Detection of Beef Suet in Butter. *Rev. Italiana Sostanze Grasse* **1983**, *60*, 557–564.

3. Kumar, A. Detection of Adulterants in Ghee. PhD Thesis, National Dairy Research Institute (Deemed University), Karnal, India, 2008; p 185.

4. Antila, V.; Luoto, K.; Antila, M. Detection of Adulteration of Milk Fat by Means of DSC. *Meijeritied Ackakausk* **1965**, *25*, 37–42.

5. Antoniani, C.; Cerutti, G. Detection of Isovaleric Acid in Fats in the Presence of Butyric Acid and Caproic Acids. *Annu. Sper. Agr.* **1954**, *8*, 801–807.

6. Arora, K. L.; Lal, D.; Seth, R.; Ram, J. Platform Test for Detection of Refined Mustard Oil Adulteration in Milk: A Short Communication. *Indian J. Dairy Sci.* **1996**, *49*, 721–723.

7. Kumar, A.; Lal, D.; Seth, R.; Sharma, R. Recent Trends in Detection of Adulteration in Milk Fat—A Review. *Indian J. Dairy Sci.* **2002**, *55*, 319–330.

8. Kumar, A.; Lal, D.; Seth, R.; Sharma, V. Turbidity Test for Detection of Liquid Paraffin in Ghee. *Indian J. Dairy Sci.* **2005**, *58*, 298.

9. Kumar, A.; Lal, D.; Seth, R., Sharma, V. Apparent Solidification Time Test for Detection of Foreign Oils and Fats Adulterated in Clarified Milk Fat, as Affected by Season and Storage. *Int. J. Dairy Technol.* **2009**, *62*, 33–38.

10. Kumar, A. Comparative Studies on Physicochemical Properties of Various Oils and Fats with a View to Detect Adulteration in Ghee. PhD Thesis, National Dairy Research Institute (Deemed University), Karnal, India, **2003**; pp 120–125.

11. Bailey, A. E. *Industrial Oil and Fat Products*; InterScience Publishers Inc.: New York, **1951**; p 216.

12. Bailey, A. V.; Harris, J. H.; Skau, E. L. Cyclopropenoid Fatty Acid Content and Fatty Acid Composition of Crude Oils from Twenty-five Varieties of Cottonseed. *J. Am. Oil Chem. Soc.* **1966**, *43*, 107–110.

13. Bartlett, J. C.; Chapman, D. G. Detection of Hydrogenated Fats in Butterfat by Measurement of Cistrans Conjugated Unsaturation. *J. Agric. Food Chem.* **1961**, *9*, 50–53.

14. Bector, B. S.; Sharma, V. *Physicochemical Characteristics of Ghee Adulterated with Palm Oil and Its Detection*; NDRI Annual Report, **2002**; pp 44–45.

15. Bhalerao, V. R.; Kummerow, F. A. Modification of the Refractive Index Method for the Detection of Foreign Fats in Dairy Products. *J. Dairy Sci.* **1956**, *39*, 947–955.

16. Bhalerao, V. R.; Kummerow, F. A. Summary of Methods for the Detection of Foreign Fats in Dairy Products. *J. Dairy Sci.* **1956**, *39*, 956–964.

17. Bhide, P. T.; Kane, J. G. Critical Temperatures of Dissolution of Ghee and Hydrogenated Oils (Vanaspati). *Indian J. Dairy Sci.* **1952**, *5*, 183–187.

18. BIS. *Handbook of Food Analysis, SP-18, Part XI- Dairy Products*; Bureau of Indian Standards: New Delhi, **1981**; pp 10–59.

19. Boghra, V. R.; Singh, S.; Sharma, R. S. Present Status of the Tests Used for the Detection of Adulterants in Ghee. *Dairy Guide* **1981**, *81*, 21–31.

20. Bolton, R. E. *Oil Fats and Fatty Foods: Their Practical Examination,* 1st ed.; Biotech Books: Delhi, **1999**; p 320.

21. Boniforti, Z. Gas Chromatographic Study of the Fatty Acids of Butter Made in Italy and Other Countries; Application to Adulteration of Commercial Butter. *Ann. Falsif. l'Expertise Chim. Toxicol.* **1962**, *55*, 255–263.

22. Bottini, E.; Campanello, F. Detection of Hydrogenated Dolphin Oil in Butter by Means of Paper Chromatography. *Ann. Sper. Agrar.* **1955**, *9*, 11–20.

23. Chen, M. H.; Bergman, C. J. A Rapid Procedure for Analyzing Rice Bran Tocopherol, Tocotrienol and γ-oryzanol Contents. *J. Food Compos. Anal.* **2005**, *18*, 139–151.

24. Christie, W. W. Cyclopropane and Cyclopropene Fatty Acids. In *Topics in Lipid Chemistry;* Gunstone, F. D., Ed.; Logos Press: London, **1970**; Vol. 1, pp 1–115.

25. Coleman, M. H. The Pancreatic Hydrolysis of Natural Fats. 3. The Influence of the Extent of Hydrolysis on Monoglyceride Composition. *J. Am. Oil Chem. Soc.* **1963**, *40*, 568–571.

26. Colombini, M.; Amelotti, G.; Vanoni, M. C. Adulteration of Butter with Beef Suet and Lard: II. GLC of Sterols and UV Spectrophotometric Analysis. *Rev. Italiana Sostanze Grasse* **1978**, *55*, 356–361.

27. Coni, E.; Pasquale, M. D.; Coppolelli, P.; Bocca, A. Detection of Animal Fats in Butter by Differential Scanning Calorimetry: A Pilot Study. *J. Am. Oil Chem. Soc.* **1994**, *71*, 807–810.

28. D'Arrigo, G. Identification of Traces of Isovaleric Acid in Butter Adulterated with Hydrogenated Dolphin Oil. *Oli Miner.* **1955**, *32*, 2–7.

29. Delforno, G. Detection of Foreign Fats in Butter Using the Crismer Test. *Latte* **1964**, *38*, 921–924.

30. DeMan, J. M. Physical Properties of Milk Fat. I. Influence of Chemical Modification. *J. Dairy Res.* **1961**, *28*, 81–86.

31. Den-Herder, P. C. Detection of Adulteration of Butter with Foreign Fats by Examination of the Sterols. *Neth. Milk Dairy J.* **1955**, *9*, 261–274.

32. Doctor, N. S.; Bajerjee, B. N.; Kothavalla, Z. R. Studies on Some Factors, Affecting the Physical and Chemical Constants of Ghee. *Indian J. Vet. Sci.* **1940**, *10*, 63–80.

33. Eckizen, A.; Deki, M. *Reports of the Central Customs Lab., No. 16*; Kobe Customs Lab.: Ikertaku, Kobe-Shi, Japan, **1976**; p 67.

34. Farag, R. S.; Abo-Raya, S. H.; Ahmed, F. A.; Hewedi, F. M.; Khalifa, H. H. Fractional Crystallization and Gas Chromatographic Analysis of Fatty Acids as a Means of Detecting Butterfat Adulteration. *J. Am. Oil Chem. Soc.* **1983**, *60*, 1665–1669.

35. Firestone, D.; Villadelmar, M. D. L. Determination of Isolated Trans-unsaturation by Infra-red Spectrophotometry. *J. Assoc. Agric. Chem.* **1961**, *44*, 459–464.

36. Fox, P. F.; McSweeney, P. L. H. Milk Lipids. In *Dairy Chemistry and Biochemistry*; Blackie Academic Professional/Chapman and Hall: New York, **1988**; pp 69–143.

37. Fox, P. F. *Advanced Dairy Chemistry*, 3rd ed.; Chapman & Hall: London, New York and Madras, **1995**; pp 1–35.

38. Francesco, F. D.; Avancini, D. The Adulteration of Butter with Trans-esterified Fats. *Latte* **1961**, *35*, 809–817.

39. Ganguli, N. C.; Jain, M. K. Ghee: Its Chemistry, Processing and Technology. *J. Dairy Sci.* **1972**, *56*, 19–25.

40. Gunstone, F. D.; Harwood, J. L.; Padley, F. B. *The Lipid Handbook*, 3rd ed.; Chapman & Hall: London, **2007**; pp 415–455.

41. Guyot, A. L. Detection of Animal Body Fats in Milk Fat by Triglyceride Gas Liquid Chromatography. *20th Int. Dairy Congr.* **1978**, *E*, 400.

42. Hamilton, R. J.; Rossel, J. B. *Analysis of Oils and Fats*; Elsevier Applied Science Publishers: London, **1986**; p 441.

43. Harper, W. J.; Armstrong, J. V. Measurement of Butyric Acid in Fat with Reference to the Detection of Substitute Fats in Dairy Products. *J. Dairy Sci.* **1954**, *37*, 481–487.

44. Homberg, E.; Bielefeld, B. Detection of Vegetable Fats in Butterfat by Gas Chromatographic Analysis of Sterols. *Z. Lebensm. Unters. Forsch.* **1979**, *169*, 464–467.

45. IDF. *Detection of Vegetable Fat in Milk Fat by Phytosteryl Acetate Test*; FIL-IDF, Boulevard Auguste Reyers 70/B, 1030 Brussels, Belgium, 1965; Vol. 32, p 5.

46. IDF. *Detection of Vegetable Fat in Milk Fat by Thin Layer Chromatography of Steryl Acetates*; FIL-IDF, Boulevard Auguste Reyers 70/B, 1030 Brussels, Belgium, 1966; Vol. 38, pp 4–6.

47. Iqbal, S.; Bhanger, M. I.; Anwar, F. Antioxidant Properties of Some Commercially Available Varieties of Rice Bran in Pakistan. *Food Chem.* **2005**, *93*, 265–272.

48. Jenness, R.; Patton, S. Milk Lipid. In *Principles of Dairy Chemistry*; John Wiley & Sons, Inc.: New York, **1969**; pp 30–72.

49. Jha, J. S. Spectrophotometric Studies of Cheuri (*Madhuca butyracea*) Fat and Ghee Mixtures. *J. Am. Oil Chem. Soc.* **1981**, *58*, 843–845.

50. Kalsi, P. S. *Dilatometeric Study of solid/Liquid Fractions of Milk Fat from Different Species*. M.Sc. Thesis, National Dairy Research Institute (Deemed University), Karnal, **1984**; pp 3–40.

51. Kane, J. G.; Ranadive, G. M. Aniline Point of Some Fats. *J. Sci. Ind. Res.* **1951**, *10*, 62–66.

52. Kapur, O. P.; Srinivasan, M.; Subrahmanyan, V. A Study in the Preparation and Properties of Hydrofuramide. *J. Sci. Ind. Res.* **1960**, *19*, 509–510.

53. Karleskind, A. *Oils and Fats (Manual): A Comprehensive Treatise*; TEC & DOC Lavoisier: France, 1996; Vol. 2, pp 53–61.

54. Keeney, M.; Bachman, K. C.; Tikriti, H. H.; King, R. L. Rapid Vitamin E Method for Detecting the Adulteration of Dairy Products with Non-coconut Vegetable Oils. *J. Dairy Sci.* **1971**, *54*, 1702–1703.

55. Kirk, R. S.; Sawyer, R. *Pearson's Composition and Analysis of Foods*; Addison-Wesley Longman, Inc.: Wiley in Harlow, Essex, New York, NY, UK, 1999; pp 45–55.

56. Konevets, V. I.; Roganova, Z. A.; Smolyanskii, A. L. Use of Infra-red Spectroscopy for Analysing Fats of Different Origin. *Izv. Vysshikh Uchebnykh Zavadenii Pischevaya Tekhnologiya* **1987**, *1*, 64–67.

57. Kuksis, A.; McCarthy, M. J. Triglyceride Gas Chromatography as a Means of Detecting Butterfat Adulteration. *J. Am. Oil Chem. Soc.* **1964**, *41*, 17–21.

58. Lakshminarayana, M.; Ramamurthy, M. K. Cow and Buffalo Milk Fat Fractions. Part III. Hydrolytic and Autoxidative Properties of Milk Fat Fractions. *Indian J. Dairy Sci.* **1986**, *39*, 251–255.

59. Lal, D.; Seth, R.; Arora, K. L.; Ram, J. Detection of Vegetable Oils in Milk. *Indian Dairym.* **1998**, *50*, 17–18.

60. Lambelet, P.; Ganguli, N. C. Detection of Pig and Buffalo Body Fat in Cow and Buffalo Ghee by Differential Scanning Calorimetry. *J. Am. Oil Chem. Soc.* **1983**, *60*, 1005–1008.

61. Lambelet, P.; Singhal, O. P.; Ganguli, N. C. Detection of Goat Body Fat in Ghee by Differential Thermal Analysis. *J. Am. Oil Chem. Soc.* **1980**, *57*, 364–366.

62. Lipp, M. Comparison of PLS, PCR and MLR for the Quantitative Determination of Foreign Oils and Fats in Butter Fats of Several European Countries by Their Triglyceride Composition. *Z. Lebensm. Unters. Fursch.* **1996**, *202*, 193–198.

63. Lipp, M. Determination of Adulteration of Butter Fat by Its Triglyceride Composition Obtained by GC. A Comparison of the Suitability of PLS & Neural Network. *Food Chem.* **1966,** *55,* 389–395.

64. Mahon, J. H.; Chapman, R. A. Detection of Adulteration of Butter with Vegetable Oils by Means of the Tocopherol Content. *Anal. Chem.* **1954,** *26,* 1195–1198.

65. Mariani, C.; Venturini, S., Fedeli, E.; Contarini, G. Detection of Refined Animal and Vegetable Fats in Adulteration of Pure Milk Fat. *J. Am. Oil Chem. Soc.* **1994,** *71,* 1381–1384.

66. Marjanovic, N.; Jankouits, I.; Turkulov, J.; Karlouic, D.; Caric, M.; Milanovic, S.; Zagorak, M. Determination of the Triglyceride Composition of Milk Fat by Gas Chromatography. *Mijekarstuo* **1984,** *34,* 67–75.

67. Markuze-Zofia, R. Z. H. Method for the Estimation of Tocopherol Content of Butter and Its Use for the Detection of Adulteration with Vegetable Fats. *Warsz* **1962,** *13,* 255–262.

68. Martin, P. G. Oils and Fats. In *Manuals of Food Quality Control. 3. Commodities;* Food & Agricultural Organization of the United Nations: Rome, **1979;** pp 41–63.

69. Krishnamurthy, M. N.; Nagaraja, K. V. Methods for Detection of Rice-bran, Mustard, Karanja Oils and Rice-bran Deoiled Cake. *Fat Sci. Technol.* **1992,** *94,* 457–458.

70. Nazir, D. J.; Magar, N. G. Tocopherol Content of Indian Butter and Its Use in Detecting Adulteration of Butterfat. *Indian J. Dairy Sci.* **1959,** *12,* 125–132.

71. Nelson, W. L. A Rapid Test for Distinguishing Between Butterfat and Fats from Plant and Other Animal Sources. *Food Technol.* **1954,** *8,* 385–386.

72. Norton, R. A. Quantification of Steryl Ferulate and p-coumarate Esters from Corn and Rice. *Lipids* **1995,** *30,* 269–274.

73. Panda, D.; Bindal, M. P. Detection of Adulteration in Ghee with Vegetable Oils Using GLC Based on a Marker Fatty Acid. *Indian J. Dairy Sci.* **1997,** *50,* 129–135.

74. Panda, D.; Bindal, M. P. Detection of Adulteration in Ghee with Animal Body Fat and Vegetable Oils Using Crystallization Test. *Indian Dairym.* **1998,** *50,* 13–16.

75. Pandey, S. N.; Suri, L. K. Cyclopropenoid Fatty Acid Content and Iodine Value of Crude Oils from Indian Cottonseed. *J. Am. Oil Chem. Soc.* **1982,** *59,* 99–101.

76. Parodi, P. W. Detection of a Synthetic Milk Fat. *Aust. J. Dairy Technol.* **1969,** *24,* 56–59.

77. Parrozzano, R., Mancinelli, F. Detection of Hydrogenated Dolphin Oils in Butter Using Paper Chromatography. *Boll. Laboratori Chim. Prov.* **1954,** *5,* 43–50.

78. PFA. *Prevention of Food Adulteration Act, 1954 & Rules, 1955 (As Amended);* Eastern Book Company: Delhi, **2009;** p 543.

79. Povolo, M.; Bonfitto, E.; Contarini, G.; Toppino, M.; Daghetta, A. Study on the Performance of Three Different Capillary Gas Chromatographic Analysis in the Evaluation of Milk Fat Purity. *J. High Resolut. Chromatogr.* **1999,** *22,* 97–102.

80. Precht, D. Quantification of Animal Body Fats and Vegetable Fats in Milk by Triglyceride Analysis. *22nd Int. Dairy Congr.* **1990,** *1,* 234.

81. Precht, D. Detection of Foreign Fat in Milk Fat. 1. Quantitative Detection of Triacylglycerol Formulae. *Z. Lebensm. Unters. Forsch.* **1992,** *194,* 1–8.

82. Priori, O. Chromatographic Detection of Hydrogenated Dolphin Oils in Market Butters. *Boll Laboratori Chim Prov.* **1995,** *6,* 45–46.

83. Provvedi, F.; Cialella, G. Gas Chromatographic Analysis of Some Food Fats. *Rev. Italia Sostanze Grasse* **1961,** *38,* 361–366.

84. Raghuveer, K. G.; Hammond, E. G. Fatty Acid Composition and Flavor of Auto-oxidized Milk Fat. *J. Dairy Sci.* **1967,** *50,* 1200–1205.

85. Rahn, O.; Sharp, P. F., Eds. *The Physics (Physik der). Milchewirtschaft*; Paul Parey: Berlin, **1928**; p 112.

86. Ramachandra, B. V.; Dastur, N. N. Application of Paper Chromatography to Differentiate Ghee from Other Fats, Part 1: Behavior of Unsaponifiable Matter on Chromatograms. *Indian J. Dairy Sci.* **1959,** *12,* 139–148.

87. Ramamurthy, M. K.; Narayanan, K. M. Hydrolytic and Auto-oxidative Properties of Buffalo and Cow Milk Fats as Influenced by Their Glyceride Structure. *Indian J. Dairy Sci.* **1974,** *27,* 227–233.

88. Ramamurthy, M. K.; Narayanan, K. M.; Bhalerao, V. R.; Dastur, V. N. A TLC Method for Detection of Adulteration of Ghee with Vegetable Fats. *Indian J. Dairy Sci.* **1967,** *20,* 11–13.

89. Rangappa, K. S.; Achaya, K. T. *Indian Dairy Products*; Asia Publishing House: Bombay, **1974**; pp 220–300.

90. Renterghem, R. V. The Triglyceride Composition of Belgian Butter in View of EU Controls on Milk Fat Purity. *Milchwissenschaft* **1997,** *52,* 79–82.

91. Roos, J. B. Methods for the Detection of Foreign Fat in Milk Fat. *Int. Dairy Fed. Annu. Bull.* **1963,** *4,* 2–54.

92. Roos, J. B.; Tuinstra, L. G. M. The Detection of Foreign Fat in Milk Fat, II: Differential Thermal Analysis of Dutch Butterfat. *Neth. Milk Dairy J.* **1969,** *23,* 37–45.

93. Sato, T.; Kawano, S.; Iwamoto, M. Detection of Foreign Fat Adulteration of Milk Fat by Near Infra-red Spectroscopic Method. *J. Dairy Sci.* **1990,** *73,* 3408–3413.

94. Schwitzer, M. K. *Margarine and Other Food Fats*; Leonardo Hill Books Ltd.: London, **1956**; pp 59–79.

95. Sharma, R.; Singhal, O. P. Physicochemical Constants of Ghee Prepared from Milk Adulterated with Foreign Fat. *Indian J. Dairy Biosci.* **1995,** *6,* 51–53.

96. Sharma, R.; Singhal, O. P. Fatty Acid Composition, Bomer Value and Opacity Profile of Ghee Prepared from Milk Adulterated with Foreign Fats. *Indian J. Dairy Sci.* **1996,** *49,* 62–67.

97. Sharma, S. K. Studies on Unsaponifiable Matter of Ghee (Clarified Butterfat) and Animal Body Fats with a View to Detect Adulteration. PhD Thesis, National Dairy Research Institute (Deemed University), Karnal, India, **1989**; p 231.

98. Sharma, V.; Lal, D.; Sharma, R. *Development of a Rapid Colorimetric Test for Detection of Vegetable Oil Adulteration in Ghee*; NDRI Annual Report, **2005**; pp 33–35.

99. Shenstone, F. S.; Vickery, J. R. Occurrence of Cyclopropene Acids in Some Plants of the Order Malvales. *Nature* **1961,** *190,* 168–169.

100. Shipe, W. F. Identification of Fats by Urea Fractions. *J. Assoc. Agric. Chem.* **1955,** *38,* 156.

101. Shipe, W. F. M53, A Modified Hydroxamate Method for Determining the Short Chain Fatty Acids in Fats. *J. Dairy Sci.* **1955,** *38,* 599.

102. Shukla, A. K; Dixit, A. K.; Singh, R. P. Detection of Adulteration in Edible Oils. *J. Oleo Sci.* **2005,** *54,* 317–324.

103. Shukla, A. K.; Johar, S. S.; Dixit, A. K.; Singh, R. P. Identification of Physically Refined Rice Bran Oil and Its Simple Detection in Other Oils. *J. Oleo Sci.* **2004,** *53,* 413–415.

104. Singhal, O. P. Studies on Ghee (Clarified Butterfat) and Animal Body Fats with a View to Detect Adulteration. PhD Thesis, Punjab University, Chandigarh, India, **1973**; pp 8–18.
105. Singhal, O. P. Adulterants and Methods for Detection. *Indian Dairym.* **1980**, *32*, 771–774.
106. FSSAI. *Lab Manual: Manual of Method of Analysis of Foods: Milk and Milk Products*; FSSAI: New Delhi, 2005; pp 107–123.
107. Singhal, O. P.; Ganguly, N. C.; Dastur, N. N. Physicochemical Properties of Different Layers of Ghee (Clarified Butterfat). *Milchwissenschaft* **1973**, *28*, 508–511.
108. Singhal, R. S.; Kulkarni, P. R.; Rege, D. V. Edible Oils and Fats. In *Handbook of Indices of Food Quality and Authenticity*; Woodhead Publishing Ltd.: Cambridge, England, 1997; pp 131–208.
109. Soliman, M. A.; Younes, N. A. Adulterated Butterfat: Fatty Acid Composition of Triglycerides and 2-monoglycerides. *J. Am. Oil Chem. Soc.* **1986**, *63*, 248–250.
110. Tappi, G.; Menziani, E. Detection of Hydrogenated Dolphin Oil in Butter. *Ann. Chim.* **1952**, *42*, 446–462.
111. Toppino, P. M.; Contarini, G.; Traversi, A. L.; Amelotti, G.; Gargano, A. Gas Chromatographic Parameters of Genuine Butter. *Riv. Italiana delle Sostanze Grasse* **1982**, *59*, 592–610.
112. Ulberth, F. Detection of Milk Fat Adulteration by Linear Discriminate Analysis of Fatty Acids. *J. Assoc. Agric. Chem.* **1994**, *77*, 1326–1334.
113. Walstra, P.; Jennes, R., Eds. *Dairy Chemistry and Physics*; John Wiley & Sons Inc.: New York, **1984**; p 431.
114. Wong, P. N.; Jenness, R. *Fundamentals of Dairy Chemistry*, 3rd ed.; Van nostrand Reinhold Company Inc.: New York, **2012**; pp 171–214.
115. Weihe, H. D. Technical Note: Interesterified Butter Oil. *J. Dairy Sci.* **1961**, *44*, 944–947.
116. Winton, A. L.; Winton, K. B. Turbidity Test for Detection of Liquid Paraffin in Ghee. *Indian J. Dairy Sci.* **1999**, *57–58*, 298.
117. Wolff, J. P. Gas Chromatography in the Control of Butter Purity. *Ann. Falsif. Expert. Chim.* **1960**, *53*, 318–325.
118. Xi, N.; Godber, J. S. Purification and Identification of Components of Gamma Oryzanol in Rice Bran Oil. *J. Agric. Food Chem.* **1999**, *47*, 2724–2728.
119. Xi, N.; Kocaogulu-Vurma, N. A.; Harper, W. J.; Rodriguez-Saona, L. E. Application of Temperature Controlled Attenuated Total Reflectance-mid-infrared (ATR-MIR) Spectroscopy for Rapid Estimation of Butter Adulteration. *Food Chem.* **2010**, *121*, 778–782.
120. Xo, K.; Kawano, S; Iwamoto, M. Detection of Foreign Fat Adulteration of Milk Fat by Near Infra-red Spectroscopic Method. *J. Dairy Sci.* **1990**, *73*, 3408–3413.
121. Xo, K.; Shuhaimi, M.; Nurrulhidayah, F.; Rohman, A.; Amin, I.; Khatib, A. Analysis of Chicken Fat as Adulterant in Butter Using Fourier Transform Infrared Spectroscopy and Chemometrics. *Grasas Aceites* **2013**, *64*(4), 349–355.

PART IV
Physical Properties of Milk and Milk Products

CHAPTER 10

PHYSICAL PARAMETERS DURING PANEER PREPARATION: A REVIEW

A. K. AGRAWAL* and GEETESH SINHA

Department of Dairy Engineering, College of Dairy Science and Food Technology Chhattisgarh Kamdhenu Vishwavidyalaya (CGKV), Raipur 492012, India

Corresponding author. E-mail: akagrawal.raipur@gmail.com

CONTENTS

ABSTRACT

Paneer (Indian cheese) is an important indigenous milk product obtained by acid coagulation of hot whole milk with subsequent drainage of whey. The coagulants most commonly used are citric acid and lactic acid, both in natural and in chemical form. Paneer is a highly nutritious and wholesome food and has great value in the diet. Good quality paneer is characterized by an acidic nutty flavor with a slight sweetish taste, a firm cohesive and spongy body, and a close, smooth texture. Textural parameters of paneer are known to have an important impact on its quality and its acceptability. The emphasis in this chapter is to summarize the textural properties of paneer and factors affecting the acceptability during manufacturing of paneer. Pressing of curd plays a vital role in developing the textural properties of paneer and the chapter covers all the details pertaining to the pressing step during manufacturing of paneer.

10.1   INTRODUCTION

Paneer is one of the acid-heated coagulated milk products. It is also called as cooking-type-acid-coagulated cheese,[6] Indian cheese,[18] and soft cheese.[11,20,34] It is extensively used for the preparation of a large number of culinary dishes (curries with peas, potatoes, spinach, etc.). It is to the vegetarians what meat is to the nonvegetarians.[16] *Paneer* resembles unripened cheese prepared from whole milk and skimmed milk.[10] *Paneer* contains nearly all proteins, fats, insoluble salts, and colloidal matter of the original milk. It also contains part of the moisture of serum of the original milk in which lactose, whey proteins, soluble salts, vitamins, and other milk components are present in proportion to their amount.

Paneer is reported to be consumed very widely in South Asian and neighboring countries. In India alone, about 5% of the milk produced is estimated to be converted into *paneer*.[19] The huge potential lies ahead for organized dairy industry for marketing of *paneer*. Its full potential can be tapped in the lines of popularization of various types of cheeses in European and American countries. Hence, there is a considerable scope to develop technologies for the industrial production of *paneer*. Successful attempt has been made to manufacture "processed *paneer*" whose production can be mechanized easily with better quality control and lower processing cost.[23] The FAO/NDDB workshop on indigenous milk products has stressed the need for developing mechanized systems for the production of indigenous milk products including *paneer*.[3]

Processes have been standardized for the production of conventional, low fat filled, fortified, and vegetable *paneer*. Process modifications have been made for *paneer* from cow and recombined/reconstituted milk.[16,17,27] But the physical parameters such as pressing of curd and chilling of pressed curd have been overlooked and therefore urgent need is required for standardization of these parameters which would be essential in designing the suitable equipment for production of *paneer*. The existing method of *paneer* production is still a batch process and confined in the unorganized sector only. The manufacture of *paneer* is limited to small scale for meeting the local demands.[6] This type of manufacturing process is vulnerable to environmental influences, permitting a very short shelf life. Pressing of curd and chilling of pressed curd are the essential and integral processes during *paneer* manufacture. These processes have been customarily treated as routine and almost no information is available on effect of their variation on recovery of solids, yield, and sensory attributes of *paneer*.

Considering the limited importance given by the investigators, the present review is undertaken to collect and document whatever information is available on the methods of applying pressure and effects of pressing and chilling processes from the methodologies described by various investigations.

10.2   PRESSING OF CURD

Paneer, essentially a harder form of *chhana*, has a characteristic texture.[25] *Paneer* resembles unripened cheese prepared from whole milk.[10] The chewiness of the *paneer* may be a factor which is synonymous to that of meat. It is clear that in *paneer*, the attributes resembling meat products would evolve with the proper combination of pressing parameters. Pressing of curd is important because paneer pressed under a high pressure or for a long time would probably be very compact lacking sponginess. On the other hand, pressing at very low pressure or for a short time may result in poor handling and storage properties owing to low cohesive strength and high moisture content.[18]

The good quality *paneer* should have firm, cohesive, spongy body, and closely knit smooth texture.[14,37] The *paneer* should be a compact body.[20] The required compactness shall come through pressing. The curd is pressed for increasing the cohesion.[18] The properties, such as cohesive strength and sponginess of *paneer* are dependent on the pressing parameters, that is, quantity of curd taken, the amount of pressure, and time of pressing. Curd is pressed to obtain proper shape and size of *paneer*, enabling it to withstand thermal stress during frying and cooking. Whey drainage[29] and matting of curd[1] are the other reasons for pressing of curd. Beside these, consolidation of curd is needed for cutting of *paneer* into pieces of regular size and shape. This facilitates its mixing with vegetables for making of culinary dishes.

10.2.1 PRESSING METHODS

10.2.1.1 IMMEDIATE PRESSING METHOD

Various investigators have prepared the *paneer* with wide variations. These methods can be broadly classified on the basis of time elapsed before start of pressing.

The standardized milk was heated at 82°C in the cheese vat and then cooled to 70°C. For coagulation, 1% citric acid solution at 70°C was added slowly to the milk and it was agitated vigorously till clean whey was separated. After coagulation, the whey was drained out and the curd was filled in rectangular hoop. Pressure was applied on top of the hoop. The pressed curd was then immersed in chilled water. The *paneer* was then removed and placed on wooden planks to allow loose water to drain for 10–15 min.[6] Same procedure was adopted with minor changes in quantity of raw materials, processes, and parameters.[4,8,9,12,13,15,21,24,28,30,31–33,38]

10.2.1.2 DELAYED PRESSING METHOD

The milk was heated to 80°C and then after lowering down the temperature, coagulant solution at 70°C was added and stirred till the coagulation was completed. The coagulated mass along with whey was transferred to a muslin cloth and hung by a thread tied between two pegs. When the dripping of whey stopped, it was transferred to a wooden hoop and pressure was applied. The pressed curd was cut into pieces and immersed in chilled water of 4–6°C for about 2 h. It was then placed on a wooden plank for 10–15 min to drain loose water and then it was wiped with dry cloth.[18,39]

10.2.2 ASSOCIATED PHENOMENON DURING PRESSING OF CURD

An attempt has been made to describe the mechanism for pressing of curd and expulsion of whey.[18] As pressure is applied, water starts to migrate readily toward the perforated wall of the press. Along with water, finer particles of curd are also transported and deposited in the outer pores. Consequently, porosity of the outer zone is reduced and further expulsion of water is restricted although sufficient moisture is present in the central zone. They reported that permeability of peripheral zone might be reduced to the extent of formation of an impervious layer and due to this phenomenon, there are chances that moisture reduction in case of higher pressure may be low.

The more or less similar observations have been reported during pressing of other intermediate dairy products. During the process of dewatering of casein curd using a pressure cell,[40] examination of pressed sample revealed that appearance near the drainage surface had changed from opaque white to translucent creamy brown. There was relatively little change in the sample appearance at the pressing surface of the sample. This phenomenon is termed as "plasticization" because of the appearance and characteristics of the translucent material formed. During the dewatering of casein curd, plasticization occurs at the drainage surface due to the pressure at this surface and due to matting of the adjacent casein molecules. Also, as the sample is compressed, the moisture is drained out and the curd particles are transported to the periphery of the sample in order to fill the voids, which result in plasticization.[36]

10.2.3 PARAMETERS FOR PRESSING OF CURD

10.2.3.1 WASHING OF CURD

The quantity of water used for washing of curd each time is about one-third of the amount of milk taken with a contact time of 10–15 min. The curd is washed with hot water (71°C) before pressing. However, other investigators did not mention the washing of curd before pressing.[1,12,30,34]

10.2.3.2 COLLECTION OF CURD FOR PRESSING

Most of the investigators gave 5–10 min residence time after coagulation and then straining of whey was done. After coagulation, the curd was allowed to remain in whey itself for 5 min for settlement.[6,11] Then the whey was drained off through a muslin cloth and the curd was subjected to pressing.[5,8,22,28,30] In another investigation, the same method was adopted but stainless steel strainer was used instead of muslin cloth.[21,32] However, the curd was collected and filled in hoops lined with strong and clean muslin cloth. Some investigators adopted the gravity drainage of whey till drippings practically stopped and then the curd was subjected to pressing.[18,39]

10.2.3.3 INITIAL THICKNESS OF CURD

The initial thickness of curd would give a clear idea about the methodology of pressing to be adopted by various investigators. Although the quantity of milk taken per batch has been reported, but unfortunately, the exact quantity of curd subjected to pressing was not mentioned in most of the studies. However, the initial quantity of curd can be estimated if the density of curd is known. The density of curd measured before pressing was 0.98 ± 0.01 g/cm^3 at $50 \pm 5\%$ moisture content.[1] For simplification, if the density of curd is 1 g/cm^3 and overfilling of curd is 10% of volume of hoop, the quantity of curd for pressing can be estimated.

The use of rectangular hoop without specifying its size was reported.[11] In a research work, the curd was filled for pressing in a rectangular box of 35 cm × 28 cm × 10 cm size.[6] It may be estimated that the quantity of curd was approximately 10.8 kg and from the hoop size, it is clear that the initial thickness of curd must have been 10 cm. In another study, the curd was transferred after stoppage of drippings of whey into a cubical wooden hoop of 15 cm × 15 cm × 15 cm size.[39] The quantity of curd must have been approximately 3.75 kg and the initial thickness of curd was 15 cm. Similarly, the squared-shaped hoop of 15 cm × 15 cm × 7.5 cm was used to accommodate 1.85 kg curd with approximately 7.5 cm initial thickness of curd.[22]

In a research work, 425 g of curd was filled into a wooden hoop of 16 cm × 8 cm × 3 cm with initial thickness of 3 cm.[13] An apparatus was developed for pressing of curd. It consisted of vertical perforated cylinder of 5.5 cm internal diameter and 12 cm height. On the basis of assumptions of overfilling and density, it showed that the amount of curd taken would have been approximately 315 g with initial curd thickness of 12 cm.[18] In a report of *paneer* preparation, the use of hoops of varying size to hold 2–10 kg curd has been mentioned.[7] For pressing, four to five *paneer* hoops were placed one over the other and 70–100 kg dead weight was placed on the top of the hoop. The size of hoop was not mentioned and therefore initial thickness of the curd in a hoop cannot be judged.

10.2.3.4   AMOUNT OF PRESSURE

In one of the pioneer work on preparation of *paneer*,[6] the curd was filled in a hoop. After closing the top and bottom of wooden hoop by planks of approximate size, a weight of 45 kg was applied on 35 cm × 28 cm top surface. This 45 kg weight on an area of 35 cm × 28 cm would create a pressure of 0.048 kg/cm^2 which was actually different from 2 kg/cm^2 approximately.[6] On the requirement of pressure, investigators seemed to have remained divided into following two specific ranges of pressures. As discussed above, the same pressure of 0.048 kg/cm^2 was applied for pressing of curd.[4,6,8,9,11,28,32,33] The use of pressure of 0.070–0.141 kg/cm^2 for pressing of curd has also been reported.[7] In another investigation, 3 kg weight was kept on a hoop of 16 cm × 8 cm surface area to create 0.023 kg/cm^2 pressure.[13]

However, the use of pressure in the range of 1–3 kg/cm^2 for pressing of curd was also reported.[11,39] In an outline of traditional method of *paneer* manufacture, it was specified that the coagulated mass was pressed with a pressure of 1–1.5 kg/cm^2. In some research, the pressure was created by centrifugal force for pressing of curd.[1]

10.2.3.5   PRESSING TEMPERATURE

During pressing, the temperature of curd is an important factor for the textural quality of *paneer.* Matting of curd is possible when the temperature of curd is greater than 60°C. From the experimentation adopted by different investigators, it is clear that the pressing was done either at coagulation or at room temperature.

After coagulation, the milk solids were allowed to settle for 5 min. During this period, one of the investigators did not permit the temperature of whey and curd to go below 63°C through circulation of hot water in the jacket of vat.[6] Then the curd was subjected to pressing. It is clear that the pressing of curd started at elevated temperature and then it cooled down as the pressing progressed.[9,28,32,33,34] However, in another study, the temperature of contents was not allowed to drop below 67°C.[22]

In other investigations, the whey was drained under the action of gravity till its dripping practically stopped.[18,39] It is obvious that this method of whey removal has brought the curd to room temperature and then at this temperature, it was subjected to pressing.

10.2.3.6   PRESSING TIME

The usual range of pressing time adopted by most of the investigators is 10–30 min. It was pressed immediately for 15 min for straining of the curd.[6] In most of the research works,[11,28,32,33] the curd was pressed for 15–20 min but in another research, the curd was pressed for 30 min.[26] In some research work on *paneer* preparation, 5 kg weight was applied for 15 min on the curd filled in a hoop.[13,22]

A general description of *paneer* making has been given,[41] where the amount of pressure was not mentioned but the pressing time of 15–20 min was mentioned. The pressing time of 15–20 min was adopted for gravity-drained curd with a higher pressure of 1–3 kg/cm².[18,39]

10.3   CHILLING OF PRESSED CURD

The practice of cooling of pressed curd blocks by dipping in chilled water has been adopted by most of the investigators. Therefore, this process is usually referred as chilling. Chilling of pressed curd blocks facilitates handling during cutting and wrapping for retail sale. Washing of the pressed curd blocks may be another purpose of chilling. The pressed curd block is usually cut into smaller pieces and then dipped in chilled water. Increase in surface area and thereby increase in moisture content and decrease in the temperature of pressed curd may be the desired objectives of chilling.

Most of the investigators have adopted the cooling of pressed curd either by dipping in chilled water or in tap water. For cooling of wrapped block of pressed curd, dipping in chilled water at 4°C has been reported.[7] Normally, producers did not bring down the temperature of *paneer* to 5–10°C.[35] As a result, when this *paneer* was transferred to a refrigerator or a cold store, it had taken quite some time to cool down to the temperature as low as 5–10°C and by that time, microorganisms may have initiated spoilage of the product.

The proper chilling of *paneer* is necessary to arrest the growth of microorganisms during its storage. It is also reported to improve the texture of *paneer*.[6] Although the process of chilling consumes the major chunk of the total time required for production of *paneer*, no work has been reported toward the reduction of time of chilling.

10.3.1 ASSOCIATED PHENOMENON DURING CHILLING OF PRESSED CURD

Increase in moisture content is the associated phenomenon along with cooling when pressed curd blocks are dipped in chilled water.[1] Diffusion of chilled water takes place through the pores into outer periphery of pressed curd resulting in the increase of moisture content. For chilling, diffusion theory can be applied for predicting the increase in moisture and determining the time for optimum moisture absorption. Considering both pressing and chilling operations together, the temperature falls down from around 65°C to approximately 5°C. During pressing, dissipation of heat occurs through the wall of the hoop and surrounding air and during chilling through water.

10.3.2 PARAMETERS OF CHILLING OF PRESSED CURD

10.3.2.1 SIZE OF PRESSED CURD

Generally, in most of the investigations, the weight of pressed curd pieces ready to be chilled was between 1 and 2 kg. Pressed curd obtained from the hoop was cut into 6–8 pieces and immersed in chilled water at 4–6°C.[6] Earlier, it is estimated that 10.8 kg curd was subjected to pressing. If removal of whey during pressing is assumed to be negligible, then the weight of pressed curd pieces would be roughly 1.3–1.8 kg. After completion of chilling, the *paneer* pieces were placed on a wooden plank for 10–15 min to drain out surface water. The cutting of the pressed curd block into pieces of 1.5 kg each and dipping in tap water for cooling were recommended.[8] Then, partially cooled *paneer* pieces were immersed in chilled water at 4–6°C for 2 h to cool them. The immersion of pressed curd pieces of 1 cm cubes in chilled water was suggested for about 2 h.[39] The usual practice of cutting of pressed curd block into ½ or 1 kg size pieces and then floating in chilled water is reported.[5]

10.3.2.2   PERIOD OF CHILLING

The time length of chilling adopted by research workers varies between 30 min and 3 h. The pressed curd block was cut into smaller pieces and was immersed in chilled water for 30 min.[26] Immersion of pressed curd in chilled water for 1 h was reported.[22] The cooling of wrapped block of pressed curd by immersing in cold water was done for 1–2 h.[7] In some investigations, immersion of pressed curd in chilled water for about 2 h was reported.[8,28,39] The possible reason for the excessive time of chilling (2–3 h) may be related to pressing. The sealing of pores at the outer periphery of pressed curd was stated.[36,40] The sealing of pores reduced the diffusion of water which in turn resulted in longer time requirement for the center of pressed curd block to reach to the temperature of chilled water.

The methodology of immersion of the pressed curd in potable tap water for cooling has been suggested till its temperature reaches to that of water and then transferred to chilled water to bring its temperature to 5–10°C. The removal of the pressed block of curd and then its immersion in tap water was done for 2 h.[32] Final washing of pressed curd with cold tap water was also reported.[29]

10.4   DEVELOPMENT OF MECHANIZED SYSTEM FOR MANUFACTURING OF *PANEER*

Continuous processes for production of *paneer* have been attempted[2] but most of other workers[6,8,12,13,32,33] mainly concentrated on the process innovations for the production of *paneer*. For the production of *paneer*, adoption of *tofu* (a Japanese product from soybean) preparation method was suggested.[20] Some prominent investigators expressed that it is really unfortunate that the simple process of pressing is not mechanized.[35] For the mechanization of *paneer* production system, suitable modifications were stressed in the hydraulic and pneumatic systems, which are available for the processing of cheese[35] with efforts from the engineers and technologists to develop such systems.

10.5 LIMITATIONS OF RESEARCH STUDIES ON PRESSING AND CHILLING PROCESSES

It can be summarized that all investigators have pressed the curd by filling in a hoop. These studies were conducted only at single pressing and chilling condition and the effect of their variations has not been investigated. During pressing, some amount of whey comes out but no observation was reported in this regard. Similarly, no efforts were made to determine the effect of amount of pressure, time of pressing and chilling, quantity of curd taken, etc. on the recovery of solids and yield of *paneer* production method. During chilling, soluble solids also dissolve into water but the investigators have neglected to make any account. The effects of size of pressed curd and temperature of chilled water on quality attributes of *paneer* have also not been attempted.

10.6 STANDARDIZED PARAMETERS OF PRESSING, CHILLING, AND LOOKING TO MANUFACTURE *PANEER*

The work by different investigators on the process of pressing and chilling was repetitive in nature and almost no basic research has been done. But these studies had reported to impart satisfactory/excellent sensory attributes to *paneer.* Those pressing or chilling parameters which were common in most of the investigations may be assumed as standardized parameters. Considering this, following parameters have been recommended. It is suggested that these can be considered for the development of mechanized method of *paneer* production.

- After coagulation, the solid should be allowed to settle down for at least 5 min.

- The temperature of curd at the time of start of pressing must be above 60°C (nearing to coagulation temperature of 70°C) and pressure adopted must be around 0.05 kg/cm².
- The quantity of curd taken may depend upon the type of equipment used for pressing but the initial thickness must not be more than 10 cm.
- Whatever may be the pressure, the time of pressing must be around 20 min. Sufficient time is required for the development of typical texture of *paneer*.
- Looking to the use of acid for coagulation and specific requirement of *paneer*, the stainless steel is a choice as material of construction because of inherent properties it possesses.
- Muslin cloth or stainless steel filter of approximately same opening size must be used for draining off the whey before pressing and during pressing.
- The weight of pressed curd pieces must not be more than 1.5 kg for dipping in chilled water.
- The temperature of water for dipping of pressed curd should be around 4–6°C.
- The duration of chilling must be equal to the time required by the center to reach the temperature of chilled water.
- After completion of chilling, surface moisture of *paneer* must be allowed to drain away and then should be wiped.

10.7 SUMMARY

This chapter recognizes the importance of production and consumption pattern of *paneer*. Traditionally, milk is first heated near to boiling, cooled to coagulation temperature, and acidified to obtain curd. The curd is pressed in a hoop and the pressed curd blocks are dipped in chilled water for 2–3 h to obtain *paneer*. On the basis of pressing, the two distinct methods adopted by investigators have been discussed. The variations in the process parameters, such as initial thickness of curd, time and temperature during pressing, amount of pressure, time of chilling, etc. were also identified. Some properties, such as cohesiveness and chewiness are developed in *paneer* due to pressing of curd and chilling of pressed curd. There are theories and associated phenomenon to understand and explain the changes during pressing and chilling. In view of the inevitable mechanization of *paneer* manufacturing, some recommendations have been made. These can be adopted as standardized parameters for further investigations and development of *paneer* manufacturing machine.

KEYWORDS

- **casein curd**
- **cheese vat**
- **chilled water**
- **chilling**
- **coagulation**
- **cooking**
- **hot water**
- **mechanization**
- **plasticization**
- **pressing**
- **sponginess**
- **tap water**
- **time of chilling**
- **time of pressing, tofu**

REFERENCES

1. Agrawal, A. K. Studies on Traditional and Centrifugal Methods of Paneer Production. Ph.D. Thesis, Indian Institute of Technology, Kharagpur, 1977; pp 1–30.
2. Aneja, V. P.; Rajorhia, G. S.; Makkar, S. K. Design and Development of a Continuous *Chhana* Making Machine. *J. Inst. Eng. (India)* **1977,** *58*(1), 1–3.
3. Anonymous. Proceedings and Recommendations of FAO/NDDB Workshop on Indigenous Milk Products. *Indian Dairym.* **1991,** *42*(2), 47–52.
4. Arora, V. K.; Gupta, S. K. Effect of Low Temperature Storage on *Paneer. Indian J. Dairy Sci.* **1980,** *33*(3), 374–380.
5. Bandyopadhyay, A. K.; Mathur, B. N. Indian Milk Products. In: *A Compendium. Dairy India;* Gupta, P. R., Ed.; Dairy India Yearbook: A-25, Priyadarshini Vihar, New Delhi, 1977; p 212.

6. Bhattacharya, D. C.; Mathur, O. N.; Srinivasan, M. R.; Samlik, O. Studies on the Method of Production and Shelf Life of *Paneer* (Cooking Type of Acid Coagulated Cottage Cheese). *J. Food Sci. Technol.* **1971**, *8*(9), 117–120.

7. Chandan, R. C. Cheeses Made by Direct Acidification. *Indian Dairym.* **1992**, *44*(3), 133–141.

8. Chawla, A. K.; Singh, S.; Kanawjia, S. K. Development of Low Fat *Paneer. Indian J. Dairy Sci.* **1985**, *38*(4), 280–283.

9. Chawla, A. K.; Singh, S.; Kanawjia, S. K. Effect of Fat Levels, Additives and Process Modifications on Composition and Quality of *Paneer* and Whey. *Asian J. Dairy Res.* **1987**, *6*(2), 87–92.

10. Chopra, S.; Mamtani, R. Say Cheese or *Paneer. Indian Dairym.* **1995**, *47*(8), 25–29.

11. De, S.; Bhattacharya, D. C.; Mathur, O. N.; Srinivasn, M. R. Production of Soft Cheese (*Paneer)* From High Acid Milk. *Indian Dairym.* **1971**, *23*(11), 224–226.

12. Dharam, P.; Garg, F. C. Utilization of Sour Butter Milk in the Production of *Paneer. Indian J. Dairy Sci.* **1989**, *42*(3), 589–594.

13. Dharam, P.; Garg, F. C. Studies on Utilization of Sweat Cream Buttermilk in the Manufacture of *Paneer. J. Food Sci. Technol.* **1989**, *26*(5), 259–264.

14. Dharam, P.; Gupta, S. K. Sensory Evaluation of Indian Milk Products. *Indian Dairym.* **1985**, *37*(10), 465–477.

15. Gupta, S. K.; Dharam, P. Suitability of Reverse Osmosis Concentrated Milk for the Manufacture of *Paneer. J. Food Sci. Technol.* **1995**, *32*(2), 166–168.

16. Kanawjia, S. K.; Roy, S. K.; Singh, S. *Paneer* Technology and Its Diversification. *Indian Dairym.* **1990**, *42*(9), 390–393.

17. Kanawjia, S. K.; Singh, S. Technological Advances in *Paneer* Making. *Indian Dairym.* **2000**, *52*(10), 45–50.

18. Kulshreshtha, M.; Agrawal, U. S.; Singh, B. P. N. Study on *Paneer* Quality in Relation to Pressing Conditions. *J. Food Sci. Technol.* **1987**, *24*(5), 239–242.

19. Mathur, B. N.; Hashizuma, K.; Musumi, S.; Nakajuwa, Y.; Watunabe, T. Traditional *Paneer* in India and Soybean Food Tofu in Japan. *Jpn. J. Dairy Food Sci.* **1986**, *35*(3), A138–A141.

20. Mathur, B. N.; Zanzad, P. N.; Rao, K. V. S. S. *Paneer* and *Tofu*: An Appraisal of Product and Process Synergies. *Indian Dairym.* **1991**, *43*(9), 407–413.

21. Pal, M. A.; Yadav, P. L. Effect of Blending Buffalo and Cow Milk on the Physicochemical and Sensory Quality of *Paneer. Indian J. Dairy Sci.* **1991**, *44*(5), 327–332.

22. Pal, M. A.; Yadav, P. L.; Sanyal, M. K. Physico-chemical and Sensory Characteristics of Low Fat *Paneer* from High Acid Heated Milk. *Indian J. Dairy Sci.* **1991**, *44*(7), 437–441.

23. Pal, M. A. Processed *Paneer*—Contemporization of the Ancient Delight. *Indian Dairym.* **2002**, *54*(10), 78–80.

24. Pant, A.; Chauhan, G. S.; Kumbhar, B. K.; Singh, S. Texture Profile Analysis of Tofu and Milk *Paneer* Before and After Deep Fat Frying. *J. Food Sci. Technol.* **1993**, *30*(6), 449–450.

25. Patil, G. R. Present Status of Traditional Dairy Products. *Indian Dairym.* **2002**, *54*(10), 35–46.

26. Pruthi, T. D.; Koul, J. L. *Paneer* from Crossbred Cow's Milk. *Indian J. Dairy Sci.* **1989**, *42*(2), 403–404.

27. Rajorhia, G. S.; Dharam Pal; Arora K. L. Quality of *Paneer* Marketed in Karnal and Delhi. *Indian J. Dairy Sci.* **1984,** *37*(1), 50–53.

28. Rao, M. N.; Rao, V. V. R.; Rao, T. J. *Paneer* from Buffalo Milk. *Indian J. Dairy Sci.* **1984,** *45*(6), 281–291.

29. Rao, K. V. S. S.; Zanzad P. N.; Mathur, B. N. *Paneer* Technology—A Review. *Indian J. Dairy Sci.* **1992,** *45*(6), 281–291.

30. Sachdeva, S.; Singh, S.; Kanawajia, S. K. Recent Developments in *Paneer* Technology. *Indian Dairym.* **1985,** *37*(11), 501–505.

31. Sachdeva, S.; Singh, S. Use of Non-conventional Coagulants in the Manufacture of *Paneer. J. Food Sci. Technol.* **1987,** *24*(6), 317–319.

32. Sachdeva, S.; Singh, S. Optimization of Processing Parameters in the Manufacture of *Paneer. J. Food Sci. Technol.* **1988,** *25*(3), 142–145.

33. Sachdeva, S.; Singh, S. Incorporation of Hydrocolloids to Improve the Yield, Solid Recovery and Quality of *Paneer. Indian Dairym.* **1988,** *41*(2), 189–193.

34. Sachdeva, S.; Singh, S. Shelf life of *Paneer* as Affected by Antimicrobial Agents, Part-1: Effect on Sensory Characteristics. *Indian Dairym.* **1990,** *43*(1), 60–63.

35. Sachdeva, S.; Singh, S. Industrial Production of *Paneer:* Innovative Approaches. *Indian Dairym.* **1995,** *47*(4), 11–14.

36. Shah, C. M.; Shah, B. P. Dewatering of Casein Curd—A Review. *Indian Dairym.* **1987,** *39*(10), 471–475.

37. Sindhu J. S.; Arora, S.; Nayak, S. K. Physico-chemical Aspects of Indigenous Dairy Products. *Indian Dairym.* **2000,** *52*(10), 51–64.

38. Singh, S.; Kanawjia, S. K. Development of Manufacturing Technique for *Paneer* from Cow Milk. *Indian J. Dairy Sci.* **1988,** *41*(3), 322–325.

39. Vishweshwaraiah, L.; Anantakrishnan, C. P. A Study on Technological Aspects of Preparing *Paneer* from Cow Milk. *Asian J. Dairy Res.* **1985,** *4*(3), 171–176.

40. Vu, J. T.; Munro, P. A. Press Dewatering of Casein Curd. *N. Z. J. Dairy Sci. Technol.* **1981,** *16*, 265–272.

41. Warner, J. N. *Principles of Dairy Processing*; Wiley Eastern Ltd.: New Delhi, 1976; p 231.

CHAPTER 11

RHEOLOGICAL AND TEXTURAL PROPERTIES OF MILK GELS

CHAKRABORTY PURBA[1], SHALINI GAUR RUDRA[2],
RAJESH KUMAR VISHWAKARMA[3], UMA SHANKER SHIVHARE[1],
SANTANU BASU[1,*], and RUPESH CHAVAN[4]

[1]*Dr. S. S. B. University Institute of Chemical Engineering and Technology, Panjab University, Chandigarh 160014, India*

[2]*Division of Food Science and Postharvest Technology, Indian Agricultural Research Institute, New Delhi 110012, India*

[3]*Central Institute of Postharvest Technology, Punjab Agricultural University, Ludhiana 141004, India*

[4]*Department of Quality Assurance, Mother Dairy Junagadh, Junagadh 362001, Gujarat, India*

Corresponding author. E-mail: santbasu@gmail.com

CONTENTS

ABSTRACT

Coagulation/gelation of milk is the most important step in the dairy industry during manufacturing of products such as cheese, dahi, yoghurt, chhana, paneer, etc. Understanding of coagulation phenomenon in milk during processing is very important to obtain a product with uniform quality and acceptability to the consumers. Coagulation of milk is the primary step that comprises the conversion of raw milk, which leads to formation of an aggregated protein network, mainly consisting of caseins. Coagulation in milk takes place due to the effect of enzymatic, heat and/ or acidic treatment and depends upon various factors such as pH, temperature, fat/protein profile of milk, and calcium content. The mechanism also varies with respect to bovine milk and buffalo milk. Coagulation or gelation properties of milk can be directly related to the composition of amino acids present and their sequences, as well as its secondary and tertiary conformation in solution. This chapter is mainly focused on gelation behavior and its impact on textural/rheological properties of different milk products.

11.1   INTRODUCTION

Food behaves like a dispersion, where particles are mostly nonspherical, flexible, interacting, and have a broad size distribution. This continuous medium contains several solutes and varied size of particles and is often non-Newtonian in nature. Fluid and viscoelastic foods consist of weak structures since its disintegration is desirable. Food gels are usually formed by combining food polymers with physical cross-linking rather than by permanent bonds.

The concept of rheology of food products, acid-induced milk gels, heat-acid-induced milk gels, enzyme-induced milk gels, rheology of yogurt, green cheese (paneer/chhana), cheese, and factors affecting the rheology of milk products are covered in this chapter.

This chapter presents an overview on the rheological and textural properties of milk gels. It emphasizes on the evaluation of texture, rheology, microstructure, and functionality; describes some important concepts; and thoroughly reviews the textural and rheological methods applied till date for the study of milk gel.

11.2 RHEOLOGY OF GELS

All food materials are viscoelastic in real time frame. The deformation timescale determines the proportion of viscous (fluid) properties to elastic (solid) properties in a material. Food materials act like "viscoelastic materials". In solids, the stress is applied and the resultant strain is measured in terms of change in dimensions or degree of displacement. For fluids, the stress always depends on the rate of change of strain with time and not on the magnitude of deformation. At times "solid" materials flow as liquids under a sustained stress. Biopolymers, on the other hand, behave usually like liquids in solution, while sometimes they form soft solids (gels). Proteins and polysaccharides are major classes of biopolymers.

Gels are composed of cross-linked polymeric molecules tangled to form interconnected networks immersed in the liquid medium. At the molecular level, gelation is the process that imparts stress resisting bulk characters (solid properties) due to the continuous framework of networks of polymer chains that extends throughout the gel phase.[2,3,32] Flory[33] proposed classification of gels based on structural criteria as:

- Well-ordered lamellar structures, including gel mesophases
- Covalent polymeric networks; completely disordered
- Polymer networks formed through physical aggregation mechanism; predominantly disordered, but with regions of local order
- Particulate, disordered structures

A "gel" may be conveniently characterized[21] by: (a) storage modulus (G') exhibiting a distinct plateau area ranging to timescales and (b) loss modulus (G'') with $G' > G''$. Upon heating or acidification of the system that gets gellified, the change from fluid to gel state with respect to temperature is detected by recording the elastic modulus (G') as a function of time. The "gel point" can be detected when G' or G'' becomes significantly greater than a preassigned threshold value.[114]

Small deformation measurements are used to analyze the rheological properties of gel, where an oscillating force is applied at a fixed frequency (ω) and a fixed maximum strain to a gel structure. The response lags the force, which is applied and is out of phase by a well-defined phase angle δ. The complex shear modulus (G^*) consists of a real (in-phase) component called the elastic modulus or storage modulus (G'), which reveals the solid-like properties of the system, and an imaginary (out-of-phase) component called the viscous modulus or loss modulus (G''), which reflects the properties of the liquid. The value of ω influences both G' and G'' and ratio of these two moduli defines the tangent of the phase angle δ. G'' is greater than G' (leading to $\delta > 45$ and $\tan \delta > 1$) when fluidlike properties are more dominant than solid-like properties. These parameters are defined as follows:

$$G^* = \sigma_0 / \gamma_0 \tag{11.1}$$

$$G' = G^* \cos \delta \tag{11.2}$$

$$G'' = G^* \sin \delta \tag{11.3}$$

$$\tan \delta = G'' / G', \tag{11.4}$$

where G^* is complex shear modulus (dimensionless), G' is elastic modulus (Pa), G'' is viscous modulus (Pa), σ_0 is the amplitude of the shear stress (Pa), γ_0 is the amplitude of the strain (Pa), and δ is phase angle.[107,131]

11.2.1 FOOD GEL RHEOLOGY

Gel formation depends on the development of "junction zones" between molecules or aggregates that confine the expansion of the network. Thus, the interconnected material network forms a continuous phase throughout the entire volume and becomes swollen with a high proportion of liquid in it.[4,51] Most food biopolymers form physical gels, structured by weak hydrogen, hydrophobic, and electrostatic interaction.[16,108,114,116] Gelling of food biopolymers is divided into "cold setting" and "heat setting," based on the gelation mechanism. In the former, gelation is induced by cooling (agarose, carrageenans, pectin, whey protein, etc.) while in the latter, gelation occurs due to heating (bovine serum albumin, myosin, etc.). On the basis of mechanical viscoelastic characteristics, food gels can again be classified into two types: strong and weak gels. Both strong and weak gels behave as solids at small deformations. However, strong gels behave as solids, while the weak gels are structured fluids at large deformations. Several researchers have identified weak gel-like behavior of food biopolymer gels and solutions.[24,35,59,74,91,96,109,115,116,133]

11.3 MILK PROTEIN GELS

Milk protein gels are aggregated gel structures. Gel formation is the crucial stage in the production of acid-based or enzyme-based gels such as yogurt, cheese, and many other dairy-based products. Several studies have been carried out to understand the gelation mechanism of milk proteins using various dairy ingredients (skim milk powder, milk protein concentrate, whey protein isolate, and sodium caseinate).[13,17,41,50]

Protein gels are normally created by the formation of cross-links or additional attractive interactions between particles in a suspension induced by acidification, enzyme action, heating, and pressure followed by the transformation of a liquid suspension to a gel. The weakness of these gel networks is particularly obvious around the sol–gel transition point, thus complicating the scope for the application of macroscopic mechanical methods. Therefore, the viscoelastic parameters of weak gels can be measured at a very small deformation and the stress–strain involved in the measurements can frequently be below the limits of accuracy of conventional rheological instruments.[19] Altering the pH of raw milk always leads to aggregation of proteins due to systematic loss of electrostatic repulsions and other physical interactions, such as hydrophobic attractions, van der Waals attractions, steric, and entropic effects.[112] Due to preheating, denaturation of whey proteins takes place and subsequent formation of micelle bound kappa casein (κ-CN)/ whey protein complexes leads to modified surface properties leading to increase in both surface hydrophobicity and apparent isoelectric point (pI) (toward the pI of whey proteins).[23,79,86] Milk protein gels are irreversible. Acidified milk gels are considered to be particle gels consisting of aggregated spherical particles forming a continuous network of clusters and strands differing in gelling properties from that of hard spherical gels with respect to the internal structure of the building blocks (casein micelles).[53,54,140]

Retention of water in the gel structure is an important functional property, which is directly related to the network structure and it is physically entrapped within the casein strands forming the gel network. Hence, the tendency for separation of whey is primarily linked to the dynamics of the network.[51,141,142] Spontaneous syneresis, referred as separation of water (or whey) from the gel without any applied external force, is a common feature in acid-induced milk gels.

Protein cross-linking enzymes for the modification of gelling methods have gained interest during past few decades. The formation and extent of enzyme-induced cross-links in protein systems depend on several factors such as the use of optimum activity and stability conditions (temperature, pH, exclusion of inhibitors, etc.) of the enzyme, morphological state of the substrate protein molecule in the reaction conditions, and availability of the targeted amino acid side chains. Proteins without a confined tertiary structure are highly prone to enzymatic cross-linking compared to densely compact globular proteins in which target amino acid residues can be embedded in the interior of the molecule and thus remain inaccessible. In this respect, caseins are good substrates for enzymatic cross-linking because of their flexible rheomorphic structure, whereas globular whey proteins in their native form are poor substrates for enzymatic catalysis.

Complete or partial denaturation of the whey proteins by chemical reduction of the disulfide bridges, exposure to alkaline pH, high temperature, or high-pressure treatments can, however, increase the extent of enzymatic cross-linking of these proteins.[27,28,57] Several theoretical approaches and models have been proposed to illustrate the gelation process. However, explicit explanation to the kinetics of network formation and growth is inadequate.[52,54]

11.3.1 CASEIN GELS

Caseins usually exist as a form of colloidal association in solution.[119] It strongly aggregates with calcium phosphate to form casein micelles of about 300 nm. These casein micelles further aggregate with whey proteins (Fig. 11.1). When the calcium phosphate is removed, the residual sodium caseinate occurs in solution mostly as a mixture of casein monomers and casein nanoparticles of 10–20 nm size. Extent of aggregation of sodium caseinate mainly depends on the relative proportions of different monomeric caseins present; pH and calcium ion concentration of the medium; temperature; and ionic strength.[7,20]

Native casein micelles are stabilized sterically by the calcium-insensitive κ-casein. In neutral pH environments, the colloidal particles are unaffected by homogenization (i.e., intense flow fields) and pasteurization (i.e., moderate heat treatment). Industrial processes lead to destabilization of the casein micelle by acidification or enzyme action and is used for the production of various products such as cheese and fermented milk. Destabilization can also be brought about by severe thermal treatment (e.g., as in sterilization) or by high-pressure treatment (at a few thousand atmospheres). Casein gels can be obtained by several coagulation methods. Acidification and enzymatic coagulation mechanisms are commonly applied. Other than these, heat-acid-induced coagulation and pressure-induced coagulation are also in use. The casein gels obtained in all the cases differ in their textural attributes as well as rheological properties. Dairy products differ in their physical properties as well as texture due to variation in coagulation mechanism. The rheology of these products is discussed later in this chapter.

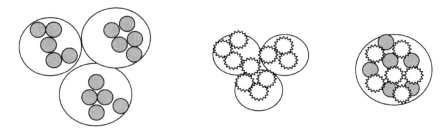

FIGURE 11.1   Structural representation of a mixture of casein micelles and whey protein aggregates.

11.4   TYPES OF MILK GELS

11.4.1   *ACID-INDUCED MILK GELS*

Two processes are followed to produce acid-induced milk gels: (a) traditional fermentation of milk by lactic acid bacteria which yields products such as yogurt, dahi and (b) direct acidification by food-grade acids such as citric acid, lactic acid or hydrochloric acid, or by addition of glucono-δ-lactone (GDL), which is hydrolyzed to gluconic acid, resulting in a reduction in pH.[78,82–85] An extensive study on the formation and properties of milk casein gels carried out by cold acidification with direct addition of hydrochloric acid and successive heating to the gelation temperature has been reported[112,113] in the literature.

Acid-induced milk gels are particulate gels and appear as porous, homogeneous material at length scales above tens of micro meter. It is well known that the main effects of lowering the milk pH are to release micellar calcium and decrease the net negative charge of the micelle. Van Hooydonk et al.[138] observed pH-dependent phenomena for both micellar casein dissociation and micelle voluminosity, which were due to the micellar calcium phosphate content and the electrostatic repulsion/attraction between charged groups. Van Vliet et al.[140] argued that acid milk gels can be considered as particle gels at the time of gel formation; but almost immediately after gel formation, they show properties similar to that of both particle and polymer gels. Polymeric gel properties develop from the fusion of casein particles. Spreadability is a sensorial parameter for protein gels with a particulate structure, which can be correlated to the distribution of broken and masticated pieces during oral processing.[137] The breakdown behavior while applying large deformation to particulate casein gels is such that fracture of the gel occurs by consuming maximum amount of energy, and thus, the structural breakdown into pieces.

11.4.1.1   YOGURT

Yogurt has been recognized since ancient periods to have positive health impact due to high protein and calcium contents besides beneficial action of its viable bacteria that compete with pathogenic bacteria for nutrients. Owing to needs of consistent diversification of food market, there has been an increase in the variants of yogurts and yogurt-based products such as reduced fat content yogurts, probiotic yogurts, fiber-enriched yogurts, yogurt shakes, drinkable yogurts, frozen yogurt, etc. The acceptability of these products by consumers is mainly based on satisfactory textural, rheological, and sensory attributes.

11.4.1.1.1   Rheology of Yogurts

Fermentation of yogurt induces irreversible changes in the properties of milk.[122] Typically, milk is fortified with dairy ingredients to produce a milk base, which is then subjected to a drastic heat treatment leading to thermal denaturation of whey proteins and their partial fixation on the β-casein micelles. Casein aggregation is hence promoted, giving stronger gels and decreasing the extent of acidification required for the association. Finally, the lactic acid production during fermentation results in destabilization of the micellar system leading to associated gelation of the proteins. As the pI of denatured proteins (pH 5.2) and casein (pH 4.6) are reached, low-energy bonds, mainly hydrophobic, are progressively developed between the proteins.[81] Slow acid development favors formation of grains in yogurt, while a coarse microstructure and low viscosity is due to short gelation time with a lower degree of casein aggregation and a looser network.[128] Textural methods such as back extrusion and penetration tests applied for evaluation of yogurts are destructive. The study of viscoelastic behavior provides important information on various aspects such as onset of gelation, extent, and strength of internal structures. Analysis of samples in the linear viscoelastic region (LVR) generates information through nondestructive manner, which although does not mimic the mouthfeel exactly as irreversible structural deformation takes place in mouth, indicates the initial experience of a consumer[62] that is predominant in the acceptability of any food.

11.4.1.1.2 Flow Behavior of Yogurts

The flow properties of stirred and set yogurts are characteristic of a non-Newtonian and weak viscoelastic fluid.[18] Primarily, the rheological characteristics of yogurt are governed by milk composition; temperature, and time of milk heat pretreatment; type and quantity of starter culture used to inoculate the milk; fermentation temperature; and storage conditions of the final product. Several researchers have studied the correlation among yogurt rheology and structure, evaluating the effects of milk heat treatment, starter culture type, incubation temperature, storage time, incorporation of fibers, probiotic cultures, exo-polysaccharide (EPS)-producing strains, etc.[15,38,111,128] Results from rheology studies reveal good correlation with the observations of the microstructural analysis. Keogh and O'Kennedy[63] studied the role of milk fat, protein, gelatin, and hydrocolloids (starch, xanthan, and locust bean gum mixture) on the rheology of stirred yogurt. The consistency index (K) and syneresis were found to be more frequently influenced by the composition than the flow behavior index (n) and the critical strain.

Herschel–Bulkley model has been widely applied to describe the shear dependency of strain in yogurts. Temporal variation of apparent viscosity indicated shear thinning characteristics of yogurt. The shear thinning mechanism for systems that are structured as biopolymer networks involves separation of aggregates or chains of biopolymers.[88] The flow behavior index decreases with increased consistency index values.[105] Further, gum solutions with a high value of "n" tended to feel slimy in the mouth. Thus, when high viscosities and a clean mouthfeel in yogurts are desired, the choice should be with a gum or hydrocolloid mixture having a low n value.

Transient shear studies of yogurts reveal a peak corresponding to a typical viscoelastic system.[15,130] For these types of systems, the power law model can be applied at long shear times using the equilibrium shear stress (σ_∞):

$$\sigma_\infty = mD^n, \tag{11.5}$$

where σ_∞ is equilibrium shear stress (Pa) at each shear rate D; m is consistency index (Pa.s), and n is a power law factor (dimensionless).

Storage time has been found to be a significant factor ($p < 0.05$) for both consistency index and the power law factor. Consistency index was found to decrease with increased storage time. Abu et al.[1] investigated flow properties of concentrated yogurt (labneh) and demonstrated thixotropic properties as evidenced by hysteresis loop. The increase in viscosity of labneh with storage time indicated more interactions in the labneh structure occurring during storage and resulting in further development of the gel structure. The non-Newtonian behavior of labneh was modeled using the two-parameter power law model:

$$\eta = m \mid \gamma \mid^{n-1} \tag{11.6}$$

The flow behavior index was decreased with storage time, which means that the deviation from Newtonian behavior ($n = 1$) increases with increasing storage time. It was concluded that rate and extent of viscosity reduction depend on both the applied shear rate and the storage time. The rate of structure breakdown is represented using following equation:

$$\left(\frac{(\eta - \eta_s)}{(\eta_0 - \eta_s)} \right)^{1-m} = (m-1)kt + 21, \tag{11.7}$$

where at $m = 2$, the transient viscosity data were satisfactorily correlated with eq 11.7. The rate constant, (k), was considered a measure of the rate of thixotropic breakdown, while the ratio of the initial to equilibrium viscosity, η_0/η_e was considered as a relative measure of the extent of thixotropy.

While investigating the effectiveness of microfluidization to produce reduced fat yogurts, Ciron et al.[15] reported a noticeable increase in viscosity for nonfat yogurts upon the use of microfluidized (Mfz) milk at 50 s[-1]. The flow profile of yogurts obtained from conventionally homogenized milk and Mfz milk were different with Mfz milk showing higher shear stress, yield stress, and greater apparent viscosity. It is thus imperative that greater shear stress is required for flow to commence and those yogurts from Mfz milk are more resistant to shearing effect. The more pronounced hysteresis effect of Mfz of heat-treated milk indicated that Mfz milk had less ability to fully recover its structure after shear-induced breakdown. Though their rheological results did not tally with textural attributes using back extrusion, they advocated that back extrusion profile would more closely relate to gel firmness (G′) and sensory firmness while rheometric studies were good indicators for sensory viscosity. In a study, on incorporation of inulin for stabilization of red capsicum carotenoids incorporated low-fat yogurts, levels of inulin higher than 4% were found to yield higher hysteresis area indicating more stable network.

In a research study by Srisuvor et al.,[129] the apparent viscosities of the probiotic low-fat banana yogurts incorporated with prebiotics, inulin, and polydextrose indicated varied inverse relationship to their increasing concentrations, with samples added with inulin recording lower apparent viscosities. Although the degree of polymerization (DP) of inulin was not specified, yet this aberration may be attributed to the fact that inulin with DP 23 gives higher viscosity and firmness to low-fat-stirred yogurts than inulin with DP of 9.[66] For stirred yogurt products, it should be recognized that mixing results in a reduction in viscosity that is only partially restored after shearing is stopped.[71,72] Recovery of the structure is called "rebodying" and always a time-dependent phenomenon. This recovery of structure also affects the apparent viscosity of yogurts.

Arshad et al.[10] reported that GDL-induced gels had only 30% recovery of the original value of the dynamic moduli even after allowing 20 h for recovery. Lee and Lucey[71] have investigated the structural breakdown in detail of the original (intact) yogurt gels that were prepared in situ in a rheometer, as well as, the rheological properties of stirred yogurts made from these gels. High preheating temperatures and low incubation temperatures in yogurt gels lead to an increase in storage modulus; apparent viscosity during structural breakdown; and yield stress, whereas a decrease in the loss tangent values permeability. Increasing milk preheating temperature and decreasing incubation temperature resulted in increased apparent viscosity along with increasing oral viscosity and sensory attributes in stirred yogurt.

11.4.1.1.3 Viscoelastic Properties of Yogurts

Oscillatory tests are widely accepted for the evaluation of rheological characteristics of yogurt.[99] The data obtained through these tests are G′, energy stored per deformation cycle during a small-angle oscillatory shear (SAOS) test, related to the stiffness of the network. Complex viscosity correlates to the cohesiveness (estimation of the amount of deformation before rupture) while tan δ reflects the viscoelastic character of the material. Full-fat milk curds significantly lowered G′ than the low-fat milk. Fat in the solid-state increases the flexibility of the casein matrix due to decrease in tan δ and thus, its gel-like behavior is increased.[122] Shortly after the formation of yogurts, the elastic or storage modulus, and acidity is low, the electrostatic interaction between casein particles is not as yet as high as it would be in aged gels.[75] However, the fracture properties of young gels can be used as an indicator of possible rearrangements. The G′–G″ crossover point in amplitude sweep runs is normally taken as the yield stress value, which showed significant correlation with the sensory "firmness" as perceived by a trained panel.[42,49] Tunick et al.[135] suggested that the inflexion of the G′ values for cheese can be linked to structure breakdown, and the inflexion point indicates the amount of stress that the sample can withstand before breakdown. The maximum gradient in the downward slope of G′ with strain percentage occurs when equilibrium between bonds rupture and reformation is established.[73] Similar behavior has been found applicable in yogurts containing soluble dietary fibers from *Pachyrhizuserosus* L. Urban, an underutilized tuberous legume crop.[106]

Sendra et al.[122] investigated the effect of incorporation of pasteurized orange fiber in yogurt on the rheological behavior of the set yogurts. The values of G′, G″, and η^* increased with fiber dose in pasteurized fiber yogurts, whereas in nonpasteurized fiber, only slight changes, not significant for all groups were observed. Low doses of 0.2 and 0.4 g/100 mL resulted in lower values for these parameters.[36] Possibly, at fiber dose lower than 4 g/L, rheological parameters decreased due to a disruptive effect of the fiber, and when fiber dose was higher than 6 g/L, rheological parameters increased. It is suggested that although the presence of fiber particles always alters yogurt structure when the fiber dose is high enough, the water absorption compensates the weakening effect of the fiber and strengthens gel structure. Yogurt deformation (γ) was found to systematically decrease with increased fiber levels.

During frequency sweep, yogurts demonstrate a predominantly elastic behavior ($G' > G''$).[102] While in true gels, the G' and G'' show a parallel behavior, and increased with the increase in frequency in yogurts. Also, the range of tan δ, 0.2–0.3 indicates yogurts to resemble an amorphous polymeric behavior, while true gels have tan δ values close to 0.01. The tan δ values have been found to decrease with increased frequency and also decrease in complex viscosity corresponds to typical shear thinning profile. Prasanna et al.[102] investigated the role of EPS producing *Bifidus* strains on the rheological properties of low-fat set yogurt. They found much higher values of the storage modulus (G') of the EPS producing strains compared to the control yogurt indicating electrostatic interactions of EPS with the caseins and whey proteins. Increased G' during storage has been widely reported owing to protein rearrangement during storage leading to higher protein–protein interactions.

Incorporation of fibers from orange, asparagus, carrot cell wall particles (CWP) leads to higher values of G', G'', and η^* in low-fat yogurts.[118] McCann et al.[87] studied the gelation kinetics of CWP added low-fat yogurts using vane geometry of rheometer. They reported shorter gelation times associated with faster rate of pH reduction with the increase of CWP concentration besides enhanced gel strength. Further, as the temperature of the set gels was lowered from 43°C to 4°C, the G' for all gels increased by about fourfold. Lucey[76] explained this phenomenon due to swelling of casein particles providing an increase in the contact area between aggregated casein particles and thus formation of a more rigid gel network. At longer storage times, yogurts with wheat, bamboo, and inulin have been found to exhibit characteristics closer to a gel,[130] where G', G'', and tan δ showed a slight variation with ω. The following weak gel model was found applicable to the dynamic curves of fiber incorporated yogurt samples:

$$G'' = a \times \omega^b \qquad\qquad (11.8)$$

$$G'' = c \times \omega^b \qquad\qquad (11.9)$$

Value of parameter "a" was 5626 Pa.s for atypical gel and 16.260 Pa.s for concentrated solutions,[131] while the reported values of a and c for yogurts lie closer to concentrated solution values. The d values have been found to vary insignificantly between 0.037 and 0.276, with most values being around 0.1. The c values are considered as dynamic consistency index.

Incorporation of cross-linked acetylated starch in yogurts has been found to strengthen the casein network through electrostatic adhesion, steric stabilization, and osmotic effect. In general, viscoelastic moduli and gel strength increase with starch concentration.[12] In addition, the introduction of acetyl group and the cross-linking group has shown to improve properties such as swelling capacity and viscosity over its native form.[103] The tan δ of set yogurt system decreases with the increase in the modified starch concentration (0–1.5%, w/w). The dynamic moduli increase with the increase of swelling power of cross-linked acetylated starch granules. An isothermal time sweep, with constant frequency and amplitude, can indicate structural changes in thixotropic samples.[104] For monitoring structure of set yogurts during storage or effect of stabilizers, the time-sweep experiment can be a useful tool. The time-sweep curves show increasing values of complex viscosity upon storage.

11.4.2 HEAT-ACID-INDUCED MILK GELS

Heating of milk at about 60°C or above leads to unfolding of the whey protein structure as well as irreversible denaturation of whey protein, which yields heat-induced whey protein aggregates (WPA).[23,127] Acidification can be divided into three stages using two light-scattering methods: a lag phase followed by a period of rapid particle growth and, finally, a dynamic equilibrium between particle growth, breakup, and shrinkage.[60] In rheological measurements of milk gel formation due to heating and acidification, there is a two-stage development of the elastic modulus around the coagulation point.[78] The primary stage of this process has been recognized as the gelation of the WPA[46] but knowledge about this "whey gel" or about the interactions between these soluble whey proteins and casein micelles are not adequate, although the rheological outcomes may suggest that these interactions do occur.[78,121] Due to acidification, these aggregates enhance the rate of casein micelle destabilization due to higher surface hydrophobicity and somewhat higher pI.[47,92,93] Aggregates are also assumed to act as bridges between casein micelles and help to strengthen the acid gel network.[5,29,46] Further coatings of the casein micelle surface by WPA (about 8%) reveal that it produced gels with considerably higher G' values than those made of only casein. Therefore, these particle interactions and mechanism of their assembly probably show the structural and mechanical properties of acid gels.[6]

Bringe and Kinsella[11] stated that hydrophobic interactions are the main driving force for protein coagulation, gel formation, and syneresis. Therefore, at low temperatures, casein gels would be weaker due to weak hydrophobic interaction. However, Lucey and Singh[80] reported that the strength of the gels, as seen from viscoelastic moduli, is greater when the temperature is lowered and that hydrophobic interactions do not play a major role in gel strength. Milk gels acidified with 1.3% (w/w) GDL at two different incubation temperatures (30°C and 40°C) and the ones incubated at lower temperature of 30°C had higher G' values than those incubated at higher temperature of 40°C.[78]

Gelation in unheated milk gels with acidification as the only coagulation method happens at about pH 4.9 and if the acidification is performed at a very high temperature, a higher gelation pH is observed.[77] Additionally, the strength of gels made from unheated milk is low. Due to the presence of dense clusters of aggregated casein particles formed from the extensive particle rearrangements during the gel formation, they have low values for the viscoelastic moduli. As a result, a lot of particles from those aggregated clusters would not be completely cross-linked during the course of the gel network formation. Furthermore, the gel strength obtained from heated milk is always higher than those made from unheated milk, because denatured WPA interact with casein micelles and behave similar to bridging materials, thus increasing the strength and number of bonds between protein particles. Gels prepared from milk heated at 80°C for 30 min prior to GDL addition had higher G' values than those made from unheated milk.[80]

A recent study on the characterization of heat-set milk protein gels, conducted by Anema et al.[8] when the gels were subjected to a temperature sweep, showed a decrease in G' with an increase in temperature. In all cases, the G′ decreased with increasing temperature. For all the samples of low-heat low-fat gels and high-heat high-fat gels, G′ decreased more evidently with increasing temperature when compared with the samples at pH 6.25–5.75.

Effect of temperature on the phase angle, where a maximum value is observed at intermediate temperatures, has been reported previously for cheese, although the point and extent of the maximum values vary a lot.[39,43,69,70,94,95,123,124,146] These protein networks are maintained through covalent bonds (linked to disulfide bonds) and through noncovalent interactions (calcium bridges, hydrophobic interactions, and hydrogen bonding). Alterations in these interactions and maintaining balance can be affected by temperature change, which leads to the rate of change of the ratio between Go and Gan leading to the maximum values of phase angle at transitional temperatures.

11.4.2.1   GREEN CHEESE (PANEER/CHHANA)

Acidified milk gels are particulate protein gels with casein as main protein contributing to the typical mechanical and sensory properties.[53] Many cheese varieties (queso blanco or Latin American white cheese, ricotta, Indian cottage cheese or paneer) worldwide can be classified as heat-acid coagulated (directly acidified) cheese.[14] Paneer or chhana (green cheese) is obtained from heat-cum-acid coagulation of casein component present in cow or buffalo milk or a combination (6%—milk fat, 9%—milk solid not fat) thereof by precipitation with sour milk, lactic acid, or citric acid. This process entraps almost all the fat, a part of denatured whey proteins and colloidal salts, as well as a part of the soluble milk solids (in proportion to the moisture content retained).[9] The body and texture of paneer must be sufficiently firm to hold its shape during cutting/slicing, yet it should be tender enough not to resist crushing during mastication, that is, the texture must be compact and smooth.

11.4.2.2   RHEOLOGY OF CHEESE

In heat-acid coagulated cheese types, incorporation of denatured whey proteins imparts certain desirable structural and textural attributes such as flow resistance and lack of elasticity. Acid gels are viscoelastic biopolymers and their rheological and textural properties can be studied with small-amplitude oscillatory shear, large-amplitude oscillatory shear (LAOS), penetration, and texture profile analysis (TPA). In this SAOS analysis, oscillatory shear strain is applied to the sample while maintaining the strain in the LVR. This technique captures the entire process of gel formation in a time frame since it is the important step in the acid-type milk gel product development. Hence, SAOS is mostly used to study the gelation phase of an acidified milk gel.[77,80] These dynamic rheological measurements have the ability to identify the process of development of gel network and distinguish between the elastic and the viscous properties. To an applied sinusoidal stress, these SAOS tests give a strain response or vice versa.

The acidified milk gels are basically casein particle gel network structure linked by weak interparticle interactions which can be effortlessly affected or destroyed by applied stress or strain used in SAOS tests, particularly when the gelation point is reached. Hence, most of the earlier rheological measurements of milk gelation are performed using very small strain data sets (<1%) and oscillating strain rates (<0.1 Hz) to avoid destruction of the gel structure.[78,139] Many studies have been performed utilizing these techniques. For examples, Kim and Kinsella[65] investigated the rheological changes during slow acid-induced gelation of milk by D-glucono-δ-lactone; Roefs et al.[112] studied the structure of acid casein gels in formation and model of gel networks; Van Marle and Zoon[139] studied the rheological properties and permeability of microbially and chemically acidified skim milk gels; Van Vliet and Keetels[141] investigated the effect of preheating of milk on the structure of acidified milk gels; Lucey et al.[79] studied the rheological properties and microstructure of acid milk gels as affected by fat content and heat treatment; Schorsch et al.[120] studied micellar casein gelation in presence of high sucrose content.

Textural properties of heat-induced acidified milk gels are affected by their microstructure. The microstructural changes in casein micelles during the acidification of milk leading to textural changes have been observed in several studies.[28,37,40] These studies focused on obtaining the microstructure of acidified milk gels with respect to texture, pH, and temperature using scanning electron microscopy (SEM), cryo-scanning electron microscopy (CSEM), and confocal laser scanning microscopy (CLSM). Gastaldi et al.[37] studied pH-induced changes during direct acidification and bacterial fermentation in casein micelles of reconstituted skim milk at a given temperature.

Five main pH ranges were proposed to observe the microstructural changes in casein micelles during milk acidification, which were: (a) the first stage of aggregation (pH 6.7–5.8), micelle cluster formation started, but the initial shape was still retained; (b) pH 5.5–5.3, casein particle deformation, stretching, and extensive coalescence; thus forming a pseudo network with an open structure; (c) pH 5.3–4.8, dense network formation and fragmentation into small units with new casein particles; (d) pH 4.8 and 4.7, casein particles fusion, dense network formation followed by a contraction and rearrangement stage, resulting in new casein particles with spherical shapes; and (e) pH 4.6, acidified milk gels formation with the casein particles aggregated into a true three-dimensional network of chains and clusters.

11.4.2.3 TEXTURAL PROPERTIES OF CHEESE

Textural properties of acidified milk gels are influenced by many factors, for example, casein concentration, pH, temperature, the history of pH and temperature, total solid contents, and processing parameters (homogenization pressure and incubation temperature).[101,145] An extremely firm texture can be obtained from a mixture containing large amount of total solids (both casein and fat) or an excessive amount of added stabilizers. On the contrary, weak texture can occur due to low total solids (fat content) in the mix, or deficient heat treatment of the milk, or low acidity, or too low gelation temperature. Lucey[77] comprehensively summarizes the effects of major processing factors on the acid coagulation of milk and properties of acid gels.

The ideal combination of deformation rate and deformation force can be used effectively for maximizing the correlation between sensory and instrumental hardness for cheese-like products. Control of processing parameters during manufacture of paneer such as temperature, pressure of press, control of pH, chilling, and freezing during storage, etc. are critical parameters affecting the textural properties of paneer. Sensory and instrumental measurements for paneer- and cheese-like products are enumerated to assess certain textural properties such as hardness, cohesiveness, adhesiveness, and springiness.[47]

Syed et al.[132] observed that the hardness was maximum for skim milk paneer when compared to cow and buffalo milk paneer. Kumari and Singh[68] reported that cow milk paneer had higher values for cohesiveness, gumminess, and chewiness than buffalo milk paneer, whereas the hardness and springiness were greater in buffalo milk paneer. However, the paneer or channa made from buffalo milk have been found to have harder and chewy texture due to a higher concentration of casein in the micelle state with bigger size, harder milk fat due to a larger proportion of high melting triglycerides in it and higher content of total and colloidal calcium.[126] Kanawjia and Singh[61] observed values of 13.2 N, 0.68, 7.8 mm, 9.8 N, and 71 N-mm, respectively, for hardness, cohesiveness, springiness, gumminess, and chewiness of paneer from buffalo milk.

Uprit and Mishra[136] investigated the effect of varying fat content in milk (0–6%) and varying proportion of soy milk (7.5 B, 0–40%) in the blend on the textural properties of fortified paneer. Based on the predicted model on the texture parameters of soy fortified paneer (SFP) as a function of fat content in milk and proportion of soy milk in blend, 3.1–3.4% fat content in milk and 14–16% soymilk blend was observed to be the optimum. This resulted in SFP (fat—11.24%, protein—24.45%) with the textural characteristics similar to that of control paneer made from 6% milk fat.

Khamrui et al.[64] studied the effect of frying, freezing, and rehydration on texture profile, and relationships between sensory and instrumental textural descriptors of paneer. Frying of paneer resulted in loss of adhesiveness and increased fracturability. Fried paneer, on rehydration by cooking in a salt solution, exhibited decreased instrumental textural values except for cohesiveness.

Shrivastava and Kumbhar[125] performed textural profile analysis on rehydrated paneer cubes. A major variation of in hardness and adhesiveness value was found, whereas springiness, cohesiveness, and chewiness had marginal variation in rehydrated paneer as compared to fresh paneer.

11.4.3 ENZYME-INDUCED MILK GELS

Gelation is a crucial step during cheese production where caseins are destabilized in the initial step for gel network formation. In enzymatic gelation, mostly rennet (chymosin enzyme) action is preferred, or acidification or combination of acid and rennet is also used.[80] At the time of renneting, chymosin excludes N-terminal of the κ-casein and hence, para-casein micelles are formed (primary phase) which lead to subparticle and particle rearrangements. The hydrophilic ends of κ-casein are released into the serum phase as caseinomacro peptide (CMP) so as to rearrange the particle clusters (Fig. 11.2). The protein system then starts to destabilize and syneresis takes place to produce gel which is the secondary phase.[141] The structured development of rennet gels leads to and regulates specific structural and rheological properties of final products.[145] Many studies with special focus on rheology and microstructural properties of rennet gels have been carried out.[34,89,90,98,147]

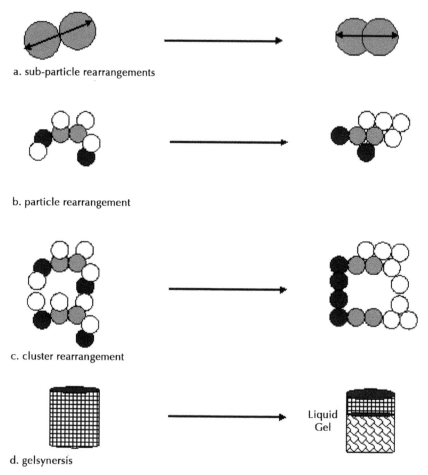

a. sub-particle rearrangements

b. particle rearrangement

c. cluster rearrangement

d. gelsynersis

FIGURE 11.2 Four types of rearrangements that can be distinguished at various levels: (a) molecules or "subparticles" rearrangement leading to particle fusion, (b) particle stage leading to interparticle movement, (c) cluster rearrangement forming strands of particles, and (d) whole gel (macroscopic view).

While studying the rheological behavior of rennet-induced milk gel, the storage modulus (G') and the maximum linear strain (γ_0) are considered as a function of time after rennet addition to monitor the changes in structural rearrangements of casein gels. The rearrangements comprise of an increase in the size of solid building blocks, fractional disappearance of fractal arrangement, and the formation of straightened strands, some of which break in due course.

At isoelectric pH, rennet gels are more porous than acid gels and have thicker protein fibers, as established by several electron microscopic studies.[142] Due to this porous structure, rennet gels are more susceptible to syneresis, leading to the production of lower moisture cheese, as compared to acid milk gels.

Hannon et al.[48] found that rennet gels are more permeable at pH 6.5 than at pH 5.2. The slower cooling rate of rennet gels helps in producing a larger number of small micelle aggregates, more junction zones, and stronger and less permeable gels. High density of casein chain cross-links leads to increase in the value of G'.

11.4.3.1   RHEOLOGY OF CHEESE

Cheese is a composite material. Its major constituents para-casein, fat, and the aqueous phase, contribute in a specific way to the structure and hence to the rheology of cheese. The para-casein matrix imparts the product solidity through the formation of three-dimensional structure. Fat is a major contributor to the temperature- and time-dependent variation in the rheological properties of cheese. Rheological characteristics of cheese are important for quality control (Table 11.1).

International Dairy Federation (IDF) has made following recommendations regarding the rheological methodology for cheese:

- A method that yields real and unequivocal rheological or fracture parameters should be used so that the results do not vary with test-piece size or with a small variation in test conditions.
- The sample preparation should be such that it does not subsequently alter the product properties.
- The type, extent, and timescale, of deformation should be in accordance with the conditions during the actual use of the cheese (e.g., eating, cutting, or storage).

- Different mechanical methods may be used for the purpose of comparison and visible changes during deformation should also be observed.

TABLE 11.1   Selected Rheological Characteristics of Cheese.

Focus area	Applications
Intermediate or final product quality control	Quality control tool; correlated with the overall texture, sensory attributes, and microstructural changes at the time of processing.
Process engineering calculations	For measuring rheological flow properties, based on instantaneous measurements of velocity profiles by means of ultrasound velocity profiling (UVP) technique with pressure difference (PD) technology.
Shelf life	Presence of moisture: decrease in hardness/firmness of cheese; initiation of spoilage.
Texture by correlation to sensory properties	Correlation between sensory and rheological properties; useful to understand cheese texture.

Davis's plastometer was developed to measure deformations in cheese, butter, etc. under compression. These include Devis's apparatus for measuring the crushing strength of cheese; a spherical compression device of Scott Blair and Coppen; Caffyn's ball compressor; and sectilometer, and several penetrometers. Some of these principles have been incorporated into commercial analytical instruments. Hoeppler consistometer is used to measure viscosity and elastic modulus of hard and semihard cheese varieties. Recently, instruments providing nondestructive dynamic measurements have been used for viscoelastic characterizations of cheese offering a better understanding of cheese texture. Several studies have correlated measurements obtained with these instruments and sensory texture properties of cheese. The correlations generally are high for hardness but low for springiness and other sensory attributes.[58]

Hard and semihard cheese varieties have been subjected to rheological measurements most frequently by uniaxial compression and to a lesser extent by deformation in tension or shear between two parallel plates. Sometimes three points bending of a cylindrical cheese sample has been employed to imitate the deformation seen while grading a cheese plug. The quantitative characterization of cheese can be achieved in terms of the compression modulus, the relative deformation at fracture, and fracture stress in a compression test. Toughness or the energy of fracture (the energy required from the onset of compression till the cheese sample fractures) is also a useful parameter. TPA using compression between parallel plates has been extensively used for cheese texture characterization but the TPA parameters have been stated to be of only a limited value in cheese texture characterization. Biaxial extensional viscosity determined by using compression between parallel discs has been found to be particularly applicable to process cheese. The so-called lubricated squeezing flow technique for determination of extensional or elongation viscosity has been used to measure the melting properties of processed and mozzarella cheese.

11.4.3.2   FACTORS AFFECTING CHEESE RHEOLOGY

Rheology and texture of cheese may be defined as a "composite sensory attribute" resulting from a combination of physical properties that are perceived by the senses of touch (including kinaesthesis and mouthfeel), sight, and hearing. It can be measured directly using a trained sensory panel. However, these are not routinely used for evaluating cheese texture due to the difficulty and cost in assembling sensory panels. Instead, cheese texture is generally measured indirectly using rheological techniques.[97] Rheology of hard or semihard cheese is mainly assessed by compression of a cylindrical or cubic cheese sample between two parallel plates of a texture analyzer.[25,26,31]

The compression may be carried out in one or two cycles (bites). Analysis of the force–displacement or stress–strain curves, often referred to as TPA enables determination of several rheological parameters (fracture stress, fracture strain, firmness, and springiness), which are related to sensory textural characteristics, such as brittleness, spreadability, hardness, and chewiness.[22,97] Everard et al.[26] showed that a three-point bending test could be used for a similar purpose. Torsion geometry involves the application of large strain shear force to a capstan-shaped sample and has been used to discriminate between various types of hard cheese.[134] Penetrometers and oscillation rheometers have been used to determine the viscosity of soft fresh cheeses and for showing the influence of the various processing steps, for example, heat treatment, homogenization, cooling, and textural properties.[67,117]

11.4.3.3 VISCOELASTIC PROPERTY OF CHEESE

The physicochemical properties include the degree of hydrolysis and hydration of the matrix *present in para-casein* and the intermolecular attraction level between *para-casein* molecules.[97] Dynamic oscillatory rheometry offers a method to indirectly evaluate the structure of cheese within a range of experimental time and temperature scales and extent of deformation. Small-amplitude oscillations are applied on the cheese sample so that the strain conditions are within the LVR where any structural breakdown is reversible within the experimental timescale.[44] If this LVR limit exceeds, the cheese structure changes and reform into a different conformation when the oscillatory strain is terminated.

Uniaxial testing is a common method for instrumental evaluation of cheese texture and to determine rheological properties of cheese.[145] It provides accurate deformation control while precisely measuring the force. Mechanical properties include deformability modulus, fracture stress, and strain, and work to fracture.[143]

A creep test provides information regarding the characteristic behavior of a viscoelastic material performed in different configurations such as compression, tension, shear, and torsion.[55,56] Frequently, the creep response of cheese is described by the Kelvin–Voigt model, Peleg model, and Maxwell model.[44,110] The texture of cheese changes constantly even after production, due to the proteolytic action of the residual enzymes. The most notable change with age decreases in fracture strain and springiness and increase in creaminess. The factors that have an effect on cheese texture during ripening are pH before and after salting and salt-in-moisture ratio. Textural changes during storage occur in two phases. In first phase, there is a rapid change in the casein network resulting in softening of cheese and in the second phase, there is a gradual proteolytic change and the protein matrix turns out to be less cohesive.

Patarin et al.[100] subjected cheese samples to increased strain percent (0.001–5%) at a constant frequency of 1 Hz. The curves demonstrated a plateau region at small strains which is followed by an LVR and then decreased. This shows the relation between the storage modulus and the loss modulus with respect to the applied strain percent which determines the degree of deformation with respect to the force applied.

Ong et al.[98] studied the dynamic time sweep analysis at a constant angular frequency (0.8 Hz) and strain (1%) to evaluate storage modulus

(G′) as the milk coagulated at renneting on the microstructure, composition, and texture of cheddar cheese. The cheese-milk was ripened to reach pH of 6.7, 6.5, 6.3, and 6.1. After the addition of rennet, gel formation was monitored by measuring G′ as a function of time. Elastic modulus values increased more rapidly in samples renneted at a lower pH. The time needed for the gel to reach a firmness value of 140 Pa was determined as the cutting time. After the gel reached its maximum G′, its firmness decreased with prolonged incubation (120 min at 33°C) in samples with pH of 6.1 and 6.3. The pH of the milk continued to decrease during incubation for these samples. This decrease in gel firmness is likely to be the result of extensive solubilization of colloidal calcium phosphate and indicates that the rheological properties of these gels are affected by the excessive acid production from the starter culture.

Vogt et al.[144] evaluated heating characteristics of cheese. cheddar and mozzarella cheese samples were heated from 20°C to 70°C at 1°C/min at an oscillating frequency of 1 Hz. The complex viscosity for all three cheese samples was found to decrease as the temperature increased. The complex viscosity of cheddar samples decreased at a higher rate than mozzarella sample.

Farahani et al.[30] studied the elastic and loss moduli of Siahmazgi cheese as a function of shear strain (oscillating frequency 0.1 Hz) at 25°C. The viscoelastic parameters including G′, G″, and tan δ were determined under various frequency range (0.1–100 Hz) in a LVR. It was observed that the lower the tan δ value (closer to 0), the less the cheese flows.

11.5 SUMMARY

Gel-based milk products are widely consumed throughout the world. Complex scientific areas such as rheology, microstructure, and functionality are being explored in recent times. This chapter presents an overview on the rheological and textural properties of milk gels. It emphasizes on the evaluation of texture, rheology, microstructure, and functionality; points out some important concepts; and thoroughly reviews the textural and rheological methods applied till date for the study of milk gel.

Knowledge of the rheological properties of milk gels can be beneficial in case of controlling quality, to understand the change in their behavior in the course of oral processing and consumption. Right choice of rheometric and textural measurement techniques considering the nature of the gel should be formulated. Various types of rheometric tests may be applied, for example, compression test, uniaxial elongation test, cutting tests, plate–plate rheometry, cone-plate rheometry, vane tests, etc.

KEYWORDS

- acid-induced milk gels
- apparent viscosity
- enzyme-induced milk gels
- flow behavior
- microstructural properties
- milk gels
- shear strain
- shear modulus
- storage modulus
- viscoelastic properties

REFERENCES

1. Abu-Jdayil, B.; Mohameed, H. Experimental and Modelling Studies of the Flow Properties of Concentrated Yogurt as Affected by the Storage Time. *J. Food Eng.* **2002,** *52,* 359–365.
2. Aguilera, J. M. Generation of Engineered Structures in Gels. In *Physical Chemistry of Foods*; Schwartzberg, H. G., Hartel, R. W., Eds.; Marcel Dekker Inc.: New York, 1992; pp 387–421.
3. Aguilera, J. M. Why Food Microstructure? *J. Food Eng.* **2005,** *67,* 3–11.
4. Aguilera, J. M.; Rademacher, B. Protein Gels. In *Proteins in Food Processing*; Yada, R. Y., Ed.; Woodhead Publishing Limited: Cambridge, 2004; pp 468–482.
5. Alexander, M.; Corredig, M.; Dalgleish, D. G. Diffusing Wave Spectroscopy of Gelling Food Systems: The Importance of the Photon Transport Mean Free Path (l^*) Parameter. *Food Hydrocoll.* **2006,** *20,* 325–331.
6. Andoyo, R.; Guyomarch, F.; Burel, A.; Famelart, M. H. Spatial Arrangements of Casein Micelles and Whey Protein Aggregate in Acid Gels: Insight on Mechanism. *Food Hydrocoll.* **2015,** *51,* 118–128.

7. Andoyo, R.; Guyomarch, F.; Cauty, C.; Famelart, M. H. Model Mixtures Evidence the Respective Roles of Whey Protein Particles and Casein Micelles During Acid Gelation. *Food Hydrocoll.* **2014,** *37,* 203–212.

8. Anema, S. G.; Lee, S. K.; Ji, Y. Characterization of Heat-set Milk Protein Gels. *Int. Dairy J.* **2016,** *54,* 10–20.

9. Arora, K. L.; Sabhiki, L.; Kanawjia, S. K. Manufacture of Paneer from Sub-standard Buffalo Milk. *Indian J. Dairy Biosci.* **1996,** *7,* 71–75.

10. Arshad, M.; Paulsson, M.; Dejmek, P. Rheology of Buildup, Breakdown and Rebodying of Acid Casein Gels. *J. Dairy Sci.* **1993,** *76,* 3310–3316.

11. Bringe, N. A.; Kinsella, J. E. Forces Involved in the Enzymatic and Acidic Coagulation of Casein Micelles. In *Developments in Food Proteins*; Hudson, B. J. F., Ed.; Elsevier Applied Science Publishers: London, 1987; pp 159–194.

12. Campo, L.; Tovar, C. Influence of the Starch Content in the Viscoelastic Properties of Surimi Gels. *J. Food Eng.* **2008,** *84,* 140–147.

13. Cavallieri, A. L. F.; Da Cunha, R. L. The Effects of Acidification Rate, pH and Ageing Time on the Acidic Cold Set Gelation of Whey Proteins. *Food Hydrocoll.* **2008,** *22*(3), 439–448.

14. Chandan, R. C.; Marin, N.; Nakrani, K. R.; Zehner, D. M. Production and Consumer Acceptance of Latin American White Cheese. *J. Dairy Sci.* **1979,** *62*(5), 691–696.

15. Ciron, C. I. E.; Gee, V. L.; Kelly, A. L.; Auty, M. A. E. Effect of Microfluidization of Heat-treated Milk on Rheology and Sensory Properties of Reduced Fat Yoghurt. *Food Hydrocoll.* **2011,** *25,* 1470–1476.

16. Clark, A. H.; Ross-Murphy, S. B. Structural and Mechanical Properties of Biopolymer Gels. *Adv. Polym. Sci.* **1987,** *87,* 57–192.

17. Cooney, M. J.; Rosenberg, M.; Shoemaker, C. F. Rheological Properties of Whey-protein Concentrate Gels. *J. Texture Stud.* **1993,** *24*(3), 325–334.

18. Cui, B.; Yan-min, L.; Cong-ping, T.; Guan-qun, W.; Gui-Hua, L. Effect of Cross-linked Acetylated Starch Content on the Structure and Stability of Set Yoghurt. *Food Hydrocoll.* **2014,** *35,* 576–582.

19. Dickinson, E. Bridging Flocculation of Sticky Hard Spheres. *J. Chem. Soc. Faraday Trans.* **1990,** *86,* 439.

20. Dickinson, E. Structure Formation in Casein-based Gels, Foams, and Emulsions. *Colloids Surfaces A: Physicochem. Eng. Asp.* **2006,** *288,* 3–11.

21. Dickinson, E.; McClements, D. J. *Advances in Food Colloids*; Blackie Academic & Professional: London, 1995; p 304.

22. Dimitreli, G.; Thomareis, A. S. Texture Evaluation of Block-type Processed Cheese as a Function of Chemical Composition and in Relation to its Apparent Viscosity. *J. Food Eng.* **2007,** *79,* 1364–1373.

23. Donato, L.; Guyomarch, F. Formation and Properties of the Whey Protein/κ-casein Complexes in Heated Skim Milk—A Review. *Dairy Sci. Technol.* **2009,** *89,* 3–29.

24. Doublier, J. L.; Launay, B.; Cuvelier, G. Viscoelastic Properties of Food Gels. In *Viscoelastic Properties of Foods*; Rao, M. A., Steffe, J. F., Eds.; E.A.F.S. Series; Elsevier Applied Science Publishers: London, U.K., 1992; pp 240–256.

25. Everard, C. D.; O'Callaghan, D. J.; O'Kennedy, B. T.; O'Donnell, C. P.; Sheehan, E. M.; Delahunty, C. M. A Three-point Bending Test for Prediction of Sensory Texture in Processed Cheese. *J. Texture Stud.* **2007,** *38,* 438–456.

26. Everard, C. D.; O'Donnell, C. P.; O'Callaghan, D. J.; Sheehan, E. M.; Delahunty, C. M.; O'Kennedy, B. T.; Howard, V. Prediction of Sensory Textural Properties from Rheological Analysis for Process Cheeses Varying in Emulsifying Salt, Protein, and Moisture Contents. *J. Sci. Food Agric.* **2007**, *87*, 641–650.

27. Færgemand, M.; Otte, J.; Qvist, K. B. Cross-linking of Whey Proteins by Enzymatic Oxidation. *J. Agric. Food Chem.* **1998**, *46*(4), 1326–1333.

28. Faergemand, M.; Otte, J.; Qvist, K. B. Emulsifying Properties of Milk Proteins Cross-linked with Microbial Transglutaminase. *Int. Dairy J.* **1998**, *8*, 715–723.

29. Famelart, M. H.; Tomazewski, J.; Piot, M.; Pezennec, S. Comprehensive Study of Acid Gelation of Heated Milk with Model Protein Systems. *Int. Dairy J.* **2004**, *14*, 313–321.

30. Farahani, G.; Ezzatpanah, H.; Abbasi, S. Characterization of Siahmazgi Cheese, an Iranian Ewe's Milk Variety: Assessment of Physico-chemical, Textural, and Rheological Specifications During Ripening. *LWT—Food Sci. Technol.* **2014**, *58*(2), 335–342.

31. Fenelon, M. A.; Guinee, T. P. Primary Proteolysis and Textural Changes During Ripening in Cheddar Cheeses Manufactured to Different Fat Contents. *Int. Dairy J.* **2000**, *10*, 151–158.

32. Flory, P. J. *Principles of Polymer Chemistry*; Cornell University Press: Ithaca, U.S.A., 1953; pp 2–16.

33. Flory, P. J. Gels and Gelling Processes: Introductory Lecture. *Faraday Discuss. Chem. Soc.* **1974**, *57*, 7.

34. Frederiksen, P. D.; Hammershoj, M.; Bakman, M.; Andersen, P. N.; Andersen, J. B.; Qvist, K. B. Variations in Coagulation Properties of Cheese Milk from Three Danish Dairy Breeds as Determined by a New Free Oscillation Rheometry-based Method. *Dairy Sci. Technol.* **2011**, *91*, 309–321.

35. Gabriele, D.; De Cindio, B.; D'Antona, P. Weak Gel Model for Foods. *Rheo. Acta* **2001**, *40*, 120–127.

36. Garcia Perez, F. J.; Lario, L.; Fernadez Loopez, J.; Barbera Sayas, E. A.; Perez Alvarez, J.; Sendra, E. Effect of Orange Fiber Addition on Yogurt Color During Fermentation and Cold Storage. *Color Res. Appl.* **2005**, *30*, 487.

37. Gastaldi, E.; Lagaude, A.; De La Fuente, B. T. Micellar Transition State in Casein Between pH 5.5 and 5.0. *J. Food Sci.* **1996**, *61*(1), 59–64.

38. Girard, M.; Schaffer-Lequart, C. Gelation and Resistance to Shearing of Fermented Milk: Role of Exopolysaccharides. *Int. Dairy J.* **2007**, *17*(6), 666–673.

39. Gliguem, H.; Lopez, C.; Michon, C.; Lesieur, P.; Ollivon, M. The Viscoelastic Properties of Processed Cheeses Depend on Their Thermal History and Fat Polymorphism. *J. Agric. Food Chem.* **2011**, *59*, 3125–3134.

40. Gras, S. L.; Onga, L.; Dagastine, R. R.; Kentish, S. E. Microstructure of Milk Gel and Cheese Curd Observed Using Cryo Scanning Electron Microscopy and Confocal Microscopy. *LWT—Food Sci. Technol.* **2011**, *44*, 1291–1302.

41. Graveland-Bikker, J. F.; Anema, S. G. Effect of Individual Whey Proteins on the Rheological Properties of Acid Gels Prepared from Heated Skim Milk. *Int. Dairy J.* **2003**, *13*(5), 401–408.

42. Guggisberg, D.; Cuthbert-Steven, J.; Piccinali, P.; BuTikofer, U.; Eberhard, P. Rheological, Microstructural and Sensory Characterization of Low-fat and Whole Milk Set Yoghurt as Influenced by Inulin Addition. *Int. Dairy J.* **2009**, *19*, 107–115.

43. Guinee, T. P.; O'Kennedy, B. T. Reducing the Level of Added Disodium Phosphate Alters the Chemical and Physical Properties of Processed Cheese. *Dairy Sci. Technol.* **2012,** *92*(5), 469–486.

44. Gunasekaran, S.; Ak, M. M. Dynamic Oscillatory Shear Testing of Foods-selected Applications. *Trends Food Sci. Technol.* **2000,** *11*(3), 115–127.

45. Guyomarćh, F.; Renan, M.; Chatriot, M.; Gamerre, V.; Famelart, M. H. Acid Gelation Properties of Heated Skim Milk as a Result of Enzymatically Induced Changes in the Micelle/Serum Distribution of the Whey Protein/κ-casein Aggregates. *J. Agric. Food Chem.* **2007,** *55*(26), 10986–10993.

46. Guyomarćh, F.; Queguiner, C.; Law, A. J. R.; Horne, D. S.; Dalgleish, D. G. Role of the Soluble and Micelle-bound Heat-induced Protein Aggregates on Network Formation in Acid Skim Milk Gels. *J. Agric. Food Chem.* **2003,** *51*(26), 7743–7750.

47. Halmos, A. L.; Foo, A. S. Texture Characteristic of Yellow Cheeses. *Food Aust.* **2002,** *54*, 459–462.

48. Hannon, J. A.; Lopez, C.; Madec, M. N.; Lortal, S. Altering Renneting pH Changes Microstructure, Cell Distribution, and Lysis of *Lactococcuslactis* AM2 in Cheese Made from Ultrafiltered Milk. *J. Dairy Sci.* **2006,** *89*, 812–823.

49. Harte, F.; Clark, S.; Barbosa-Cánovas, G. V. Yield Stress for Initial Firmness Determination on Yogurt. *J. Food Eng.* **2007,** *80*(3), 990–995.

50. Hashizume, K.; Sato, T. Gel-forming Characteristics of Milk-proteins 2. Roles of Sulfhydryl-groups and Disulfide Bonds. *J. Dairy Sci.* **1998,** *71*(6), 1447–1454.

51. Hermansson, A. M. Structuring Water by Gelation. In *Food Materials Science*; Aguilera, J. M., Lillford, P. J., Eds.; Springer: New York, 2008; pp 255–280.

52. Horne, D. S. Casein Interactions: Casting Light on the Black Boxes, the Structure in Dairy Products. *Int. Dairy J.* **1998,** *8*, 171–177.

53. Horne, D. S. Formation and Structure of Acidified Milk Gels. *Int. Dairy J.* **1999,** *9*, 261–268.

54. Horne, D. S. Casein Micelles as Hard Spheres: Limitations of the Model in Acidified Gel Formation. *Colloids Surfaces A: Physicochem. Eng. Asp.* **2003,** *213*(2–3), 255–263.

55. Huang, C. Y.; Soltz, M. A.; Kopacz, M.; Mow, V. C.; Ateshian, G. A. Experimental Verification of the Roles of Intrinsic Matrix Viscoelasticity and Tension-compression Nonlinearity in the Biphasic Response of Cartilage. *J. Biomech. Eng.* **2003,** *125*(1), 84–93.

56. Hung, V.; Morita, N. Effects of Granule Sizes on Physicochemical Properties of Cross-linked and Acetylated Wheat Starches. *Starch/Stärke* **2005,** *57*, 413–420.

57. Huppertz, T.; de Kruif, C. G. Ethanol Stability of Casein Micelles Cross-linked with Transglutaminase. *Int. Dairy J.* **2006,** *17*, 436–441.

58. Hwang, C. H.; Gunasekaran, S. Measuring Crumbliness of Some Commercial Queso Fresco-type Latin American Cheeses. *Milchwissenschaft* **2001,** *56*(8), 446–450.

59. Ikeda, S.; Nishinari, K. Weak Gel-type Rheological Properties of Aqueous Dispersions of Non-aggregated κ-carrageenan Helices. *J. Agric. Food Chem.* **2001,** *49*(9), 4436–4441.

60. Jablonka, M. S.; Munro, P. A. Particle Size Distribution and Calcium Content of Batch-precipitated Acid Casein Curd: Effect of Precipitation Temperature and pH. *J. Dairy Res.* **1985,** *52*, 419–424.

61. Kanawjia, S. K.; Singh, S. Sensory and Textural Changes in Paneer During Storage. *Buffalo J.* **1996**, *12*(3), 329–334.

62. Kealy, T. Application of Liquid and Solid Rheological Technologies to the Textural Characterisation of Semisolid Foods. *Food Res. Int.* **2006**, *39*(3), 265–276.

63. Keogh, M. K.; O'Kennedy, B. T. Rheology of Stirred Yogurt as Affected by Added Milk Fat, Protein, and Hydrocolloids. *J. Food Sci.* **1998**, *63*(1), 108–112.

64. Khamrui, K.; Dutta, S.; Dave, K. Effect of Frying, Freezing, and Rehydration on Texture Profile of Paneer and Relationships Between its Instrumental and Sensory Textural Attributes. *Milchwissenschaft* **2004**, *59*, 640–644.

65. Kim, B. Y.; Kinsella, J. E. Rheological Changes During Slow Acid-induced Gelation of Milk By D-glucono-δ-lactone. *J. Food Sci.* **1989**, *54*(4), 894–898.

66. Kip, P.; Meyer, D.; Jellema, R. H. Inulins Improve Sensoric and Textural Properties of Low-fat Yoghurts. *Int. Dairy J.* **2006**, *16*(9), 1098–1103.

67. Korolczuk, J.; Mahaut, M. Studies on Acid Cheese Texture by a Computerized Constant Speed, Cone Penetrometer. *Lait* **1988**, *68*, 349–362.

68. Kumari, S.; Singh, G. Textural Characteristics of Channa and Paneer Made from Cow and Buffalo Milk. *Beverage Food World* **1992**, *19*(3), 20–21.

69. Lee, S. K.; Anema, S. G. The Effect of the pH at Cooking on the Properties of Processed Cheese Spreads Containing Whey Proteins. *Food Chem.* **2009**, *115*(4), 1373–1380.

70. Lee, S. K.; Huss, M.; Klostermeyer, H.; Anema, S. G. The Effect of Pre-denatured Whey Proteins on the Textural and Micro-structural Properties of Model Processed Cheese Spreads. *Int. Dairy J.* **2013**, *32*, 79–88.

71. Lee, W. J.; Lucey, J. A. Impact of Gelation Conditions and Structural Breakdown on the Physical and Sensory Properties of Stirred Yogurts. *J. Dairy Sci.* **2006**, *89*(7), 2374–2385.

72. Lee, W. J.; Lucey, J. A. Formation and Physical Properties of Yogurt. *Asian Australas. J. Anim. Sci.* **2010**, *23*(9), 1127–1136.

73. Lobato, C. C.; Rodríguez, E.; Sandoval-Castilla, O.; Vernon-Carter, E. J.; Alvarez-Ramirez, J. Reduced-fat White Fresh Cheese-like Products Obtained from $W_1/O/W_2$ Multiple Emulsions: Viscoelastic and High-resolution Image Analyses. *Food Res. Int.* **2006**, *39*, 678–685.

74. Löfgren, C.; Walkenström, P.; Hermansson, A. M. Microstructure and Rheological Behavior of Pure and Mixed Pectin Gels. *Biomacromolecules* **2002**, *3*(6), 1144–1153.

75. Lucey, J. A. The Relationship Between Rheological Parameters and Whey Separation in Acid Milk Gels. *Food Hydrocoll.* **2001**, *15*, 603–608.

76. Lucey, J. A. Acid and Acid/Heat Coagulated Cheese. In *Encyclopedia of Dairy Sciences*; Roginski, H., Fox, P. F., Fuquay, J. W., Eds.; Academic Press: London, U.K., 2002; pp 350–356.

77. Lucey, J. A. Formation, Structural Properties and Rheology of Acid-coagulated Milk Gels. In *Cheese: Chemistry, Physics, and Microbiology*; 3rd ed.; Fox, P. F., McSweeney, P. L. H., Eds.; Elsevier Applied Science: Barking, Essex, 2004; pp 105–135.

78. Lucey, J. A.; Singh, H. Formation and Physical Properties of Acid Milk Gels: A Review. *Food Res. Int.* **1998**, *30*, 529–542.

79. Lucey, J. A.; Munro, P. A.; Singh, H. Whey Separation in Acid Skim Milk Gels Made with Glucono-δ-lactone: Effects of Heat Treatment and Gelation Temperature. *J. Texture Stud.* **1998**, *29*, 413–426.

80. Lucey, J. A.; Singh, H. Acid Coagulation of Milk. In *Advanced Dairy Chemistry, Vol. 2: Proteins Chemistry*, 2nd ed.; Fox, P. F., McSweeney, P. L. H., Eds.; Aspen Publishers: Gaithersburg, 2003; pp 997–1021.

81. Lucey, J. A.; Munro, P. A.; Singh, H. Effects of Heat Treatment and Whey Protein Addition on the Rheological Properties and Structure of Acid Skim Milk Gels. *Int. Dairy J.* **1999**, *9*, 275–279.

82. Lucey, J. A.; Johnson, M. E.; Horne, D. S. Invited Review: Perspectives on the Basis of the Rheology and Texture Properties of Cheese. *J. Dairy Sci.* **2003**, *86*, 2725–2743.

83. Lucey, J. A.; Munro, P. A.; Singh, H. Rheological Properties and Microstructure of Acid Milk Gels as Affected by Fat Content and Heat Treatment. *J. Food Sci.* **1998**, *63*, 660–664.

84. Lucey, J. A.; Tamehana, M.; Singh, H.; Munro, P. A. A Comparison of the Formation, Rheological Properties and Microstructure of Acid Milk Gels Made with a Bacterial Culture or Glucono-δ-lactone. *Food Res. Int.* **1998**, *31*, 147–155.

85. Lucey, J. A.; Van Vliet, T.; Grolle, K.; Geurts, T.; Walstra, P. Properties of Acid Gels Made from Na-caseinate with Glucono-δ-lactone, II: Syneresis, Permeability and Microstructural Properties. *Int. Dairy J.* **1997**, *7*, 389–397.

86. Lucey, J. A.; Van Vliet, T.; Grolle, K.; Geurts, T.; Walstra, P. Properties of Acid Casein Gels Made by Acidification with Glucono-δ-lactone, 1: Rheological Properties. *Int. Dairy J.* **1997**, *7*, 381–388.

87. McCann, T. H.; Fabre, F.; Day, Li. Microstructure, Rheology, and Storage Stability of Low-fat Yoghurt Structured by Carrot Cell Wall Particles. *Food Res. Int.* **2011**, *44*, 884–892.

88. McClements, D. J. *Food Emulsions: Principles, Practice, and Techniques*; CRC Press: Boca Raton, FL, 1999; pp 1–54.

89. Mellema, M.; Van Opheusden, J. H. J.; Van Vilet, T. Categorization of Rheological Scaling Models for Particle Gels Applied to Casein Gels. *J. Rheol.* **2002**, *46*(1), 11.

90. Mellema, M.; Van Vilet, T.; Van Opheusden, J. H. J. Categorization of Rheological Scaling Models for Particle Gels. In *Proceedings of the 2nd International Symposium on Food Rheology and Structure*; Fischer, P., Marti, I., Windhab, E. J., Eds.; ETH Zurich: Zurich, 2000; pp 1–18.

91. Mleko, S.; Foegeding, E. A. pH-induced Aggregation and Weak Gel Formation of Whey Protein Polymers. *J. Food Sci.* **2000**, *65*(1), 139–143.

92. Morand, M.; Dekkari, A.; Guyomarćh, F.; Famelart, M. H. Increasing the Hydrophobicity of the Heat-induced Whey Protein Complexes improves the Acid Gelation of Skim Milk. *Int. Dairy J.* **2012**, *25*,103–111.

93. Morand, M.; Legland, D.; Guyomarćh, F.; Famelart, M. H. Changing the Isoelectric Point of the Heat-induced Whey Protein Complexes Affects the Acid Gelation of Skim Milk. *Int. Dairy J.* **2012**, *23*, 9–17.

94. Mounsey, J. S.; O'Riordan, E. D. Empirical and Dynamic Rheological Data Correlation to Characterize Melt Characteristics of Imitation Cheese. *J. Food Sci.* **1999**, *64*, 701–703.

95. Mounsey, J. S.; O'Kennedy, B. T.; Kelly, P. M. Effect of the Aggregation-state of Whey Protein-based Ingredients on Processed Cheese Functionality. *Milchwiss. Milk Sci. Int.* **2007**, *62*, 44–47.

96. Ng, T. S. K.; McKinley, G. H. Power-law Gels at Finite Strains. *J. Rheol.* **2008**, *52*(2), 417–449.

97. O'Callaghan, D. J.; Guinee, T. P. Rheology and Texture of Cheese. In *Cheese: Chemistry, Physics and Microbiology, Volume 1: General Aspects*, 3rd ed.; Fox, P. F., McSweeney, P. L. H., Cogan, T. M., Guinee, T. P., Eds.; Elsevier Academic Press: Amsterdam, 2004; pp 511–540.

98. Ong, L.; Dagastine, R. R.; Kentish, S. E.; Gras, S. L. The Effect of pH at Renneting on the Microstructure, Composition and Texture of Cheddar Cheese. *Food Res. Int.* **2012**, *48*(1), 119–130.

99. Ozer, B. H.; Grandison, A. S.; Bell, A. E., Robinson, R. K. Rheological Properties of Concentrated Yoghurt (Labneh). *J. Texture Stud.* **1997**, *29*(1), 67–79.

100. Patarin, J.; Galliard, H.; Magnin, A.; Goldschmidt, B. Vane and Plate–plate Rheometry of Cheeses Under Oscillations and Large Strains: A Comparative Study and Experimental Conditions Analysis. *Int. Dairy J.* **2014**, *38*(1), 24–30.

101. Phadungath, C. *A Study of Structure Development in Cream Cheese and the Impact of Processing Conditions on Cheese Texture, and Sensory Properties*. M.S. Thesis University of Wisconsin: Madison, 2003; pp 8–23.

102. Prasanna, P. H. P.; Grandison, A. S.; Charalampopoulos, D. Microbiological, Chemical and Rheological Properties of Low Fat Set Yoghurt Produced with Exopolysaccharide (EPS) Producing *Bifidobacterium* Strains. *Food Res. Int.* **2013**, *51*(1), 15–22.

103. Raina, C. S.; Singh, S.; Bawa, A. S.; Saxena, D. C. Some Characteristics of Acetylated, Cross-linked and Dual Modified Indian Rice Starches. *Eur. Food Res. Technol.* **2006**, *223*, 561–570.

104. Ramaswamy, H. S.; Chen, C. R.; Navneet, S. Comparison of Viscoelastic Properties of Set and Stirred Yogurts Made from High Pressure and Thermally Treated Milks. *Int. J. Food Prop.* **2015**, *18*(7), 1513–1523.

105. Ramaswamy, S.; Basak, S. Pectin and Raspberry Concentrate Effects on the Rheology of Stirred Commercial Yogurt. *J. Food Sci.* **1992**, *57*(2), 357–360.

106. Ramírez-Santiago, C.; Ramos-Solis, L.; Lobato-Calleros, C.; Peña-Valdivia, C.; Vernon-Carter, E. J.; Alvarez-Ramírez, J. Enrichment of Stirred Yogurt with Soluble Dietary Fiber from *Pachyrhizuserosus* L. Urban: Effect on Syneresis, Microstructure and Rheological Properties. *J. Food Eng.* **2010**, *101*, 229–235.

107. Rao, M. A. *Rheology of Fluids and Semisolid Foods*; Aspen Publishers, Inc.: Maryland, USA, 1999; pp 1–125.

108. Rao, M. A. Phase Transitions, Food Texture and Structure. In *Texture in Food: Volume 1: Semi-solid Foods*; Mckenna, B. M., Ed.; Woodhead Publishing: Cambridge, U.K., 2003; pp 36–62.

109. Rao, M. A.; Cooley, H. J. Rheological Behavior of Tomato Pastes in Steady and Dynamic Shear. *J. Texture Stud.* **1992**, *23*, 415–425.

110. Rao, M. A.; Rizvi, S. S. H.; Datta, A. K.; Ahmed, J. *Engineering Properties of Foods*; CRC Press: Boca Raton, FL, 2014; pp 41–91.

111. Renan, M.; Guyomarćh, F.; Arnoult-Delest, V.; Paquet, D.; Brule, G.; Famelart, M. H. Rheological Properties of Stirred Yoghurt as Affected by Gel pH on Stirring, Storage Temperature and pH Changes After Stirring. *Int. Dairy J.* **2009**, *19*(3), 142–148.

112. Roefs, S. P. F. M.; Van Vliet, T. Structure of Acid Casein Gels 2. Dynamic Measurements and Type of Interaction Forces. *Colloids Surf.* **1990**, *50*, 161–175.

113. Roefs, S. P. F. M.; de Groot-Mostert, A. E. A.; Van Vliet, T. Structure of Acid Casein Gels 1. Formation and Model of Gel Network. *Colloids Surf.* **1990**, *50*, 141–159.
114. Ross-Murphy, S. B. Rheological Characterization of Polymer Gels and Networks. *Polym. Gels Netw.* **1994**, *2*(3–4), 229–237.
115. Ross-Murphy, S. B. Structure–Property Relationships in Food Biopolymer Gels and Solutions. *J. Rheol.* **1995**, *39*, 1451–1463.
116. Ross-Murphy, S. B. Rheological Characterization of Gels. *J. Texture Stud.* **1995**, *26*, 391–400.
117. Sanchez, C.; Beauregard, J. L.; Chassagne, M. H.; Bimhenet, J. J.; Hardy, J. Effects of Processing on Rheology and Structure of Double Cream Cheese. *Food Res. Int.* **1996**, *28*, 541–552.
118. Sanz, T.; Salvador, A.; Jimenez, A.; Fiszman, S. Yogurt Enhancement with Functional Asparagus Fiber, Effect of Fiber Extraction Method on Rheological Properties, Color, and Sensory Acceptance. *Eur. Food Res. Technol.* **2008**, *227*, 1515–1521.
119. Schmidt, D. G. *Developments in Dairy Chemistry*; Applied Science Publisher: New York, 1982; Vol. 1, pp 61–86.
120. Schorsch, C.; Jones, M. G.; Norton, I. T. Micellar Casein Gelation at High Sucrose Content. *J. Dairy Sci.* **2002**, *85*, 3155–3163.
121. Schorsch, C.; Wilkins, D. K.; Jones, M. G.; Norton, I. T. Gelation of Casein–Whey Mixtures: Effects of Heating Whey Proteins Alone or in the Presence of Casein Micelles. *J. Dairy Res.* **2001**, *68*(3), 471–481.
122. Sendra, E.; Kuri, V.; Fernández-López, L.; Sayas-Barberá, E.; Navarro, C.; Pérez-Alvarez, P. A. Viscoelastic Properties of Orange Fiber Enriched Yogurt as a Function of Fiber Dose, Size, and Thermal Treatment. *LWT—Food Sci. Technol.* **2010**, *43*(4), 708–714.
123. Shirashoji, N.; Jaeggi, J. J.; Lucey, J. A. Effect of Trisodium Citrate Concentration and Cooking Time on the Physicochemical Properties of Pasteurized Process Cheese. *J. Dairy Sci.* **2006**, *89*, 15–28.
124. Shirashoji, N.; Jaeggi, J. J.; Lucey, J. A. Effect of Sodium Hexametaphosphate Concentration and Cooking Time on the Physicochemical Properties of Pasteurized Process Cheese. *J. Dairy Sci.* **2010**, *93*, 2827–2837.
125. Shrivastava, S.; Kumbhar, B. K. Textural Profile Analysis of Paneer Dried with Low Pressure Superheated Steam. *J. Food Sci. Technol.* **2010**, *47*, 355–357.
126. Sindhu, J. S. Suitability of Buffalo Milk for Products Manufacturing. *Indian Dairym.* **1996**, *48*, 41–47.
127. Singh, H.; Creamer, L. K. Denaturation, Aggregation and Heat Stability of Milk Protein During the Manufacture of Skim Milk Powder. *J. Dairy Res.* **1991**, *58*, 269–283.
128. Sodini, I.; Lucas, A.; Tisier, J. P.; Corrieu, G. Physical Properties and Microstructure of Yogurts Supplemented with Milk Protein Hydrolysates. *Int. Dairy J.* **2005**, *15*(1), 29–35.
129. Srisuvor, N.; Chinprahast, N.; Prakitchaiwattana, C.; Subhimaros, S. Effects of Inulin and Polydextrose on Physicochemical and Sensory Properties of Low-fat Set Yoghurt with Probiotic-cultured Banana Purée. *LWT—Food Sci. Technol.* **2013**, *51*, 30–36.
130. Staffolo, M. D.; Bertola, N.; Martino, M.; Bevilacqua, A. Influence of Dietary Fiber Addition on Sensory and Rheological Properties of Yogurt. *Int. Dairy J.* **2004**, *14*, 263–268.

131. Steffe, J. F. *Rheological Methods in Foods Process Engineering*, 2nd ed.; Freeman Press, East Lansing: Michigan, U.S.A., 1996; pp 1–118.

132. Syed, H. M.; Rathi, S. D.; Jadhav, S. A. Studies on Quality of Paneer. *J. Food Sci. Technol.* **1992,** *29*, 117–118.

133. Tunick, M. H. Rheology of *Dairy* Foods that *Gel*, Stretch, and Fracture. *J. Dairy Sci.* **2000,** *83*(8), 1892–1898.

134. Tunick, M. H.; Van Hekken, D. L. Torsion Gelometry of Cheese. *J. Dairy Sci.* **2002,** *85*(11), 2743–2749.

135. Tunick, M. H.; Nolan, E. J.; Shieh, J. J.; Basch, J. J.; Thompson, M. P.; Maleeff, B. E. Cheddar and Cheshire Rheology. *J. Dairy Sci.* **1990,** *73*, 1671–1675.

136. Uprit, S.; Mishra, H. N. Instrumental Textural Profile Analysis of Soy Fortified Pressed Chilled Acid Coagulated Curd (paneer). *Int. J. Food Prop.* **2004,** *7*(3), 367–378.

137. Van den Berg, L.; Carolas, A. L.; Van Vliet, T.; Van der Linden, E.; Van Boekel, M. A. J. S.; Van de Velde, F. Energy Storage Controls Crumbly Perception in Whey Proteins/Polysaccharide Mixed Gels. *Food Hydrocoll.* **2008,** *22*(7), 1404–1417.

138. Van Hooydonk, A. C. M.; Hagedoorn, H. G.; Boerrigter, I. J. The Effect of Various Cations on the Renneting of Milk. *Neth. Milk Dairy J.* **1986,** *40*(4), 369–390.

139. Van Marle, M. E.; Zoon, P. Permeability and Rheological Properties of Microbially and Chemically Acidified Skim-milk Gels. *Neth. Milk Dairy J.* **1995,** *49*, 47–65.

140. Van Vliet, T.; Walstra, P. Water in Casein Gels; How to Keep it Out or Keep it in. *J. Food Eng.* **1994,** *22*, 75–88.

141. Van Vliet, T.; Keetels, C. J. A. M. Effect of Pre-heating of Milk on the Structure of Acidified Milk Gels. *Neth. Milk Dairy J.* **1995,** *49*, 27–35.

142. Van Vliet, T.; Roefs, S. P. F. M.; Van Den Bijgaart, H. J. C. M.; de Groot-Mostert, A. E. A.; Walstra, P. Structure of Casein Gels Made by Combined Acidification and Rennet Action. *Neth. Milk Dairy J.* **1990,** *44*, 159–188.

143. Velmurugan, R.; Gupta, N. K.; Solaimurugan, S.; Elayaperumal, A. The Effect of Stitching on FRP Cylindrical Shells Under Axial Compression. *Int. J. Impact Eng.* **2004,** *30*(8), 923–938.

144. Vogt, S. J.; Smith, J. R.; Seymour, J. D.; Carr, A. J.; Golding, M. D.; Codd, S. L. Assessment of the Changes in the Structure and Component Mobility of Mozzarella and Cheddar Cheese During Heating. *J. Food Eng.* **2015,** *150*, 35–43.

145. Walstra, P. Relation Between Structure and Texture of Cultured Milk Products. In *Texture of Fermented Milk Products and Dairy Desserts*; International Dairy Federation: Brussels, 1997, pp 9–15.

146. Wang, F.; Zhang, X.; Luo, J.; Guo, H.; Zeng, S. S.; Ren, F. Effect of Proteolysis and Calcium Equilibrium on Functional Properties of Natural Cheddar Cheese During Ripening and the Resultant Processed Cheese. *J. Food Sci.* **2011,** *76*, E248–E253.

147. Zoon, P.; Van Vliet, T.; Walstra, P. Rheological Properties of Rennet-induced Skim Milk Gels 1: Introduction. *Neth. Milk Dairy J.* **1988,** *42*, 249–269.

CHAPTER 12

TEXTURAL ANALYSES OF DAIRY PRODUCTS

AKANKSHA WADEHRA[1,*], PRASAD S. PATIL[2],
SHAIK ABDUL HUSSAIN[1], ASHISH KUMAR SINGH[1],
and RUPESH S. CHAVAN[3]

[1]Lab No. 133, Dairy Technology Division, National Dairy Research Institute (NDRI), Karnal 132001, Haryana, India

[2]Lab No. 112, NCDC Lab, Dairy Microbiology Division, National Dairy Research Institute (NDRI), Karnal 132001, Haryana, India

[3]Department of Quality Assurance, Mother Dairy, Junagadh 362001, Gujarat, India

*Corresponding author. E-mail: smartakanksha@gmail.com

CONTENTS

ABSTRACT

Texture is the attribute of a substance resulting from a combination of physical properties and perceived by the senses of touch (including kinesthesis and mouth feel), sight, and hearing. Textural properties of milk products are considered for determining ideal processing parameters during manufacturing; for monitoring, maintaining, and improving the consistency and quality; for increasing the consumer acceptability of the finished or semifinished product; for selection of equipment's used for processing and packing; for selecting right form and material of packaging for products; for determining the serving conditions for semicooked milk and milk products. Considering the importance of textural properties, the following chapter deals with the importance of texture in respect to milk and milk products. The various principles and methods of texture measurements used for evaluating the textural properties of dairy products are also dealt in detail for every individual milk product.

12.1   INTRODUCTION

Texture, which comes from Latin *textura* (*cloth*), is a vocable used to describe the crosslinking style of weave threads. It now defines disposition and arrangement for the different parts of a system. It was not until the 1960s that texture began to be considered important in evaluating food quality. Since then, a number of definitions have been proposed that are synthesized in the corresponding ISO standard 5492 (1992). In the ISO standard, texture is defined as, "All the mechanical, geometrical and surface attributes of a product perceptible by means of mechanical, tactile and, where appropriate, visual and auditory receptors." Texture can be evaluated by sensory qualities of dairy products and the evaluation of texture is subjective based on opinions of a panel. On the other hand, the texture evaluation based on chemical composition and the physical properties of a certain dairy product is objective.

The evaluation of texture is a multifaceted, dynamic process that comprises visual observation of the product surface, product behavior in response to previous handling, and integration of sensations experienced in the mouth during mastication and further swallowing. The human brain compiles all of these, and a unique sensation is built up.

Consideration is given to the rheological and mechanical properties of foods. In the first place, it should be taken into account that during mastication, foods are broken down into small pieces. Size, shape, and surface roughness of the resultant particles, as well as their capacity to absorb water from saliva, are not rheological properties. Less direct connections can be established between rheology and the textural perceptions linked to both visual and auditory receptors. Depending on the nature of the food considered, human perception of different attributes by different senses may lend more or less importance to the integrated texture sensation perceived. In general, attributes perceived in the mouth are the most important. Even when considering only those aspects of food texture directly related to the perception of rheological or mechanical properties, the entire sensation could be modified by interactions with other sensory attributes, such as flavor or color, and by certain hedonic associations that are hard to remove. Therefore, even if a rheological or mechanical method is available to explain a high proportion of the variability observed in some of the textural characteristics, it will never afford sufficient information to evaluate texture as a whole.

In summary, "It is only the human being that can perceive, describe and quantify texture, according to Szczesniak."[22] According to the ISO standard 5492 (1992), sensory analysis is the "examination of organoleptic attributes of a product by the sense organs."

This chapter discusses the importance, principle, and methods of texture measurements in the evaluation of dairy products.

12.2 MEASUREMENTS OF FOOD TEXTURE

Due to the technological advancements in the second half of the 20th century, along with the development of sensory methods to evaluate texture, there has been significant research studies aimed at describing the perception mechanisms involved in assessing the textures of liquid, solid, and semisolid foods. Today, sufficient tools are available to study the sensory evaluation of food texture. There are different methods of food texture measurement that will be presented in this chapter.

The principle of a texture measurement system is to deform the sample in predefined controlled conditions and simultaneous measurement of its response.[24] The graphical presentation of force response of the sample against the time/distance is shown on the digital display moving up or down depending on the mode of compression using a probe or tension using grips. Force created during these movements is manipulated to recreate conditions that foods are exposed to when we eat them or process them. These forces are a function of the properties of the sample and the parameters of the test method (Table 12.1). The texture of the product can be analyzed using either single-bite compression (Fig. 12.1) or double-bite compression (Fig. 12.2), depending upon the textural properties to be measured. Table 12.2 shows parameters measured during texture profile analysis of foods

TABLE 12.1   Human Perception of Different Attributes by Different Senses.

Nature of the attribute	Examples
Auditory	Sound intensity during mastication (crunchy, crispy foods)
Tactile, nonoral	Resistance to deformation (fruits and bread)
	Resistance to cut with a knife (meats)
	Resistance to cut with a spoon (dairy desserts)
Tactile, oral	Resistance to mastication (solid foods)
	Resistance to displacement in mouth (liquid foods)
	Structural characteristics (fibrousness, granularity, flouriness, etc.)
	In mouth movements (liquids and solids)
Visual	Color (fruits and vegetables)
	Dropping rate (liquids)

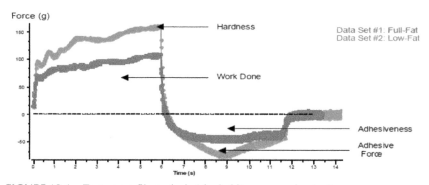

FIGURE 12.1   Texture profile analysis (single-bite compression test).

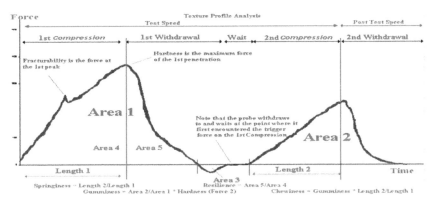

FIGURE 12.2 Texture profile analysis (double-bite compression) curve.

In general, a texture analyzer from various suppliers may be available with following accessories, which are used for instrumental measurement of texture of various solids and semisolid foods:

- Ball probes
- Cone probes
- Cylinder probes
- Compression plate
- Gel bloom kit
- Dough preparation set
- Kramer type shear cell
- Three point bend food jig
- Back extrusion cell
- Ottawa forward extrusion cell
- Warner–Bratzler shear blade set
- Spaghetti/noodle test fixture
- Puncture probe set
- Knife edge probe set
- Gluten dough extensibility jig
- Confectionery accessories kit
- Butter cutting jig

TABLE 12.2   Parameters Measured During Texture Profile Analysis of Foods.

Parameters	Sensorial definition	Instrumental definition
Adhesiveness (A)	The work necessary to overcome the attractive forces between the surface of the food and the surface of other materials with which the food comes into contact (e.g., tongue, teeth, palate). Work required to pull food away from a surface	The negative area for the first bite, representing the work necessary to pull compression probe away from sample
Brittleness (B)	Force at which a material fractures. Related to the primary parameters of hardness and cohesiveness, where brittle materials have low cohesiveness. Not all foods fracture and thus value may relate to hardness if only single peak is present. Brittle foods are never adhesive	The first significant break in the first compression cycle
Chewiness (Ch)	Energy required to chew a solid food product to a state where it is ready for swallowing. Attribute is difficult to quantify precisely due to complexities of mastication (shear, compression, tearing, and penetration)	Calculated parameter: Gumminess × springiness (or hardness × cohesiveness × springiness)
Cohesiveness (Co)	The strength of internal bonds making up the body of the product (greater the value, greater the cohesiveness)	The ratio of positive force during the second to that of the first compression cycle (downward strokes only)
Gumminess (G)	Energy required to disintegrate a semi-solid food product to a steady state for swallowing. Related to foods with low hardness levels	Calculated parameter: Hardness × cohesiveness
Hardness (H)	Force required to compress a food between molars. Defined as force necessary to attain given deformation	Peak force of the first compression cycle
Springiness	Rate at which a deformed material goes back to its original condition after deforming force is removed	Height that the food recovers during the time that elapses between the end of the first bite and the start of the second bite

12.3   TEXTURE MEASUREMENT TECHNIQUES FOR MILK AND MILK PRODUCTS

Texture measurement techniques can be grouped as either subjective or instrumental. The subjective measurements or sensory evaluations are made by the trained taste panel. The instrumental methods can be broadly grouped under the following three categories: empirical, imitative, and fundamental.

12.3.1 EMPIRICAL METHODS

These are based on the subjection of cheese to a stress or strain that results in visual fracture, that is, permanent deformation. In these tests, the test conditions are arbitrary and the aim is to obtain a number that gives a vague indication of the textural characteristics (e.g., its hardness). Empirical tests are relatively simple procedures that typically measure a force on a sample and the accompanying deflection. These methods rely completely on test parameters, such as sample volume, shape, and testing speed. The penetrometer, puncture test, and ball compressor tests are good examples of empirical measurements.

12.3.2 IMITATIVE METHODS

These may also be called semifundamental methods that are used to make mechanical measurements with little control of experimental variables (e.g., probe type and size, product shape, etc.). These attempt to mechanically mimic the sensory evaluation of human evaluators. In fact, when the instrumental test used mimics the action of the human assessor, more accurate models of food texture attributes could be developed. The test results from imitative tests are analyzed and correlated to sensory perceptions of taste panels without valid structural and molecular level reasoning. Therefore, the test results, at best, serve as relative measures of textural attributes of products tested. Nonetheless, the imitative methods are perhaps the largest group of instrumented texture measurement methods. The widely adopted texture profile analysis (TPA) belongs to this group.

12.3.3 FUNDAMENTAL METHODS

These employ valid rheological test techniques, and the data are analyzed using well-defined rheological, structural, and molecular theories. The fundamental test methods also yield results that are independent of test instrument. Some nondestructive methods have also being developed to measure rheology of milk and milk products.

12.4   TEXTURAL TESTS FOR CHEESE

Cheese is the product which is mostly subjected for textural analysis among all of the milk products. The textural tests involve subjecting a cheese sample to a stress or strain by various techniques (e.g., inserting a penetrometer). The different types of tests include:

- Compression tests, where the extent of compression under a constant load for a specified time is measured (e.g., ball compressor test).
- Penetration tests, where the force required to insert a probe a given distance into the cheese or, alternatively, the depth of penetration by a probe under a fixed load for a given time is measured (e.g., penetrometer test).
- Cutting tests, measure the resistance to the passage of a knife or a wire through a cheese (e.g., Cherry and Burrell curd tension meter)
- Curd tests, measure the characteristics of the curd. Various tests come under this, which measures firmness, pitching point, etc. of the curd.

12.4.1   BALL COMPRESSOR TEST

The early devices for evaluating the hardness of cheese involved compression by a ball, in an instrument known as the ball compressor (Fig. 12.3), where deformation resulting from applying a fixed force for a specified time was measured.[21] The action is stimulated by a thumb pressing against cheese when making a sensory evaluation of the product. The ball compressor test measures the depth of indentation after a given time made by a small ball or hemisphere when placed under a given load (stress) on the cheese surface. The depth of penetration has been used directly as an index of firmness. Alternatively, by making a number of simple assumptions, testers may use it to calculate a modulus, analogous to an elastic modulus G, given by the equation:

$$G = 3M[16(RD^3)]^{1/2} \qquad (12.1)$$

where, M is the applied force and R and D are the radius and depth of the indentation, respectively.

FIGURE 12.3 Ball penetrometer test being performed on cheese slice.

This method is a nondestructive means of testing and does not require an operator to undergo any specialized training. Also, it is used on the whole cheese, whereas almost every other rheological test requires that a sample be cut from the whole block. The ball compressor has the merits of cheapness and simplicity, but the time taken to obtain a representative reading limits its use to the research laboratory. Ball compressors, penetrometers, and curd tension meters have been used as empirically measuring tools for rheological attributes of the cheese in cheese plants for many years.[21] These equipments are not precision instruments because of the arbitrary test conditions,[23] and the test results are difficult to compare with those from more rigorous experiments.[16]

12.4.2 PENETROMETER TEST

The penetrometer test is one other empirical tests, and is known for its simplicity and not quite nondestructive, but very nearly so, since it only requires that a needle be driven into the body of the cheese; no separate sampling is required. It measures the depth to which a penetrometer (e.g., needle or cone or rod) can be forced into a cheese under a constant stress. As the needle or cone penetrates the cheese, the cheese in its path is fractured and forced apart. The progress of the penetrometer is retarded to an extent that depends on the hardness of the cheese in its path, the adhesion of the cheese to its surface (which increases with the depth of penetration into the cheese), and its surface area of contact with the cheese (regulated by the thickness of the needle or angle of the cone used). Eventually, the retardation stresses become equal to the applied stress and penetration ceases. In case of cone penetrometer, the penetration depth at rest (h) is used to calculate an "apparent yield stress," σ_{app}:

$$\sigma_{app} = \frac{Mg}{\pi h^2 \tan^2\left(\dfrac{\alpha}{2}\right)} \tag{12.2}$$

where, M is the cone mass; g is the acceleration due to gravity; α is the cone angle; H is the cone height; A is the cone angle; and h is the cone penetration depth.

For a given cone of known mass and cone angle, the equation simplifies in terms of just h, that is, $\sigma_{app} = k/h^2$, where, k is a constant for the particular cone. Cone probes with various angles (e.g., 20–90°) are available to be used with commercial instruments. This test, which is used to provide an index of hardness (i.e., resistance of a surface to penetration), is suitable for closed textured cheese such as Gouda and mozzarella, which are macroscopically homogeneous. Conversely, it is unsuitable for open-textured cheese with small mechanical openings or eyes (e.g., Tilsit and Gruyere) or cheese that is macroscopically nonuniform owing to the presence of chip boundaries (e.g., Cheddar).

Hennequin and Hardy[8] used a cylindrical probe (5 mm diameter at a speed of 10 mm/min to a depth of 10 mm) to penetrate soft cheese (e.g., Camembert, Coulommier, and Munster). They found that the force at 10 mm penetration gave a high correlation with sensory firmness ($r = 0.94$, $n = 19$). They concluded that the technique is suitable as a rapid method for texture measurement in soft cheese. Breuil and Meullenet[4] found a significant correlation between measurements obtained using a cone penetrometer (30°), or a 2 mm needle, and textural characteristics of a wide range of commercial cheeses (e.g., Colby, Edam, Cheddar, mozzarella, and cream cheese) as measured by a sensory panel.

12.4.3 TEXTURE PROFILE ANALYSIS

This test imitates the grinding action of the jaw; it is performed by subjecting a specimen to two-step compression. The first compression step, known as the "first bite," is followed by a second compression, the "second bite." This is to stimulate the first two bites taken during chewing of the food.[6] The two compression steps may be separated by an optional wait time. A typical TPA test performed using a universal testing machine (UTM) would generate a force–time profile. The textural parameters determined from TPA force–time (or deformation) curve are: hardness, cohesiveness, adhesiveness, chewiness, gumminess, springiness, resilience, and fracturability.

Hardness is the force necessary to attain a given deformation. Fracturability is the force at significant break in the curve on the first bite (originally known as "brittleness"). Cohesiveness (no units) is the strength of the internal bonds making up the product. Adhesiveness (A) is the work necessary to overcome the attractive forces between the surface of the food and the surface of other materials with which the food comes into contact.

Gumminess is the energy needed to disintegrate a semisolid food until it is ready for swallowing. Chewiness (Ch) is the energy needed to chew a solid food until it is ready for swallowing. Springiness is the distance recovered by the sample during the time between the end of first bite and the start of second bite (originally known as "elasticity," rate at which a deformed material goes back to its undeformed condition after the deforming force is removed). Resilience (no units) is the measure of how well a product "fights to regain its original position."[6]

12.4.4 WIRE CUTTING TEST

The wire cutting test is used to determine firmness, consistency, and fracture properties of cheese. The same test can also be used for butter and ice cream blocks (Fig. 12.4A). Typically, a wire is pushed into the material until a steady state cutting force is achieved. A typical force–displacement curve for a wire cutting test consists of two phases. In the first phase, the wire indents into the material. At some point during the indentation, the material starts to fracture and a steady state cutting phase proceeds. It has been stated that the force F required for cutting is proportional to the wire diameter (d) and that there is a constant component arising from G_c, the fracture toughness. The proportionality factor with the diameter was then determined from the deformation energy and the friction. A linear equation between the cutting force and wire diameter is:

$$\frac{F}{b} = G_c + (1+\mu)\sigma_y d \qquad (12.3)$$

where, b is the width of the sample, σ_y is the characteristic stress, and μ is the coefficient of friction.

(A)   (B)   (C)

FIGURE 12.4   Wire-cutting test on butter (A); multiple puncture probe test on butter (B); extrusion test on a dairy product (C).

12.4.5 MULTIPLE PUNCTURE PROBE TEST

This test allows food manufacturers to test nonuniform products containing particulates of different size, shape, structure, and levels of hardness to provide repeatable results. Such products have nonhomogeneous textures due to the presence of, for example, dried fruits, vegetables, or nuts or consist of different layers to provide more interest to the consumer.[11] By penetrating the product in several areas at the same time (Fig. 12.4B), the multiple puncture probe produces an averaging effect and is therefore more representative. This test is generally used for hard variety of cheeses.

12.4.6   BACK EXTRUSION RIG

The back extrusion rig offers an invaluable measure of product consistency. A sample container of the test matter is compressed by a disc plunger attached to the texture analyzer, which extrudes the product up and around the edge of the disc (Fig. 12.4C). The effort taken to do this is measured using the software and results give an indication of the viscosity. Such a measure relates to the body or flowing nature of the product, its mouth coating potential, and spoonability. Achieving the desired texture of semi-solid product is therefore possible, and can be monitored from batch to batch, and throughout the shelf life of the product. This is generally used for cheese sauces.

12.5   SPREADABILITY OF MILK AND MILK PRODUCTS

Spreadability of margarine and butter is an important aspect of the consumer acceptability of these products. Spreadability, in pragmatic terms, is the ease with which a spread can be applied in a thin, even layer to, for instance, bread. Firmness or hardness maybe measured by the force required to obtain a given deformation or by the amount of deformation under a given force (Fig. 12.5). Although spreadability is also a deformation under an external load, yet it is a more dynamic property. Measurements of

FIGURE 12.5   Cheese spreadability test.

firmness and spreadability are usually correlated. However, the relationship is rarely perfect, and is partly a function of work softening. Margarine, for instance, work softens (when spread on bread) more easily than butter, which allows it to be more spreadable even when hardness values are initially equal.

12.5.1   FACTORS AFFECTING SPREADABILITY

Many people are interested in measuring the spreadability of food products such as butter, spreads, peanut butter, margarines, cheese, and cream cheese. These products are often very shear sensitive and are difficult to consistently prepare for testing. Penetration and compression style tests are simple methods that give results of sample hardness. Hardness measurements, however, even with cone probes, are not always good discriminators of spreadability. The TTC spreadability fixture is an attachment which measures the ease with which a product, such as margarine or table spread, can be applied in a thin, even layer. It comprises a male 90° cone probe and five precisely matched female perspex cone-shaped holders. The material is either deposited or allowed to set up in the lower cone holders in advance of testing or is filled with a spatula and the surface then leveled. Excessive work is not introduced into the product and different styles of filling only affect the early part of the test. The fixture comes with five replicable female cone sample holders which can be filled in advance of testing and then easily locked into the base holder precisely centered under the matching upper cone probe. The sample holders can be stored in refrigerators or frozen environments, or they can be left ambient before testing. The important action that the test is designed to measure spreadability occurs in the later stage of the test. The cone-shaped holders offer no locations into which the product can be packed or compressed, so the product is forced to flow outward at 45° between the male and female cone surfaces, the ease of which indicates the degree of spreadability. The probe withdrawal may also offer some insights into a product's adhesive characteristics.

12.5.2   DESCRIPTION OF TEST AND RESULTS

During the test, the force is seen to increase up until the point of maximum penetration depth of the cone probe. This peak force value (maximum force to shear) can be taken as "firmness" at the specified depth. A firmer sample shows a correspondingly larger area which represents the total amount of force (or otherwise referred to as "work of shear") required to perform the shearing process. Both of these values have been shown to rank samples in the same order of spreadability (and firmness), but for some samples one may prove to be more suitable than the other. The probe then proceeds to withdraw from the sample and any adhesive characteristics are indicated by a negative force region on the curve.

Knowledge of the rheological properties of a semisolid food, such as margarine, is important in process design, quality control, and the development of new products. Soft or tub margarine is mainly used as a spread for toast and sandwiches. Soft margarines have to be stored under refrigeration and should be spreadable when taken from the refrigerator. The spreadability out of the refrigerator and at room temperature should be even and smooth with no syneresis and separation.

12.6 TEXTURE OF YOGURT

The textural characteristics of yogurt are generally studied in the cup, using a spoon, or in the mouth by sensory methods. The most common sensory attributes relating to yogurt texture are thickness/viscosity, smoothness (opposite to lumpiness, graininess, and grittiness), and sliminess (or ropiness). Determination of the yogurt texture usually comprises sensory, structure, and rheology analyses.[18] Generally, textural properties of yogurt can be assessed using following methods:

- Instrumental analysis: rheology and texture
  - Small-amplitude oscillatory rheology
  - Large-amplitude oscillatory shear
  - Penetrometry
  - TPA
  - Viscometry
- Instrumental analysis: structure
  - Visual observation of gel microstructure and indirect evaluation of network homogeneity by measurements of water-holding capacity (WHC)
  - Scanning or transmission electron microscopy

- Confocal scanning laser microscopy
- Permeability of drainage test

• Sensory analysis
  - Smoothness: lumpiness, graininess, grittiness, sliminess, or ropiness

12.6.1 SET YOGURT TEXTURE

In case of set yogurt, the assessment of the product must be approached in a manner that does not destroy the delicate coagulum. The falling sphere technique[14] can be used, but the most convenient test involves the use of the penetrometer. The texture profile analyzer may also be used to improve the repeatability of the measurements. A sensory analysis is also used to measure the texture of set yogurts. For example, physical features or faults, namely lumpiness or the presence of nodules will be detected during sensory analysis.

12.6.2 FACTORS AFFECTING TEXTURAL PROPERTIES OF YOGURT

Parameters influencing the yogurt texture include fortification level and materials used, stabilizers type and usage levels, fat content and homogenization conditions, milk heat-treatment conditions, starter culture [type, rate of acid development, and production of exopolysaccharide (EPS)], incubation temperature (influences growth of starter cultures), cooling conditions, handling of product postmanufacture (e.g., physical and temperature), gel aggregation, bond strength, and pH at breaking. The gel strength of yogurt is related to the total effects of chemical interactions. During denaturation β-lactoglobulin interacts the binding of β-lactoglobulin with the κ-casein on the casein micelle surface by disulfide bridging which is responsible for the increase of gel strength and viscosity of yogurt. The physical properties of yogurt gels including gel stiffness and permeability, rearrangement of protein particles in gel network, and structure breakdown of stirred-type yogurts are important factors that influence the physical and structural properties of yogurts. Deep understanding of the mechanisms involved in the formation of texture in yogurts and impact of the processing conditions on texture development helps to improve the quality of yogurt.

12.6.3 TEXTURE OF STIRRED AND FLUID YOGURT

A range of methods are available to measure the viscosity of stirred and liquid yogurts. However, the choice of method is really a matter of operator preference. Some subjective measurement techniques employed to measure textural quality of stirred and liquid yogurts include:

- Scooping a sample of yogurt onto the back of a spoon, and then gently inclining the spoon downwards—the rate at which the yogurt drips from the spoon is a reflection of its viscosity; the same technique will also reveal any irregularities in the coagulum.
- Inserting a plastic teaspoon into a typical retail carton of yogurt—if the spoon remains upright, the product has an acceptable, spoonable viscosity.

Even though, these techniques are subjective in nature, they offer some guidance to experienced operators/analysts to test the quality of the product. These techniques are not reproducible. Several reproducible techniques are also available to test the textural characteristics of stirred and fluid yogurts. A rotating cylinder can be used to measure the viscosity of yogurt.[5] The rotating cylinder is tilted until the yogurt began to pour; the angle necessary to initiate flow can be taken as a measure of product viscosity. Bottazzi[3] described a method, where time taken for a standard metal sphere to descend a certain distance through a prescribed volume of yogurt can provide information on its viscosity.

The flow rate of yogurt through funnels of prescribed orifice sizes to measure their viscosity was proposed by Posthumus.[15] In the Posthumus funnel, the time taken for the yogurt—surface to pass between the starting point and the centrally located pointer gives a measure of the viscosity of the product.

The time taken for a yogurt sample of known volume to flow down an inclined plane, with or without weirs, to get an idea of its viscosity was proposed by Hilker.[9] However, the most universally accepted approach is that employing a rotational viscometer, or the torsion wire apparatus.

Another empirical method used to measure the rheological property of stirred yogurt is known as the Bostwick consistometer. The unit resembles a rectangular channel made from stainless steel and fitted with removable slot or door. The consistometer can be used on-site in the production area where a sample of yogurt is placed in the slotted compartment; the door is removed and the rate of flow per time is measured on a scale. The thinner or low viscous yogurt tends to flow faster. The ease of operation makes the rotational viscometer a popular choice, and once the type of spindle and its speed of rotation have been established for a given product, comparison between successive batches presents few problems.

However, some researchers argue that these tests do not reflect the true nature of the product, since the shearing effect of the spindle destroys the integrity of the coagulum. Consequently, it has been suggested that as stirred yogurt is a viscoelastic material, that is, it has some of the properties of a viscous liquid and some of an elastic solid,[12,13] dynamic oscillatory testing would be more appropriate.[19] The Rheotech international controlled-stress rheometer is one of the typical instruments that can be used to make dynamic measurements. However, the disadvantages of this approach are that the equipment is expensive in contrast to the cost of a viscometer, and that taking the measurements can be technically demanding, and hence for routine operation in quality control, it is a system that is unlikely to find much application.

Important approaches to modify the texture of yogurt and decrease textural defects include:

- Use of novel stabilizers and various types of milk-derived ingredients
- Use of various types of membrane concentration methods
- Specific cultures (e.g., producing a specific EPS, type, and content)
- Enzymatic crosslinking of milk proteins, (e.g., transglutaminase)
- Use of high hydrostatic pressure (e.g., >200 MPa) to the milk to cause denaturation of whey proteins or of the yogurt to prevent post acidification
- Use of very high-pressure homogenization
- Increasing gel stiffness (elastic modulus) and yield stress

Higher levels (above 2%) of inulin and guar gum significantly improve textural characteristics of yogurt.

12.7   TEXTURE OF ICE CREAM

While there are many different kinds and flavors of ice cream, it is important to distinguish between two types: those with particulates, for example, cookie dough pieces, chocolate chips, fruit chunks, and those without.

When testing a uniform ice cream such as French vanilla or plain choco-late, Warner–Bratzler blade may be used (Fig. 12.6).

FIGURE 12.6   Warner–Bratzler blade probe for determining the hardness of ice cream.

Quantifying the firmness/hardness of ice cream containing particulates can be challenging. The number, size, shape, and distribution of particu-lates are usually randomly distributed within each container. While it is relatively easy to quantify the firmness of the ice cream matrix on its own using standard penetration probes, it doesn't work as well on some vari-ants such as mint chip, chocolate chip, and cookie dough. For this, multiple puncture probe is generally recommended. The use of a multiple puncture probe for a nonuniform sample often increases reproducibility over single puncture probes.

12.8   TEXTURE OF BUTTER

For quality control purposes, instrumental evaluations of firmness and spreadability of edible fats are usually performed by using empirical rheological methods. These are rapid, arbitrary easy-to-perform and single-point measurement techniques, which correlate well with sensory assessments of the desired properties. For firmness and firmness and spreadability determinations frequently applied methods include wire sectility testing, orifice extrusion, constant speed, as well as, constant load penetration with varying devices, and rotational rheometry. Magnitude estimation experiments performed only recently showed that the sectility force in wire cutting experiments and the apparent yield value in constant load penetrations closely correlate with sensory evaluations of firmness and spreadability, respectively.[17]

The wire sectility method (Fig. 12.4C) has been standardized for cutting speed (6 mm/min), specimen shape and size (cube, 25 mm edge length), and temperature (15°C). Besides influence of sample size and tempera-ture, special attention should be paid to effects of sectility speed on the test results. From the sensory point of view it can be assumed that the wire sectility method is almost imitative, that is, subjects assess firmness on the basis of the force they apply to cut the sample with a knife, which is precisely the underlying principle of sectility tests. This, however, butter cutter is used for the determination of butter firmness according to the ISP 16305. In this test a stainless steel wire cuts through a defined cube of butter. This test is very dependent on temperature and is performed as a standard method in a hot water bath. This is also used for the determina-tion of the cutting strength of cheese, eggs, vegetables, and fruit.

12.9   TEXTURE OF KHOA/PANEER/PEDA

The textural profile analysis of peda samples were performed on the texture analyzer (TA.XT *plus,* Stable Micro Systems, United Kingdom) and data were analyzed using software *Exponent Lite* fitted with a 5 kg load cell. TPA was done to characterize the various textural attributes such as hard-ness, adhesiveness, springiness, cohesiveness, gumminess, and chewiness of the product. The samples were cut into 1 cm^3 size and subjected to texture analyzer. Exponent was carried out by monoaxial compression of 5 mm height with 5 g trigger force that generated plot of force (*g*) versus time (*s*). 75 mm (P/75) compression platen probe was used to measure texture of set peda samples at a temperature of 25 ± 1°C. The force distance curve was obtained for two-bite compression cycle with 1.0 mm/s test speed with 200 pps data acquisition rate and the pre- and posttest speed were set up at 1.0 and 5.0 mm/s, respectively.[2]

In another study, the frozen samples of gundpak (a traditional milk product of Nepal) were equilibrated at room temperature (25°C). The texture profiles of samples were measured by making uniform size sample with a circular diameter (46 mm) and height (19 mm) with a stainless steel ring and average weight was 40 ± 2 g per piece of the sample. The hardness of samples was measured by an Instron Universal Texture Testing machine (Model LR5K, Llyods Instruments, United Kingdom). The product was compressed by 50% in two bites. The load cell of 5 KN at a cross head speed of 100 mm/min was used and the force of 1 KN was applied for compression of the samples.[1]

In one study, the paneer samples were cut into cube shape (10 mm × 10 mm × 10 mm) and were warmed to 25°C in a temperature-controlled cabinet for 1–2 h and the textural studies were carried out at the same temperature. The central temperature of the samples was measured by a digital thermocouple. The operating conditions were cross-head pretest speed of 5.0 mm/s, and test speed of 5.0 mm/s. Posttest speed of 10 mm/s and 5 s interval between successive bites were employed for 50% compression. The texture profile parameters were determined by the TPA curve as reported by Koca and Metin[10] and Halmos et al.[7]

The textural properties were evaluated using the TA.XT plus Texture analyzer of the Stable Micro System equipped with 25 kg load cell. The analyzer is linked to a computer that recorded the data via a software program. Paneer sample of length 1 cm^3 was cut from the central portion of tofu cake with a stainless steel cutter. The stainless steel probe of 5 mm diameter with a flat end was used to determine the textural properties with settings.[20]

Test modes are: compression; pretest speed: 1 mm/s; test speed: 2 mm/s; posttest speed: 2 mm/s; target mode: distance; distance: 3 mm; two counts.

12.10 SUMMARY

Food texture is a collection of sensory features perceived during eating and handling of a food. Each texture property reflects only one particular aspect of the eating (sensory) experience of food. There are a number of tests used by sensory scientists to determine product acceptability with a special reference to individual characteristics such as texture and areas of potential improvement. Moskowitz developed some novel mathematical business-oriented methods of texture optimization. Texture selected for a specific product should be compatible with the image the product is intended to convey. During eating, an individual sensory feature changes its intensity and accordingly there will be a changing profile of the dominating sensory feature. This dynamic process is directly linked to or caused by the changing length scale of food particles and the length scale involved in the deformation process that controls a materials mechanical response.

KEYWORDS

- brittleness
- chewiness
- cohesiveness
- food texture
- hardness
- penetrometer
- texture profile analysis
- universal testing machine
- wire-cutting test

REFERENCES

1. Acharya, P. P.; Kharel, G. P.; Chetana, R. Physicochemical and Microbiological Characteristics of Gundpak—A Traditional Milk Product of Nepal. *J. Food Res.* **2015,** *4,* 30–37.
2. Banjare, K.; Kumar, M.; Goel, B. K.; Uprit, S. Studies on Chemical, Textural and Sensory Characteristics of Market and Laboratory Peda Samples Manufactured in Raipur City of Chhattisgarh. *Orient J. Chem.* **2015,** *31,* 231–238.
3. Brennan, S. C.; Tudorica, M. C. Carbohydrate-based Fat Replacers in the Modification of the Rheological, Textural and Sensory Quality of Yogurt: Comparative Study of the Utilisation of Barley Beta-glucan, Guar Gum and Inulin. *Int. J. Food Sci. Technol.* **2008,** *43,* 824–833.
4. Breuil, P.; Meullenet, J. F. A Comparison of Three Instrumental Tests for Predicting Sensory Texture Profiles of Cheese. *J. Texture Stud.* **2001,** *32,* 41–55.
5. Davis, J. G. Laboratory Control of Yogurt. *Dairy Ind.* **1970,** *35,* 139–144.
6. Gunasekaran, S.; Mehmet, A. K. M. *Cheese Rheology and Texture*; CRC Press: Boca Raton, FL, 2002; p 512.

7. Halmos, A. L.; Pollard, A.; Sherkat, F.; Seuret, F.; Seuret, M. G. Natural Cheddar Cheese Texture Variation as a Result of Milk Seasonality. *J. Texture Stud.* **2003**, *34*, 21–40.

8. Hennequin, D.; Hardy, J. Instrumental and Sensory Evaluation of Some Textural Properties of Soft Cheese. *Int. Dairy J.* **1993**, *3*, 635–647.

9. Hilker, L. D. A Method for Measuring the Body of Cultured Cream. *J. Dairy Sci.* **1947**, *30*, 161–164.

10. Koca, N.; Metin, M. Textural, Melting and Sensory Properties of Low-fat Fresh Cheese Produced by Using Fat Replacers. *Int. Dairy J.* **2004**, *14*, 365–373.

11. Moskowitz, H. H.; Jacobs, B. E. *Consumer Evaluation and Optimization of Food Texture—Instrumental and Sensory Measurement*; Marcel Dekker Inc.: New York, 1987; pp 293–328.

12. Ozer, B. H. Rheological Properties of Labneh (Concentrated Yogurt). PhD Thesis, The University of Reading, UK, 1997; pp 12–30.

13. Ozer, B. H.; Robinson, R. K.; Grandison, A. S.; Bell, A. Gelation Properties of Milk Concentrated by Different Techniques. *Int. Dairy J.* **1998**, *8*, 793–799.

14. Pette, J. M.; Lolkeme. Yogurts and Yogurt Cultures. *Netherlands Milk Dairy J.* **1950**, *4*, 261–267.

15. Posthumus, G. New Approach to Viscosity Measurement. *Off. Organ FNZ* **1954**, *6*, 46–55.

16. Rao, V. N. M.; Skinner, G. E. Rheological Properties of Solid Foods. In *Engineering Properties of Foods;* Rao, M. A., Rizvi, S. S. H., Eds.; Marcel Dekker: New York, NY, 1986; pp 215–254.

17. Rohm, H.; Ulberth, F. Use of Magnitude Estimation in Sensory Texture Analysis of Butter. *J. Texture Stud.* **1989**, *20*, 409–418.

18. Sodini, I.; Remeuf, F.; Haddad, S.; Corrieu, G. The Relative Effect of Milk Base, Starter, and Process on Yogurt Texture: A Review. *Crit. Rev. Food Sci. Nutr.* **2004**, *44*, 113–137.

19. Steventon, A. J.; Parkinson, C. J.; Fryer, P. J.; Bottomley, R. C. The Rheology of Yogurt. In *Rheology of Food, Pharmaceutical and Biological Materials with General Rheology*; Carter, R. E., Ed.; Elsevier: London, 1990; pp 197–210.

20. Supekar, A. L.; Narwade, S. G.; Syed, I.; Kadam, R. P. Effect of Goat Milk Fortification on Chemical, Microbial and Sensorial Quality Characteristics of Paneer. *Res. J. Anim. Husb. Dairy Sci.* **2014**, *5*, 131–135.

21. Szczesniak, A. S. Classification of Textural Characteristics. *J. Food Sci.* **1963**, *28*, 385–389.

22. Szczesniak, A. S. Correlating Sensory with Instrumental Texture Measurements—An Overview of Recent Developments. *J. Texture Stud.* **1987**, *18*, 1–15.

23. Voisey, P. W. Instrumental Measurements of Food Texture. In *Rheology and Texture in Food Quality*; deMan, J. M., Voisey, P. W., Rasper, V. F., Stanley, D. W., Eds.; AVI: Westport, CT, 1976; pp 79–141.

24. Wilkinson, C.; Dijksterhuis, G. B.; Minekus, M. From Food Structure to Texture. *Trends Food Sci. Technol.* **2000**, *11*, 442–450.

GLOSSARY OF TECHNICAL TERMS

3A Sanitary Standards have been developed for a variety of equipment used in the dairy industry, as well as some equipment used in egg processing. The founding organizations of 3A represent equipment manufacturers, dairy processors, and regulatory officials. 3A Standards are used as a reference under the Grade A Pasteurized Milk Ordinance (PMO), the official regulatory document for the National Conference on Interstate Milk Shipments (NCIMS).

Adhesiveness is the work necessary to overcome the attractive forces between the surfaces of the sample and the other materials with which the sample comes in contact. It is the negative force area for the first bite curve.

Anthocyanins are versatile and plentiful flavonoid pigments found in red/purplish fruits and vegetables, including purple cabbage, beets, blueberries, cherries, raspberries, and purple grapes. Within the plant, they serve as key antioxidants and pigments contributing to the coloration of flowers.

Antibiotics refer to substances produced by microorganisms that act against another microorganism. Thus, antibiotics do not include antimicrobial substances that are synthetic (sulfonamides and quinolones), or semisynthetic (methicillin and amoxicillin), or those which come from plants (quercetin and alkaloids), or animals (lysozyme).

Antimicrobial refers to all agents that act against microbial organisms. It includes all agents that act against all types of microorganisms—bacteria (antibacterial), viruses (antiviral), fungi (antifungal), and protozoa (antiprotozoal).

Apparent particle density or envelope density is the mass per unit volume of a particle, excluding the open pores but including the closed pores. Gas or liquid displacement methods like gas or liquid pycnometry are adopted for the measurement of apparent particle density.

Apparent solidification time (AST) value means fat sample in test tube to apparently become solidified at 18°C.

Apparent viscosity is the measurement of the resistance to flow or the fluidity of a material and is the ratio of shear stress to shear rate.

Bacteriostatic concentration is the lowest concentration at which bacteria fail to grow in broth, but are cultured when broth is plated onto agar.

Biosynthesis (de novo synthesis) means, when a specific compound is synthesized with the help of metabolizing cells is called biosynthesis.

Biotechnology refers to the exploitation of biological processes for industrial and other purposes is called biotechnology.

Biotransformation refers to the modification of chemical substrate with the help of microbial cells..

Bömer value refers to the sum of the melting point of saturated triglycerides (isolated by diethyl ether method) and twice the difference between this melting point and that of the fatty acids obtained after the saponification of these triglycerides.

Bulk density, also known as apparent or packing density, is a measure of the mass of milk powder, which occupies a fixed volume. It is dependent on particle density, particle internal porosity, and arrangement of particles in the container.

Carotene refers to yellow-red pigments (lipochromes) widely distributed in plants and animals, notably in carrots, and closely related in structure to the xanthophylls and lycopenes and to the open-chain squalene; of particular interest in that they include precursors of the vitamins A. Chemically, they consist of eight isoprene units in a symmetric chain with the two isoprenes at each end cyclized, forming either α-carotene or β-carotene.

Casein is a typical milk protein, exists in milk as particles called casein micelles, which are made up of calcium phosphate and casein components. Buffalo milk is richer in total protein, particularly the casein and whey proteins.

Caseinomacro peptide is the enzymatic breakdown product of para κ-casein having hydrophilic nature due to glyco-conjugation to peptides. Caseinomacropeptide is responsible for increased efficiency of digestion, prevention of neonate hypersensitivity to ingested proteins, and inhibition of gastric pathogens.

Cheese ripening is a series of complex physical, chemical, and microbiological changes in cheese, leads to formation of distinct flavor, taste, body, and texture.

Chewiness refers to the energy required to masticate a food to a state ready for swallowing. It is a product of hardness, cohesiveness, and springiness.

Chhana (curd) is a solid product formed by heat/acid precipitation of buffalo milk during preparation of paneer. Chhana obtained from cow milk is an intermediate product used for preparation of a wide variety of traditional milk-based sweets. Colostrum refers to the first drawn secretion after parturition differs greatly from normal milk. It has very high serum protein contents. Milk is lacteal secretion practically free from colostrum obtained from healthy animals. It must contain not less than 3.5% fat and 8.5% SNF.

CIP means a totally automatic cleaning sequence with no manual involvement.

Cleaning-out-of-place (COP) is defined as a method of cleaning equipment items by removing them from their operational area and taking them to a designated cleaning station for cleaning. It requires dismantling an apparatus, washing it in a central washing area using an automated system, and checking it at reassembly.

Cohesiveness refers to the extent to which a material can be deformed before it ruptures. It is the ratio of the area under the second peak to that under the first peak.

Complete liquification time (CLT) test means time required to complete melting of milk fat at 44°C.

Complex shear modulus (G^*) includes both storage and loss moduli values and is an indicator of the strength of a gel.

Critical temperature of dissolution (CTD) is the temperature at which turbidity appears on gradual cooling of the fat dissolved in a warm solvent or solvent mixture.

Crohn's disease is a long-term condition that causes inflammation of the lining of the digestive system. Inflammation can affect any part of the digestive system, from the mouth to the back passage, but most commonly occurs in the last section of the small intestine (ileum) or the large intestine (colon).

Crystallization time test means the suitable quantity of milk fat dissolved in solvent and recording the onset of crystallization of milk fat at 17°C.

Decaffeination is the removal of caffeine from coffee beans, cocoa, and other caffeine-containing materials. Decaffeinated drinks contain typically

1–2% of the original caffeine content, and sometimes as much as 20%. Decaffeinated products are commonly termed decaf.

Dispersibility refers to the ability of a milk powder to separate into individual particles when dispersed in water with gentle mixing is an important consideration in industrial settings. It is the ease with which lumps and agglomerates of powder fall apart in the water. It is expressed as the percentage of the solids dissolved.

Encapsulation is a process to entrap one substance (active agent) within another substance (wall material). The encapsulated substance, except active agent, can be called the core, fill, active, internal, or payload phase. The substance that is encapsulating is often called the coating, membrane, shell, capsule, carrier material, external phase, or matrix.

Essential oils: The International Organization for Standardization (ISO) in their Vocabulary of Natural Materials (ISO/D1S9235.2) defines an essential oil as a product made by distillation with either water or steam or by mechanical processing of citrus rinds or by dry distillation of natural materials. Following the distillation, the essential oil is physically separated from the water phase.

First-order kinetics is the progress of reaction which totally depends on the concentration of one reactant.

Flavor is the sum of those characteristics of any material taken in the mouth, perceived principally by the senses of taste, and smell, and also the general pain and tactile receptors in the mouth, as received and interpreted by the brain.

Flowability of a powder refers to the ease with which the powder particles move with respect to each another, that is the resistance to flow.

Foam formation is important in the development of the texture of foods such as ice cream, mousse, whipped topping, meringues, and even coffee (espresso coffee). While most foams are formed at low temperature, there is also interest in the foaming properties of milk at high temperature, for example, in the foams produced by steam injection for hot milk-based beverages.

Fouling is the accumulation and formation of unwanted materials on the surfaces of processing equipment, which can seriously deteriorate the capacity of the surface to transfer heat under the temperature difference conditions for which it was designed.

Fractionation of milk fat refers to the milk fat that is subjected to a specific temperature/time profile to allow a portion of milk fat to crystallize.

Fracture force is a force with which a product fractures. It is the force at the first significant break fracture in the first bite curve.

Glass transition temperature (Tg) is defined as the temperature at which an amorphous system changes from a glassy to a rubbery state.

Gumminess is the energy required to masticate a sample to a state ready for swallowing. It is a product of hardness and cohesiveness multiplied by 100.

Hardness is the maximum force that is exerted on the sample to attain a given deformation that is the highest point of the peak in the first bite curve at 25% compression.

High-pressure processing (HPP) is a novel processing technique during which the product is treated in a vessel of suitable strength at a high pressure, generally in the range of 100–1000 MPa.

Homogenization is a mechanical process which converts large fat globules into small particles to uniformly distribute fat throughout the milk.

Hurdle technology is a combination of preservation method used to kill pathogens in food products thereby ensuring safe consumption and extended shelf life.

Hygroscopicity is a measure of the water absorption by a powder. It is often measured by passing air of a known humidity level (usually 80% at 20°C) over a powder until equilibrium is reached, then measuring the weight gain of the powder. Powders which absorb much moisture may cake during storage.

Immersion cleaning is the type of cleaning in which the parts to be cleaned are placed in the cleaning solutions to come in contact with the entire surface of the parts. Immersion cleaning is preferred for parts that must be placed in baskets and for processes requiring a long soaking time because of the type of contamination to be removed or the shape of the parts to be cleaned.

Inflammatory bowel disease involves chronic inflammation of all or part of your digestive tract. IBD primarily includes ulcerative colitis and Crohn's disease.

ISO 14159: 2002 This International Standard specifies hygiene requirements of machines and provides information for the intended use to be provided by the manufacturer. It applies to all types of machines and associated equipment used in applications where hygiene risks to the consumer of the product can occur.

Leaf protein concentrate (LPC) is prepared by grinding young leaves to a pulp, pressing the paste, and then isolating a liquid fraction containing protein by filter or centrifuge. Herbaceous plants and legumes, such as clover and lucerne, produce higher yields of protein concentrate than perennial grasses.

Loss modulus (G'') is the component out of phase with the strain or viscous behavior.

Maillard reaction is defined as the reaction between amino acids and reducing sugars present in milk which leads to brown color to milk and milk products during high-heat treatment. Its desirability depends on the type of milk and milk product.

Manual cleaning is the universal practice and the efficiency is accomplished by training the cleaning operators, ensuring the exact method of cleaning in the manual cleaning SOP, validating the method from different operators, and verifying the procedure with an interval of time.

Membrane distillation (MD) is a thermally_based procedure used for the separation of vapor molecules which can penetrate the microporous membrane layer.

Milk fat crystallization is the formation of crystals, when milk fats are cooled down from a high temperature (completely melted fat) to temperatures below the melting point.

Minimum bactericidal concentration (MBC) is a concentration where 99.9% or more of the initial inoculum is killed.

Minimum inhibitory concentration (MIC) is a lowest concentration resulting in maintenance or reduction of inoculum viability.

Particle porosity is defined as the fraction of air or void space over the total bed volume. It is affected by factors such as mechanical compaction, particle size (particle size distribution), moisture, temperature, chemical nature of each constituent, processing conditions, moisture, and temperature during storage. The changes are created due to increase in adhesiveness, variation in the mass due to water sorption or evaporation, or due to the phase change of fatty components with temperature.

Phase angle (δ) is directly related to the energy lost per cycle divided by the energy stored per cycle. Phase angles can vary from 0° to 90°, with 0° indicating an ideal solid material (Hookean solid), and 90° indicating an ideal viscous material (Newtonian fluid).

PLC is a digital computer widely used for automation of industrial electromechanical processes, such as control of automated factory assembly lines, robotics, or batch process controls.

Prebiotics are generally defined as nondigestible food ingredients that beneficially affect the host by selectively stimulating the growth and/or activity of one or a limited number of bacterial species already established in the colon, and thus in effect improve host health.

Probiotics refer to live microorganisms that, when administered in adequate amounts, confer a health benefit on the host.

Proteolysis is termed for the enzymatic breakdown of proteins.

Pulsed electric field (PEF) is a nonthermal method of food processing that uses short bursts of electricity for microbial inactivation and causes minimal or no detrimental effect on food quality attributes. High-intensity pulsed electric field processing involves the application of pulses of high voltage (typically 20–80 kV/cm) for short time periods (less than 1 s) to fluid foods placed between two electrodes.

Pulsed light (PL) is a method which involves the use of intense and short-duration pulses of broad spectrum "white light" (ultraviolet to the near infrared region).

Radapertization process is not recommended for most foods as it involves treating the product to levels of radiation of approximately 30 kGy.

Radicidation treatment is used to destroy pathogenic organisms on the food product and is able to kill vegetative cells. The treatment dose for this treatment ranges from 2.5 to 5.0 kGy.

Radurization is used to treat foods that are high in pH and which have high water activity with a treatment dose of approximately 1 kGy.

Recombined milk refers to the product obtained when butter oil, skim milk powder, and water are combined in the correct proportions to yield fluid milk.

Reconstituted milk refers to milk prepared by dispersing whole milk powder in water approximately in the proportion of 1 part powder to the 7–8 parts water.

Refractive index refers to the property which concerns the degree of bending of light waves passing through a liquid or transparent solid is a characteristic for the particular liquid or solid.

Rennet coagulation is defined as the coagulation of milk which induced by the proteolytic action of rennet. This process includes two phase, in first phase the chymosin enzyme breaks κ-casein at Phe_{105}–Met_{106} position, and in second stage, destabilization and coagulation of milk proteins takes place.

Rheology is the science of deformation and flow of matter. It is the study of the manner in which a material responds to applied stress or strain.

Saponification refers to a reaction of fat with alkali and forms soap.

Sensory analysis is examination of organoleptic attributes of a product by the sense organs.

Serum is defined as milk plasma minus casein micelle. Removal of the micelles from skim milk by clotting with rennet yields the liquid called whey.

Shear modulus is the constant of proportionality used to relate shear stress with shear strain.

Shear strain occurs when stress is applied parallel to the surface of the material.

Shear stress (τ) is defined as a frictional force per unit area.

Sinkability is the ability of powder particles to overcome the surface tension of water and sink into the water, after passing through the surface. It is expressed as milligram of powder that sink/min/cm^2 surface area.

Skimmed milk is separated milk but fat separation is not completed.

Solidification point is the temperature at which fat shows the first sign of appearance of solid phase on cooling.

Solubility/insolubility index is an important feature of milk powders. Poorly soluble powders can cause processing difficulties and can result in economic losses. Solubility is a measure of the final condition to which the constituents of the powder can be brought into solution or stable suspension.

Springiness is the height that the sample recovers between the first and second compression on the removal of the deforming forces.

Standardized milk is milk whose fat and/or solid not fat content has been adjusted to a certain predetermined level (fat: 4.5%, solid not fat: 8.5%). The standardization can be done by partially skimming the fat in the milk or by admixture with fresh or reconstituted skim milk in proper proportions.

Sticky point temperature (T_s) is defined as the value at which a powdery material will start caking. Since the stickiness of dried fruit powders normally develops once the transition from glassy to rubbery took place, the sticky point temperature is always higher than the glass transition temperature. Furthermore, sticky point temperature can be increased by increasing T_g.

Stiffness is the ratio of crushing work done to the area under the compression part of the curve.

Storage modulus (G') indicates the degree of elastic behavior in a material.

Supercritical fluid extraction (SFE) is defined as separation of chemicals, flavors from the products such as coffee, tea, hops, herbs, and spices which are mixed with supercritical fluid to form a mobile phase. In this process, the mobile phase is subjected to pressures and temperatures near or above the critical point for the purpose of enhancing the mobile phase solvating power.

Supercritical fluid is any substance at a temperature and pressure above its critical point, where distinct liquid and gas phases do not exist. It can effuse through solids like a gas, and dissolve materials like a liquid.

Synbiotic is used when a product contains both probiotics and prebiotics. Because the word alludes to synergism, this term should be reserved for products in which the prebiotic compound selectively favors the probiotic compound.

Syneresis is the release of whey from the surface of fermented milk gels.

Texture refers to all the mechanical, geometrical, and surface attributes of a product perceptible by means of mechanical, tactile, and, where appropriate, visual, and auditory receptors.

Titre value of milk fat represents the highest temperature reached when the liberated water insoluble fatty acids are crystallized under arbitrarily controlled conditions.

Tofu is prepared from combinations of soya milk and skim milk, or soya milk and buffalo milk. In tofu manufacture, hexameta phosphate or polyphosphate can be added to delay the coagulation of soya milk until it has been charged into containers.

Transmission electron microscopy is a microscopy technique in which a beam of electrons is passed through an ultrathin specimen, to form an image from the interaction of the electrons transmitted through the specimen. The image is magnified and focused onto an imaging device, such as a fluorescent screen, on a layer of photographic film, or to be detected by a sensor such as a charge-coupled device.

Triglyceride refers to ester of fatty acids with glycerol.

True particle density is the ratio of the mass of the particle to the volume of the particle, excluding open and closed pores.

Ulcerative colitis is a disease that causes inflammation and sores, called ulcers, in the lining of the rectum and colon. It is one of a group of diseases called inflammatory bowel disease.

Ultrafiltration is a membrane process and the driving force of which is pressure and is suitable for the separation of large molecules and colloidal substances from liquids.

Ultrasound is a form of energy generated by sound waves of frequencies that are too high to be detected by human ear, that is, above 16 kHz.

Ultraviolet processing involves the use of radiation from the ultraviolet region of the electromagnetic spectrum.

Unsaponifiable matter refers to that portion of oil/fat which could not react with alkali.

Vanaspati oil refers to vegetable oil which is hydrogenated and hardened.

Wettability is a measure of the ability of a powder to absorb water on the surface, to be wetted, and to penetrate the surface of still water. Contact angles indicate the degree of wetting when a solid and liquid interact. The lower is the contact angle, the greater is the wetting. It defines the potential for a powder to wet and absorb water at a given temperature.

INDEX